White Gold

Karl Froschauer

White Gold:
Hydroelectric Power in Canada

UBCPress / Vancouver

Printed in Canada on acid-free paper ∞

ISBN 0-7748-0708-3 (hardcover)
ISBN 0-7748-0709-1 (paperback)

Canadian Cataloguing in Publication Data

Froschauer, Karl, 1946
   White gold
Includes bibliographical references and index.
ISBN 0-7748-0708-3 (bound); ISBN 0-7748-0709-1 (pbk.)
1. Electric utilities – Canada. 2. Water-power – Canada. I. Title.
HD9685.C32F76   1999   333.91′415′0971   C99-910638-4

This book has been published with a grant from the Humanities and Social
Sciences Federation of Canada, using funds provided by the Social Sciences and
Humanities Research Council of Canada.

UBC Press acknowledges the financial support of the Government of Canada
through the Book Publishing Industry Development Program (BPIDP) for our
publishing activities.
Canadä

We also gratefully acknowledge the ongoing support to our publishing program
from the Canada Council for the Arts and the British Columbia Arts Council.

Set in Stone by Brenda and Neil West, BN Typographics West
Printed and bound in Canada by Friesens
Cartographer: Eric Leinberger
Indexer: Annette Lorek

UBC Press
University of British Columbia
6344 Memorial Road
Vancouver, BC V6T 1Z2
(604) 822-5959
Fax: 1-800-668-0821
E-mail: info@ubcpress.ubc.ca
www.ubcpress.ubc.ca

Contents

Figures, Tables, and Appendices

Appendices

Preface

In the last four decades of the twentieth century, many Canadians had their lives and efforts subordinated to the political and economic imperative of damming rivers to produce hydroelectricity for industry and export. These included Native peoples, environmentalists, and others who opposed Canada's dam-building rush, and also the thousands of construction workers, technical employees, and engineers whose careers were broken when the rush ended in the 1990s. The political and economic leaders responsible for Canada's hydroelectric development policies often dismissed opponents' views in favour of their own claims that provincial electrical systems needed to expand for local industrial transformations and that electricity exports are in Canada's national interest. At the time, the 25,000 employees now laid off by Ontario Hydro, Hydro-Québec, and BC Hydro were often convinced by those claims. A closer examination of the claims reveals, however, that something went wrong with the initial privatizations of hydro resources, with the kind of industrial development the addition of new hydroelectric projects was expected to bring, with the timing and size of the provinces' electrical supply expansion, and with the national and regional initiatives to link these major projects to a trans-Canada power grid.

In probing the relationship between hydro expansion, industrial diversification, and the weakening of national power grid initiatives through export of power, I have had to conduct my research at both the federal and provincial levels. This was because the Canadian Constitution assigns the authority to develop natural resources, including hydroelectric power, to the provinces, but gives jurisdiction over trade to the federal government without the authority to unilaterally enforce its policies; thus the federal government must try to achieve consensus among the provinces regarding interprovincial trade and exports to the US.

Federal Cabinet records, engineering and economic assessments, federal-provincial conferences and negotiations, and studies of provincial utilities in provinces owning major water-power resources (e.g., waterfalls, canyons, river narrows) reveal that, in the 1960s, 1970s, and 1980s, it would have been possible to establish either a national network or several regional power grids, which would have resulted in more integrated service to the country. However, provincial resistance to federal plans, the inability and unwillingness of provinces to accept or form an extra-provincial authority to regulate the transport of electricity across provincial transmission systems, the preference of provinces to export electricity to the US, and the strategies of provinces to develop hydro for local economic purposes undermined such initiatives.

In my examination of the causes of the failure of these initiatives, I consider the views of a number of political economists. Thomas Hughes, for example, sees the cause of failure in the particular mix of private and public ownership and the often narrow profit and political interest of both utilities and governments. I also employ the insights of the new Canadian political economists, such as Janine Brodie (national policies and regional outcomes), Michel Duquette (centralized energy policies and defensive provincial continentalism), Laura Macdonald (free trade and differential continental integration), and Rianne Mahon (locations of resistance in an era of fragmenting state sovereignty), to examine Canada's national, regional, and continental power system integration in the context of specific political practices. Furthermore, I also cite research by engineering firms to show that (1) during the initial planning stage of various hydro mega-projects in the 1960s, the use of Canada's provincial electricity surplus capacities could have been coordinated to benefit both neighbouring provinces and the country as a whole by optimizing plant operations interprovincially and thereby reducing their environmental, social, and fiscal impacts; but that (2) provincial pursuit of exports and hydro-related industrial policies, as well as federal continental bias, undermined initiatives in that direction.

Other new Canadian political economists, including Mel Watkins, Wallace Clement, Glen Williams, Gordon Laxer, and Neil Bradford, have shown that Canada's peculiar industrial development processes vary regionally and split into at least three paths: the emergence of manufacturing entrepreneurs from within local, provincial, and national communities; the establishment of manufacturing in branch plants by foreign direct investment; and export-oriented commodity extraction for manufacturing elsewhere. Provincial reliance on foreign, often global, corporations for economic growth makes the planning of infrastructures, such as hydro-electric systems, for local industry difficult. An historic case-in-point, as noted by H.V. Nelles, was that of the Niagara Falls development in Ontario, where the province had to buy back the water rights and power plants from private American interests who failed to provide electricity for Ontario communities and industries because of their export of power to manufacturers on the American side (see Chapter 3).

More recent hydro projects, such as Churchill Falls (Labrador), James Bay, Nelson River, and Peace River, are in northern regions of the country that have only since the 1950s been opened for extensive development. The policy of accelerating dam-building in these areas during the 1960s and 1970s was intended, in part, to further the cooperative interconnection of the new northern generating facilities with a national power network, to benefit Canada as a whole, and to diversify its industrial growth. But the Liberal federal government of Prime Minister Lester B. Pearson weakened this strategy by encouraging early dam construction for power exports. Subsequent electricity exports evolved according to federal and provincial building-for-export agendas. These agendas were legitimized by the National Energy Board (NEB) and by the federal Cabinet but were denied initially by provincial utility officials until the 1980s and 1990s, because it was uncertain whether official utility mandates allowed the building of hydroelectric facilities exclusively for provincial markets or also for export.

Because hydro expansion programs occurred with provincial predominance in national, continental, and (with respect to attracting industries) global contexts, the required multiprovincial inquiry into, and interpretations of, varied outcomes presents a theoretical and methodological challenge. In Chapter 1, I provide academic readers with general insights from new Canadian political economists that allow a better understanding of hydroelectric development and the related politics within Canadian national, transborder regional, and North American continental spaces. To provide a more specific understanding of provincial cases, I also introduce in Chapter 1 an analytical guide to the inquiry into the five expansions of provincial hydro systems presented in Chapters 3, 4, 5, 6, and 7. Although theoretical insights are integrated throughout this book, I have kept discussion of theoretical literature to a minimum and replaced academic jargon, where possible, with more common terms.

I reviewed documents and records in both provincial and national archives, in utility archives, in the Centre de recherche en développement industriel et technologique in Montréal, and in libraries and industrial development departments as well as in some regulatory agencies. I reviewed most manuscripts, correspondence, and contracts with industry at their locations in Québec City, Montréal, Ottawa, Toronto, Winnipeg, Vancouver, and Victoria. Historic collections consulted include papers from Sir Adam Beck (Ontario Hydro), the records of Prime Ministers John Diefenbaker's and Lester B. Pearson's federal Cabinet discussions concerning the national power network, the records of the Churchill Falls (Labrador) Corporation, transcripts of the public discourse in export hearings, the papers of Edward Schreyer, the records of the James Bay Development Company, submissions to provincial commissions of inquiry, and the collections held by provincial utilities and in the National Archives in Ottawa. I have integrated all this historical and other material in the main chapters of the book in order to demonstrate the tensions over national and regional power-system integration and to illustrate in five provincial case studies the repetitive historical pattern in the development of hydro power, industry, and exports.

To permit a better understanding of archival records and other sources, I also conducted twenty unstructured interviews with industrial development officers, utility economists, procurement officers, utility officials, and people who have researched the relationship between hydro development and industry. Until 1984, manufacturers used to report their electricity consumption to Statistics Canada, which provided evidence for determining whether new supplies of power stimulated new manufacturing activities. Since then, Statistics Canada has ended such surveys, and so the numerical quantification of this relationship cannot be extended. Nevertheless, the decline in manufacturing in Canada in the 1980s and 1990s suggests that new supplies of electricity have not led to new manufacturing in the last decade.

In this book I do not attempt to describe the entire history of Canada's federal, provincial, or continental electricity policies or of its major utilities; rather, after discussing Canada's proposed national and regional power networks, I describe five provincial hydroelectric developments and confine my analysis to specific time frames and to key development patterns that recur in each case. In my Niagara Falls discussion, I give a brief history of the pre-1900 period and of the first three decades of the twentieth century. The more recent period of major hydroelectric system expansions (with the exception of the proposed Conawapa project on the Nelson River) started in the early 1960s and ended in 1972 in Labrador, in 1984 in British Columbia, in 1992 in Manitoba, and in 1994 (with the exception of Québec's and Newfoundland's 1998 proposal to develop the Lower Churchill project) in Québec. Within those periods, I concentrate on power projects at Churchill Falls (Labrador), in the James Bay region (Québec), on the Nelson River (Manitoba), and on the Peace River (British Columbia) (Figure 1.1). I trace the influence these projects had on expected provincial industrial transformation until the mid-1980s and examine their role in power exports and in the deintegration of Canadian public power systems in order to integrate them in the late 1990s with US power systems. In each provincial case, I briefly review the historical pattern of initially releasing public water power for private development and then reclaiming these privatized hydro-power resources, often with completed power facilities, for public ownership. Finally, I argue that now, at the dawn of the twenty-first century, in keeping with the neo-liberal economic policies of governments in the 1990s, hydroelectric resources are in danger of being re-privatized.

I present archival material and secondary sources in order to demonstrate that political issues prevented the development of a national power grid or regional grids and that certain development patterns are repeated in one or more provincial hydro cases (e.g., privatization reversals, expected but unrealized hydro-related industrial development, reported electricity surpluses, and the implications of exporting these surpluses to the US for utility restructuring in Canada). Overall, my research findings support the claim that, instead of providing the infrastructure for a national power grid and serving as a force for indigenous secondary industry, the provincial expansions of Canada's hydro resources have merely fostered continued dependence on branch-plant industrial development and staples export and have created vast surpluses of electricity for continental, rather than national, use.

Acknowledgments

Many people have given their time to make this manuscript possible, and I thank them for their help. Patricia Marchak, whose book *Green Gold* was never far from my mind when writing this manuscript, was the first to encourage me to examine the influence of hydro projects on industrial development. For sharing their analytical insights, for their continued interest, and for their supportive comments when this work was still an emerging thesis, I express my gratitude to Wallace Clement, Rianne Mahon, and John Harp of Carleton University and to June Corman of Brock University. To my friend Shahid Alvi, for the long discussions, theoretical debates, and motivational support that helped advance this work, I also express my gratitude. Without the assistance from archivists, librarians, and utility personnel in Québec, Montréal, Ottawa, Toronto, Winnipeg, and Vancouver, this book would have little substance. Anonymous readers for the Humanities and Social Sciences Federation of Canada and UBC Press have provided a thorough review. I am grateful to Gary Teeple for his encouragement when I most needed it. For the financial support from the Aid to Scholarly Publishing Programme of the Humanities and Social Sciences Federation I express my gratitude. My thankfulness extends also to Jean Wilson, senior editor of UBC Press, for her continuous support. John Marriott, Jenny Lawn, and Holly Keller-Brohman deserve credit for their helpful suggestions on improving my prose and the orderly presentation of ideas. For enduring and supporting me while I took time from them, I thank my family, Josefine, Toby, and Tina.

Abbreviations

AC	Allis-Chalmers
Alcoa	Aluminum Company of America
AMPCO	Association of Major Power Consumers in Ontario
BCFP	British Columbia Forest Products
BCUC	British Columbia Utilities Commission
BPA	Bonneville Power Authority
Brinco	British Newfoundland Corporation
Brinex	British Newfoundland Exploration
CAC	Canadian Allis-Chalmers
CFE	Comisión Federal de Electricidad
CFLCo	Churchill Falls (Labrador) Corporation
CGE	Canadian General Electric Company
CREDIT	Centre de recherche en développement industriel et technologique
CWES	Canadian Westinghouse
DEW	Dominion Engineering Works Limited
E.U.	États-Unis (United States)
EE	English Electric
FERC	US Federal Energy Regulatory Commission
FTA	Canada-United States Free Trade Agreement
FUJI	Fuji
HITA	Hitachi Ltd.
HVAC	High Voltage Alternating Current
HVDC	High Voltage Direct Current
Inco	International Nickel Company
IPACE	Interprovincial Advisory Council on Energy
ISI	Import Substitution Industrialization
JBDC	James Bay Development Company (French acronym SDBJ)
JBEC	James Bay Energy Corporation (French acronym SEBJ)
JI	John Inglis

LG1-4	La Grande River Generating Stations, numbers 1 to 4
LMW	Leningrad Metal Works
MAPP	Mid-Continent Area Power Pool
MEC	Maritime Energy Corporation
MIL	Marine Industries Limited
MITI	Mitsubishi
MITS	Mitsui
MP	Member of Parliament
MP&L	Minnesota Power and Light Company
MTS	Manitoba Telephone System
NA&NR	Northern Affairs and National Resources
NAFTA	North American Free Trade Agreement
NDP	New Democratic Party
NEB	National Energy Board
NEP	National Energy Policy
NEU	New England Utilities
NEYC	Neyrpic
NSP	Northern States Power Company
OPG	Ontario Power Generation
OHSC	Ontario Hydro Services Company
PASNY	Power Authority of the State of New York
Powerex	British Columbia Power Exchange Corporation
PWW	Pelton Water Wheel
RTG	Regional Transmission Groups
SMS	S. Morgan Smith
SPEC	Society for the Promotion of Environmental Conservation
TOBA	Toshiba
VEW	Vancouver Engineering Works
VIW	Vancouver Iron Works
YMCA	Young Men's Christian Association

1

Introduction:
Federal and Provincial Power

When explorers travelled toward and reached Niagara Falls, they
told themselves, they had arrived at the threshold of a new world
of unimaginable wonder, beauty and power. It was the beginning
of the dream of North America, a dream of which Niagara Falls
was the first and greatest symbol. Anything they believed would
be possible for those who could control the power this torrent of
dropping water represented.[1]

Niagara Falls – that indescribable, glistening, rushing dream of
water and light, falling, and falling, and falling. Niagara Falls was
the crucible of American industrialization where hydroelectric
power made the automated assembly line possible.[2]

– CBC, 'The Falls,' from the radio program *Ideas*, n.d.

The tremendous flow of Niagara Falls is divided between Canada and the
US by the international boundary. By 1907 American utilities diverted
Canadian flow through their powerhouses on the Ontario side of the Falls,
directing the electricity produced in Canada to industries in the State of
New York. In small southern Ontario towns, casket makers, mill owners,
cigar box manufacturers, carriage makers, and furniture factory owners
(most still using steam engines to run their machinery) watched as
Ontario's power drove electric motors in modern factories in industrial
parks across the border, and they looked with envy at the economic success of American industries perched on the cliffs on the US side of the
Falls. This sight convinced small Ontario manufacturers, some of whom
were also politicians, to try to mobilize their provincial and federal governments to repatriate the water rights and electricity from these US-
owned powerhouses. Soon this slogan was heard: 'Power exported is
power lost.'

With too many railways already built, federal and provincial land sur-
veyors, international speculators, and some developers of dubious
integrity soon mapped the mighty northern rivers in Manitoba, British
Columbia, Québec, and Labrador – rivers that propelled their tremendous
cascades of white gold into deep rock pools and through narrow canyons.
Decades later, free-flowing rivers were blocked behind barriers of concrete
retaining walls and earth-fill dams, creating large impounded lakes for

power production. The surface of such expansive reservoirs drowned forests, some containing the burial sites of Canada's Aboriginal peoples, and carried barges heaped with logs or minerals to gigantic mills and to roads on barely cleared shorelines. Trapped salmon no longer leaped up the rapids to spawn, and some caribou drowned in unseasonable, utility-regulated river flows. Although some concerned citizens had resisted these developments, their protests were overwhelmed by the rhetoric of progress. Soon electricity could be transported great distances to light skyscrapers, melt metals, grind pulp, and facilitate chemical reactions. However, the price of this electrification was often, even from the very beginning at Niagara Falls, the poisoning of some of the world's largest freshwater lakes, the discharge of effluence from pulp mills into the sea, and the leaking of acid from mines into rivers.

By the 1960s, Canadian engineers could transport electricity over greater distances: from northern rivers to southern cities, for thousands of miles from east to west across the Prairies, and to the south across the US border. Just as farmers had saved money by connecting farm houses to shared electric generators, some Canadian prime ministers and provincial premiers believed that, like the railways in the previous century, a national power grid could hold the country together. They proposed saving capital expenditures by interconnecting provincial power grids and large hydro-electric projects on a national scale. Many provincial premiers, already overseers of public power commissions, thought they could not go wrong in booming, postwar North America by rushing to build mega-dams with mega-loans from their American neighbours. They believed that they could sell electricity to power-hungry customers south of the border and that transnational companies would invest in setting up manufacturing plants, using cheap power and cheap resources in their provinces. More construction, mill, and factory workers would then be employed, and the benefits from this economic growth would easily allow them to pay off the interest and mega-bills.

Soon, coloured photographs of mega-dams, with the world's largest tur-bines, 500-megawatt (million watt) generators, and control rooms with push-button consoles, created the impression that new manufacturing would soon develop. Adversaries of this rush to build dams were told that they were holding up progress and asked if they would like to return to wood stoves and candlelight. Then in the 1970s, when oil exporters in the Middle East increased their prices and oil companies in Canada predicted that non-renewable oil and gas would run out in a dozen years, some pre-miers dreamed that, by damming more rivers, they could produce for export renewable hydroelectricity that would turn their provinces into Kuwaits of the North. There were some opponents of electricity exports to the US, who thought it better to phase out some of the polluting fossil-fuel plants, avoid nuclear plants, and supply neighbouring provinces with surplus power within a system of regionally integrated transmission networks. American presidents, however, continued to be interested in Canadian renewable energy, and prime ministers and political leaders in Manitoba, Québec, and British Columbia thought that, by signing trade agreements with their southern neighbours, the nation and their provinces could profit from the sale of surplus electricity.

Hydro development policy in several provinces shows, in fact, that surplus generating capacities developed for export and industry had unexpected repercussions. In the last four decades of the twentieth century, the damming of provincial rivers to attract global industry and to permit exports has resulted in the construction of more hydroelectric plant capacity than either the nation or the provinces need.[3] Under trade and access reciprocity, exporting the surplus electricity generated in these extra-capacity plants to the US can lead to both the possible access by US power marketers to Canadian customers and the required restructuring of Canadian utilities (that is, the deintegration of most vertically integrated provincial power utilities).[4] Although in the free trade period from 1988 to 1996 US customers bought, on average, only 6 percent of the total electricity generated in Canada, maintaining or increasing this small percentage of export sales continues to have an influence disproportionate to its size on utility managers' approaches to power planning and utility organization in Canada.[5] More significantly, Canadian industrial customers, not US-based customers, consumed the largest proportion – about 40 percent – of all electricity generated in Canada in the mid-1990s.[6] Therefore, in the examination of hydro-related development patterns, the relationship between industrial development and hydro development becomes a priority. In previous decades, industrial customers demanded that provincial governments intervene in the market to increase public generating capacity and provide power at or below cost; but in the late 1990s, advocates of industrial customers have proposed that industry be allowed to abandon public power supply and buy electricity from private suppliers in a continentalized electricity market.[7]

Not only these outcomes are problematic, however. In recent decades, the accelerated damming of rivers to generate hydroelectricity resulted in surplus capacities which in turn contributed to project cancellations, controversies, and criticisms. In the 1980s and 1990s, provincial governments cancelled planned power projects in British Columbia, Manitoba, Ontario, and Québec. In fact, the Parti Québécois government's cancellation of the $13 billion Great Whale hydro project in 1994 virtually ended three decades of provincial hydro system expansions in most of Canada.[8] Furthermore, critics have measured the problematic impacts of these projects in terms of the environment (river flow changes, fishing and wildlife harvesting losses, and flooding), debt ($90 billion in long-term utility debt in 1996),[9] social disruption (dislocation of First Nations communities and large-scale layoffs of 25,000 technical utility employees), and 'unplanned' power surpluses. Other critics focused on how British Columbia's new electricity trading arrangements with the western US required the deintegration of BC Hydro into separate business units (for the purposes of generation, transmission, distribution, and export) and argued that deintegration, happening in western provinces, could result in the private 'grab' of corporate elements of public utilities.[10]

Similar to western utilities, radical changes in the internal structure and mode of operation of Ontario Hydro have also occurred. The Ontario Energy Competition Act (Bill 35) of October 1998 resulted in the breakup of Ontario Hydro on 1 April 1999 into a generating company, a transmission and distribution services company (Ontario Hydro Services Company, OHSC), and a transmission network operating unit.[11] Ronald Osborne, Ontario Hydro's president and chief executive officer, plans to reduce the provincial market share of Ontario Power Generating (which, effective 1 April 1999, owns all of Ontario Hydro's 78 generating stations with 30,000+ MW) from 85 percent to 35 percent by the year 2010, while placing no limits on exports over upgraded OHSC-owned transmission lines beyond the province and to the $300 billion per year US market.[12] As will be discussed in more detail below, this deintegration process and the preparation for possible privatization coincide with 'orders' to meet US transmission regulations. As the early history of Niagara Falls shows, major Ontario hydroelectric resources had been privatized before Ontario Hydro became a public company in 1906, so that one wonders why advocates of privatization do not ask why utilities became publicly owned in the first place, an issue that will be addressed in Chapters 3 to 7 of this book.

The efforts of the hydro pioneers at Niagara Falls in the early twentieth century were upstaged in the 1970s by the rush to build some of the world's largest power plants at Churchill Falls, in the James Bay region, and on the Nelson and Peace Rivers. These accelerated provincial developments appear to have proceeded with little coordination between provinces and have left a legacy of debt, uncertain export markets, employee layoffs, as well as adverse impacts on Aboriginal communities. Some Canadians, however, did envisage an ecologically less detrimental, a financially less burdensome, and a politically more integrative alternative to overbuilding of provincial hydroelectric systems. I will show in Chapter 2 that, as early as 1961, engineers, leaders in government, utility planners, and political economists had advocated unifying provincial grids into a national network; that is, they envisaged a large, cooperatively controlled electric power system embracing interconnecting power plants and loads (consumer demands) of varied characteristics.[13] Notwithstanding, since then other politicians, civil servants, and private investors, all favouring provincial power systems serving industrial development and continental integration, overrode these proposals.

So far, I have introduced some of the problematic outcomes of Canada's major hydroelectric schemes when built for industry and export, but I have not offered any theoretical insights that might help in understanding hydroelectric development in a wider spatial context. Political and economic decision making within provinces was not the only influence on hydroelectric development; influences external to provinces also played a major role. Therefore, I will present the theoretical discussion in a spatial sequence by relating major expansions of hydroelectric systems (e.g., Churchill Falls) to a context that is international, national, regional, and provincial. Because provinces play the major policy role, however, I will provide an analytical framework for examining recurrent patterns in hydro expansion schemes that also allows for cross-provincial comparisons.

The New Political Economy of Hydro Power

Wallace Clement argues that, in the broadest terms, 'the place of the "external" in Canada's "internal" development has long been at the core of all the traditions within Canadian political economy.'[14] Political economists examining Canada's economic growth at the national, regional, and local levels agree that, historically, both international and domestic relations shape the spatial distribution of development in particular territories. Such relations have resulted in Canada's industrialization being characterized by pervasive foreign direct investment, the dominant features of which are a branch-plant structure in central Canada and predominant export-oriented resource extraction in the regions. To understand this development, a number of political economists have examined the past practices and decisions of economic and political leaders. Although Canadian provinces can grant licences to extract natural resources conditional on their transformation in local manufacturing and finished goods industries and, possibly, increase sustainability of renewable resources, the actions of Canadian political leaders have led, instead, only to the provision of infrastructure for the development of export-based staple extraction.[15] To extract such staples, Canada must provide, and often overbuilds, transportation and other infrastructure:

> Staples are the key policy instrument for export-led growth; the state assumes infrastructure costs (such as railways, ports, pipelines, and hydro projects) and foreign debts. To pay the debt, more staple extraction is required, and fewer manufacturing conditions are applied. Short-term solutions appear to work because the construction phase is labour-intensive

and welcomed by local capitalists, especially construction firms, but in the longer term they produce structures of dependence and debt ... and leave few jobs behind.[16]

At times, such infrastructure programs have been part of national development policies that also have come to serve regional, provincial, and continental interests but have done little to more evenly spread economic growth in Canada.

National, Regional, and Continental Expansions

Periodically, national policies have had a distributive function in Canada's industrial development. Political economists, such as Janine Brodie, argue that such national policies have played a considerable role in the uneven spatial distribution of industrial benefits in Canada. The nineteenth-century national policy (1867-85), with the trans-Canada railway as the infrastructure binding the country together, 'forged an east-west economy, leaving in its wake an industrialized centre, a deindustrialized east and a primary [resource]-exporting western hinterland.'[17] Focusing less on this spatial distribution of development, other political economists presume that railway infrastructure and electricity networks would together yield both industrial and national benefits. For instance, Hugh Aitken thinks that industrial growth effects produced by the all-Canadian railway system would be repeated if 'cheap Canadian industrial power' were used to build the nation.[18] That was also the basic assumption of the initiative taken by Prime Minister John Diefenbaker's Cabinet toward building a national power network (Chapter 2). As an analytical concept, however, Canada's first national policy (i.e., that centred on the transcontinental railway) is of limited value in the second half of the twentieth century, especially for decentralized energy infrastructure development (in which regional politics play a major part) and the hoped for industrial development.

Canada's Constitution grants provincial governments the jurisdiction to plan, construct, and operate hydroelectric systems within their provinces (a right insisted upon particularly by Québec, but also by other provinces), and it grants the federal government only restricted legislative authority over electricity transportation across provincial boundaries. Therefore, to achieve a national power grid, the provinces either would have to grant authority to the federal government or create an extra-provincial agency to transport electricity across interprovincial power lines. Also, in Canada's 'internal' development, the 'external,' such as the reliance on US capital and the preference of political leaders and policy makers for continental integration, is always present to make binding the country together through national infrastructure-centred policies a complex undertaking. Among the decision makers, both continentalists and nationalists took turns at different periods of Canadian hydroelectric development at attempting to influence policy. Others devised a national-continental policy, such as the revision of Diefenbaker's national grid policy by the Pearson Cabinet, which could be read as supportive of either position (see Chapter 2).

An analyst's historical account of the political background of electrical-system expansion will reflect his or her interpretive framework. Using a political economy approach, Thomas Hughes, in his 1974 essay 'Technology as a Force of Change in History: The Efforts to Form a Unified Electric Power System in Weimar Germany,' found that the clear technical and economic advantages of a national electrical system were not pursued in Germany because of the divergent interests of regional governments and private owners.[19] Ten years later, his interpretive framework changed, and he incorporated concepts from goal-oriented systems theory, including those of sociologists such as Talcott Parsons.[20] Whereas I argue that repetitive patterns ('privatization reversals,' for example) are discernable, Hughes, in *Networks of Power,* argues that utility systems do not evolve according to a strict pattern and proposes the concept of 'style' to describe organizational aspects of regional hydroelectric systems.[21] Canadian utility historians, such as Christopher Armstrong and H.V. Nelles in *Monopoly's Moment,* use similar concepts to analyze 'organizational styles in Canadian electricity supply' and to determine the variety of factors that shaped them before the 1930s.[22] Brodie, on the other hand, emphasizes the importance of regionalism, that is, the great importance of issues of *where* people live and *how* economic development, state activity, and political power are distributed across geographic space.[23] Brodie's insight is helpful in analyzing federal and provincial clashes over the proposed national and regional benefits of hydroelectric development. The organizational form of utilities (e.g., whether they are public or private), which Armstrong and Hughes stress in their later work, is only one component of my analysis; related industrial linkages, overbuilding, and the implications of exports are others. Hughes's earlier political economy perspective on the German power grid conflicts is more compatible with my own view of why attempts to integrate such grids often fail. Going beyond Hughes's interpretation, I contend that the possibility of establishing a national power network in Canada did exist, but that initiatives to integrate provincial power systems were undermined by provincial resistance to federal control and by biases towards continentalization at both levels of government.

Janine Brodie maintains that among Canadian regions (e.g., the West, the East, and the Maritimes) conflicting political and social tensions arise that are deeply embedded in Canada's collective historical experience.[24] The divisive political conflict around issues of territorial autonomy and the distribution of resources across geographical space that she identifies are evident in differences over 'where' mega-projects are developed, 'where' surplus is available and allocated regionally, and 'where' authority should be held. The 'where' questions remain central to interprovincial conflicts over finding regional power-grid compromises: Where will the electricity-generating facility be built (at Limestone, Lepreau, or Gull Island)? Which province would get most of the construction jobs? Where would the electricity be sold – to the neighbouring provinces, to the US, or to both? Where would the extra-provincial authority to direct the planning, coordination, and administration of a regional power network be located? These issues surfaced in the clashes between politicians favouring politicized power grids within each province and those speaking for the formation of regional or national power grids (Chapter 2).

Moreover, as Michel Duquette observes, a particular political tension emerges in Canada if nationalizing policies pertaining to energy development go too 'far through forced centralization and eventually constitutional reform[:] whatever their rationale, they will be perceived by the periphery as internal colonialism.'[25] He suggests that, as a defensive reaction, provincial policy makers in the Canadian periphery 'may try to "escape" economically what they believe to be the heavy burden of federalism by overwhelmingly committing their industrial strategy to the logic of foreign markets.'[26] This logic of commitment to foreign markets (also evident at the federal level) and of distancing themselves from national electricity policy goals is evident in how the provinces look to neighbouring American states rather than to other provinces for long-term trade arrangements. Thus, as Garth Stevenson observes, the commitment to transboundary regional arrangements for transporting energy (e.g., gas pipelines and power transmission lines) becomes stronger than does the commitment to national arrangements.

Stevenson claims that the nineteenth-century east-west centralization of infrastructural benefits from the national railway was only temporary and that, during the twentieth century, infrastructure actually had a decentralizing effect. He stresses that, since the 1930s, transboundary, regionally installed infrastructures (e.g., highways, airline routes, and communications networks) opened up several *continental* axes that have become interconnected with the American regional markets via the provinces.[27] For example, Stevenson sees increased integration of energy infrastructures (e.g., oil and gas pipelines, and reservoir storage in British Columbia for US-based electricity generation on the Columbia River) as yet another example of how twentieth-century infrastructural development directly served continentalism. In the matter of power use and distribution, decisions in the twentieth century about Canada's hydroelectric infrastructure expansions were generally made with provincial interests taking precedence over national interests.

Historically, the politics and policies of foreign countries have also influenced the Canadian economy and the formation of policy within Canada. As Mel Watkins argues, in some English-speaking countries in the 1980s, free-market policies had triumphed – 'in Britain under Margaret Thatcher and in the United States under Ronald Reagan. With the election of the first Mulroney government in 1984 these policies came to Canada.'[28] As Wallace Clement and Glen Williams observe, 'from the American point of view, energy became the main resource to be coveted in the 1980s.'[29] Elected on 17 September 1984, the Mulroney Conservatives, with support from the private oil sector, dismantled Trudeau's National Energy Policy (NEP) and opened the door to making energy a continentalized commodity. This process was strengthened when, 'in March 1985, Prime Minister Mulroney and President Reagan agreed at the Quebec City Summit to work towards opening access by their respective countries' [to each other's] energy markets by reducing barriers to trade.'[30] By the mid-1980s, Canada was ready to accept the US neo-liberal agenda of free trade in energy, including electricity.[31] The free trade agreements that followed also promoted neo-liberal restructuring (including deregulation, deintegration of utilities, minimal government intervention, and opening investment opportunities for new suppliers) and continental integration in the energy and electricity sectors at an administrative level, developments which will make it difficult for future governments to reverse these processes.[32] The continuing evolution of domestic political forces in favour of continental integration[33] ensures the inclusion of electricity in free trade agreements.

Ricardo Grinspun and Robert Kreklewich argue that the Canada-US Free Trade Agreement (FTA) and the North American Free Trade Agreement (NAFTA, which also includes Mexico) 'can serve as "conditioning frameworks" to promote continental integration and the consolidation of neoliberal restructuring.'[34] In this way, the weaker nation state's bureaucracies and regulatory institutions, often with the consent of its administrators, can be penetrated by American administrative rules or orders (which is what has happened with the adoption of the US Federal Energy Regulatory Commission's orders for some provincial transmission systems, as will be discussed in succeeding chapters).[35] With agreements such as the FTA and NAFTA, the relative fortunes of regions increasingly depend on how they integrate their trade into the continental market.[36]

Although provincial transmission infrastructure is not fully part of the 'zone' that the Canadian economy constitutes 'within the US economy' (to use Williams's description), since 1996 provincial power lines have increasingly been open for wholesale trade in a south-north direction (a reversal of the mostly north-south export pattern), and the business of transporting electricity is increasingly becoming part of that 'zone.' Provinces, in the transitional phase, can either remain electricity hinterlands for US utilities organized in regional transmission groups or become an emerging market for US electricity suppliers. So far, these insights allow us to understand hydroelectric developments spatially in their national, regional, and continental outcomes. In analyzing specific provincial cases, concepts from the new political economists that address economic growth issues will be integrated into a specific analytical framework in order to uncover and illuminate the Canadian experience of building power projects for industry and for export.

Analysis of Provincial Expansions

Until now, the analysis of hydro development has been characterized by individual case studies, and few analyses employ theoretical insights from the new Canadian political economists. In an attempt to develop an analytical framework that will allow for the analysis of several provincial case studies, both in the formulation of the research questions and in the explanation of the findings, I use an approach that draws on concepts taken from the new Canadian political economists as well as from compatible perspectives.

The research upon which this book is based critically examines selected provincial hydro expansions as 'powerful agents' for hydro-related industrial development. One of the first major hydro developments in Canada occurred at Niagara Falls in 1906, but it was not until the 1960s that construction accelerated elsewhere, producing the largest hydroelectric projects in Canadian history: Churchill Falls (5,429 MW, Labrador), La Grande 2 (5,328 MW, James Bay, Québec), La Grande 4 (2,535 MW, James Bay, Québec), Gordon M. Shrum (2,426 MW, Peace River, BC), and the hydro stations on the Nelson River in Manitoba (3,932 MW) (Figure 1.1). These large projects became part of major hydroelectric expansions, only certain aspects of which are analyzed: their development under private or state ownership; their use as industrial infrastructure; their contribution to overbuilding provincial systems; and the direction of their power surpluses towards continental, rather than national, use.

Public or Private Ownership?

In Canada, many major developments were advanced under private ownership before coming under public ownership. What conditions gave rise to the development of infrastructure expansion first under private ownership and then under public ownership? Given historic and economic conditions, I emphasize that whether an infrastructure, such as hydro plants and the transmission system, was operated as a capitalist enterprise or as a public enterprise depended on whether private entrepreneurs found its provision sufficiently profitable.[37] In Canada, where provinces initially allocated the water rights for canyons and waterfalls to private owners,[38]

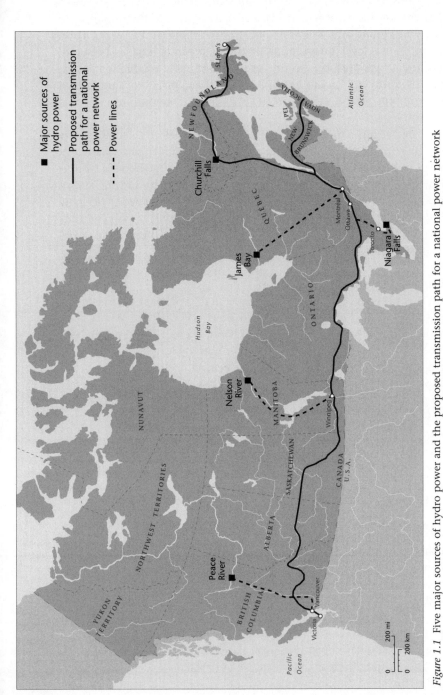

Figure 1.1 Five major sources of hydro power and the proposed transmission path for a national power network

Source: Walter Dinsdale, Minister, Department of Northern Affairs and National Resources, 'Memorandum to the Cabinet: Long-Distance Power Transmission,' Ottawa, 6 December 1961; Ingledow & Associates, Consulting Engineers, 'National Power Network Stage II Assessment,' vol. I (Vancouver: T. Ingledow, February 1967), p.1; David Cass-Beggs, 'Economic Feasibility of Trans-Canada Electrical Interconnection,' March 1960; and Chapter 2 in this book.

private companies often failed to bridge large transmission gaps or to develop the necessary electrical infrastructure, especially in small towns and in the less lucrative regional power markets within the provinces of Québec, Newfoundland, Manitoba, British Columbia, and Ontario. Thus, at the provincial level, intervention occurred, in part at least, as a result of the failure of private firms to provide less profitable production inputs (e.g., providing electricity in less industrialized regions) even though such missing imputs were necessary in order to give momentum to industrial expansion.

Once ownership returned to the provinces, hydroelectric facilities were more likely to be overbuilt rather than underbuilt. Projects were constructed not solely in response to demand, but also in the hope of *creating new* demand through accelerated industrial development and the associated accumulation process. Provincial governments went beyond bridging the hydro-infrastructural gaps and integrating provincial systems; they tried to make electricity production a key feature of their industrial development strategies.

Hydro and Industrial Development
Politicians who have supported the expansion of provincial hydro systems since the 1960s include Premiers Robert Bourassa (Québec), W.A.C. Bennett (BC), Joseph Smallwood (Newfoundland), René Lévesque (Québec), and Edward Schreyer (Manitoba). Together with hydro planners, they set out to realize their ambitions of making electrical utilities cornerstones of their provincial economies. The utilities, they thought, would not merely generate and distribute electricity to households and businesses, but would also become politicized enterprises and development companies that would advance the growth of secondary industries.

Why did so many premiers and utility executives assume that the availability of new hydroelectricity would foster industrial transformation? Their belief stems from Eurocentric notions about the relationship between energy and industry. Many analysts of Canada's social and economic development have relied on the industrial history of Europe as a 'blueprint' for Canadian industrial development, assuming that key features of the Industrial Revolution in European societies could be replicated in Canada. The combination of new forms of industrial energy (such as steam engines), the innovations in new technology and the factory system, the industrial capitalist class emerging mainly from within the community (whether regional or national), and the availability of labour helped to transform pre-industrial societies such as Britain, Germany, and France into fully developed industrial nations. Many Canadian analysts have ignored the difference between the emergence of Canadian ownership structures and the emergence of European ownership structures. Also ignored was the transition from steam power to hydroelectric power, which, in particular, has led to the perpetuation of false analogies. Canadian politicians and developers assumed that hydroelectric power which, produced from falling water and discovered 100 years after the invention of steam power, would generate industrial development in Canada, much as steam power had in Britain, despite the fact that Canada, unlike Britain, was reliant on foreign entrepreneurs. It was thought that towns with waterfalls suitable for electricity generation would attract industrial innovators who would use sophisticated machinery powered by electric motors to produce consumer goods.

Politicians, utility executives, and analysts claimed, and continued to claim throughout the twentieth century, that hydroelectric development was 'a revolutionary force' that would lead the provinces and the country into a 'new type of economic development.'[39] In 1899, T.C. Keefer, a professional engineer and transportation philosopher, thought the development of abundant electrical power would start a 'new industrialism,' as it had on the US side of Niagara Falls, and that it would 'deliver the dominion from its "hewer of wood" servitude to American industry and its bondage to American coal.'[40] As recently as 1987, the chair of Manitoba Hydro, Marc Eliesen, claimed that 'low cost electricity supply maintains and attracts new investment to the province, helping to encourage a stable, diversified economy.'[41] Even in 1997 the BC government initiated a 'power for jobs' scheme whereby the government proposed to offer cheap power for energy-hungry industries in return for the creation of employment. It also wanted to sell the surplus capacity that would be available from the Columbia River Treaty's downstream benefits (eventually 1,400 MW per year) to industrial customers.[42]

Canadian political economists, Harold Innis and John Dales, also believed that hydroelectric power would generate industrial development. Innis, writing in the 1930s, claimed that the public control and ownership of Niagara Falls water power was simply the outcome of previous political decisions: 'The control of water power has been placed by the British North America Act, and later decisions of the Privy Council, under the provinces. Consequently, the development of government ownership in relation to hydro-electric power has been undertaken by the provinces.'[43] Once hydro power was available, Innis had no doubt that it would be a powerful agent in the development of manufacturing linkages. For Innis, it was 'beyond question' that the Ontario Hydro-Electric Commission had been 'responsible for the rapid increase in the industrialization of Ontario.'[44] He thought that 'the dangers of over-expansion [had] been avoided through a strong, healthy criticism and effective checks in the form of commissions of investigation appointed by the government.'[45] Innis appears to have subscribed to the popular 1930s notion that the availability of hydroelectricity would lead to the creation of consumer goods industries. John Dales, writing in 1957, argued that the 'power station succeeded the railway as the main development agency' in Canada and that diversification had been most successful where power companies had to develop their own local industrial markets.[46] He claimed that 'in such situations, hydroelectric development has been a powerful agent in the promotion of a twentieth-century industrial revolution in Canada – an agent, that is to say, in the transition from an economy based on the exploitation and exportation [sic] of a few natural resources to an economy which is more diversified, which depends to a larger extent on manufacturing industry, and which is more self-reliant.'[47] Basing his claim on a comprehensive analysis of the industrial development resulting from the installation of regional power systems in Québec (1898 to 1940), he did not hesitate to extend his provincial thesis to the entire country.

What Dales overlooks is that the interests of foreign industrialists owning businesses in Canada differed from the interests of indigenous industrialists in Britain during the Industrial Revolution. Furthermore, in Canada, because of Crown ownership of natural resources, the provincial and federal governments played a more important role in industrial development than did the British government during the Industrial Revolution. Canada, historically, has relied on the expertise of industrialists or developers from the US (Niagara Falls), Britain (Churchill Falls), Sweden (Peace River), and other countries to finance and establish power and industrial plants. Rather than creating new industries, however, many of these invited industrial firms have simply extracted natural resources from Canada for their factories elsewhere.

The forms and failures of industrial development stemming from hydro development raise questions about Dales's thesis and support my argument that the development of Canada's large waterfalls did not bring about the transition he envisaged – a qualitative transformation from resource-based semi-processing to self-reliant manufacturing. Nor will hydro development bring about such a transition in the future, given the persistence of a continentalized power policy and a national industrial development policy that emphasizes 'industrialization by invitation.' Rather than bringing an 'Industrial Revolution,' hydro development has, at best, fostered dependent (branch plant) industrialization and extended Canada's dependence on the energy-intensive production of staples (i.e., semi-processed resource exports).[48]

Hydroelectricity that is exported to the US is often conceptualized as a staple. Defined as a staple, hydroelectric power is seen as a natural resource product that has been extracted from falling water through the use of hydro-mechanical technology for the purpose of export to the manufacturing sector in the US. Is this an adequate definition? Should electricity be understood as a staple (exported to the US) or should it be understood as an infrastructure for manufacturing and staple (e.g., pulp and metal ingot) processing in Canada? There are three possible ways of looking at hydroelectricity: (1) as an infrastructure for Canadian industries; (2) as energy input to extend staple production in Canada; and (3) as a staple for export to the US.[49] Which of these is most adequate?

An examination of actual electricity use eases the choice. As the product of power plants, electricity is used in a variety of staple-processing facilities (e.g., pulp and smelting) and in manufacturing industries (e.g., automobile production). In addition, power is produced in provincial plants for export to the US. In the major expansionary periods under study (between the 1960s and 1990s) electricity exports to the US (up to 1996) averaged only 6 percent of the total electricity generated in Canada (and fluctuated between 3.5 percent and 8.4 percent in the 1990s), whereas the consumption by domestic industries averaged about 40 percent of the total electricity generated in the period between 1960 and 1994.[50] Given that, since the 1960s, more than seven times the amount of electricity was consumed by industry in Canada than was exported to the US, it is appropriate to analyze hydroelectricity primarily as an infrastructure for Canadian industries (including staple industries) and only secondarily as a staple for export to the US.

Linkages to Industry

To determine more precisely how hydro schemes (developed by both public and private firms) have failed to advance industrial diversification, we can examine each provincial case for an array of new hydro-related production linkages.[51] Where hydro schemes lead to new investments in input supplying facilities, such as turbines, generators, and electrical apparatus manufacturing plants, 'backward linkages' are formed (Figure 1.2); where they lead to 'new investments in output-using facilities' (i.e., the use of new supplies of electricity to manufacture finished products from natural resources) 'forward linkages' result.[52] Besides such production linkages, more indirect linkages are created when, by applying Albert Hirschman's insight, provincial profits from energy projects are invested to subsidize energy consumption in manufacturing facilities. When this investment occurs the energy infrastructure fosters fiscal linkages, a concept that, in this book, pertains only to the results of manufacturing subsidies on the Island of Newfoundland which are financed by Churchill Falls revenues.[53] These processes, however, do not operate on their own. The type of industrial linkages related to hydro development result from courses of action taken by officials in private firms and provincial governments. When provincial government officials rely on indigenous entrepreneurs, then an independent path of growth ensues; when officials rely on foreign entrepreneurs (or foreign direct investment), then they pursue a dependent path of development.[54]

Backward linkages are formed when, through the imposition of nationalistic manufacturing policies, provincial governments make the procurement of technical equipment conditional upon its manufacture within the province.[55] This outcome is often achieved by substituting imported hydroelectric equipment with domestic equipment. This process is akin to the import substitution industrialization (ISI) process identified by Glen Williams, and usually constitutes a dependently formed linkage.[56] Whether such government-sought backward linkages are created outside Ontario and Québec will depend on the strength of nationalistic procurement practices in other provinces, as well as on their technological capacity. As I will show in Chapter 4, the insistence upon a 'buy-Québec' policy by Hydro-Québec executives and the James Bay Energy Corporation during the 1960s and 1970s did strengthen both the private- and government-owned turbine and generator industry.

Forward linkages are fostered when hydro schemes lead manufacturers to invest in facilities that use new hydroelectricity to diversify products made from natural resources. When virgin resources were allocated to private owners, some provincial governments, as proprietors, were able to impose manufacturing conditions on these owners in natural-resource

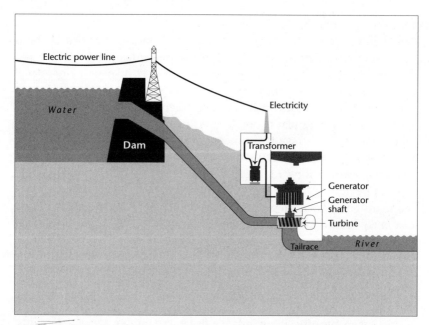

Figure 1.2 Hydroelectric generating station with turbine and generator detail
Source: 'Electric Power Generation,' *The Canadian Encyclopedia* (Edmonton: Hurtig, 1988), p. 680.
Permission to use this diagram, courtesy Ontario Power Generation, 1999.

harvesting leases and, thus, to encourage forward linkages.[57] Where such forward linkages are developed through 'industrialization by invitation' programs that target firms already vertically integrated with manufacturing centres elsewhere, the linkages are unlikely to become diversified.[58] Such firms often only minimally process natural resources for export to their factories. If foreign firms do not integrate their operations in Canada, then the forward linkages using cheap power and natural resources will continue to be of the semi-processed goods variety (such as wood pulp or mineral concentrates).

Overbuilding the Infrastructure
Since the 1970s, provincial governments have been increasingly unable to plan and coordinate provincial hydro and industrial development. At times, provinces attempted to plan an economy using electricity and its related infrastructure (including the storage reservoir) for invited industries, only to have foreign-controlled industrial firms thwart the process. As Claus Offe's research indicates, matching the size and timing of infrastructure built for specific uses with the changing requirements of profit-oriented production in private industry poses severe difficulties in capitalist societies whose industrialists and investment emerge from within.[59] In Canada, this situation is aggravated by a reliance on investors from abroad. When too much reliance is placed on industrialists who are expected to arrive but may not, or when multinational firms show little loyalty to production in a particular country or province, then the ability to plan efficient capacity and the timing of hydroelectric infrastructure becomes difficult. The US ownership of power plants at Niagara Falls earlier in the twentieth century meant that the Ontario government lost the opportunity to direct its power to Ontario industries. By inviting the British Newfoundland Corporation to Churchill Falls, Newfoundland lost decision-making powers over its electricity exports, plant ownership, and power repatriation. When provincial utilities (e.g., BC Hydro and Ontario Hydro) built plants in anticipation of firm contracts and plant expansions promised by investors from abroad, companies (mostly foreign-owned) could and did cancel their commitments in accord with the production decisions of their parent firms. These cancellations aggravated the growth of the province's electricity surplus. Furthermore, encouraged by the federal government, provinces built a surplus of plants and then exported power to the US. As will be shown in the chapters below, developing hydro for exports to the US does not, however, guarantee a market.

Directing Surpluses
In Canada, the planning and building of hydro projects comes under provincial jurisdiction, whereas continental and national (interprovincial)

trade is to a large extent the responsibility of the federal government.[60] Theoretically, the existence of a provincial surplus would allow the federal government, the National Energy Board (NEB), and the provinces to choose between two paths of action: (1) using surplus as a national industrial input that would require a national or regional power network, interprovincial sharing of surplus and transmission facilities, and the blocking or reduction of exports; and (2) using surplus as a provincially produced continental export commodity that would require US interconnection, no enforced interprovincial sharing of surpluses and transmission, and the authorization of exports.

The second (i.e., the continental) option in directing surpluses was predominant in the thinking of government officials from the 1960s to the 1980s, in part, because public hearings and decisions by the NEB and the federal Cabinet supported export licences and international power lines. This regulatory process was weighted towards continentalist interests, especially since the 1990 NEB Act amendments have made hearings an unlikely occurrence. Such hearings, albeit an attempt to include the *demos* – the people – and so to add democratic legitimacy to export decisions, tend to be subject to the kind of preferred and pre-selected imperatives that Rianne Mahon identifies in her 1977 analysis of unequal structures of representation.[61] Similarly, Offe argues that through the pre-selection of intervenors (potential opponents) and the pre-selection of imperatives (e.g., to grant export permits), the critical effect of such hearings is undermined, and all that remains in the Canadian case are scientifically supported commercial glorifications of Canadian benefits from continental integration.[62] In such cases, official decisions are based on narrow, pre-selected, export-oriented parameters, with the result that problematic planning matters brought forward by the public can be, and frequently are, neglected.[63] Thus the NEB and the federal government are not neutral mediators between export planning interests (utilities, provinces, and the public) and those adversely affected by exports (neighbouring provinces and opponents bringing forward social, environmental, or economic arguments); rather, their decisions favour the continental integration of energy infrastructure. Furthermore, with energy trade reciprocity in North America increasing, some of the sovereignty over energy and electricity has shifted to US government departments. In such instances, when analytical categories such as state sovereignty are crumbling, Mahon adapts Edward Soja's redefinition of space as 'a configuration of differentiated and hierarchically organized locales' in which centres of power at the local, national, and supranational levels interact in a complex and dynamic way.[64] With sovereignty over sections of the economy shifting to new public administrative and/or private corporate levels, or both, this new perspective on hierarchical space, according to Mahon, 'helps open up new strategic horizons by rendering visible the multiplicity of sites of action that have simultaneously to be considered' for resistance.[65]

In summary, I will employ Claus Offe's 'criterion of non-profitability'[66] regarding infrastructure provision in order to examine such patterns of government intervention as the privatization and reappropriation of waterfalls. In so doing I will demonstrate that state intervention in Canadian provinces changes from merely allocating natural resources (including water-power rights) to intervening in the production of hydroelectricity so as to create private accumulation conditions in a number of regions and industries.[67] Mel Watkins and Albert Hirschman provide an understanding of 'backward,' 'forward,' and 'fiscal' linkages from resources to industries; Glen Williams's concept of 'import substitution industrialization' is suitable for analyzing dependent development within home markets;[68] and Armstrong and Nelles find this dependency characterized by utilities imitating not only US business structures, but also relying on US technology.[69] Furthermore, H.V. Nelles's 'manufacturing conditions' can be applied to assess the degree of provincial regulatory power over the processing of natural resources; and Tom Naylor's conception of 'industrialization by invitation,' along with Philippe Faucher's and Kevin Fitzgibbins's work on the 'nationalistic' procurement practices utilized by provincial governments, allow further spatial refinements in the analysis of development. This analytical framework, derived from proponents of the new political economy probing relations between the state, natural resources, industrialization, and foreign ownership, gives rise to four key research questions:

(1) Why did private hydro development become subject to state intervention? More specifically, why, in several cases, were rights to extract power from waterfalls and to generate, transmit, and distribute power from them first privatized and then reappropriated by provincial governments?

(2) To what degree have hydro schemes developed by provincial governments and private firms advanced industrial linkages within the provinces?

(3) What contributed to the building of too many power plants and the resulting provincial energy surpluses? Was it the continentalist planning by the provinces, the failure by invited manufacturers to use the energy, or some other combination of factors?

(4) How did the Canadian government legitimate continental, rather than national, sharing of surplus energy from provincial hydro projects?

As part of my analysis, these four research questions will be applied to each of the case studies (Chapters 3 to 7).

Prospectus of Chapters
I present my findings systematically in order to show that Canadian provincial hydro expansions, when required to serve a policy of 'industrialization by invitation' and to favour temporary exports to the US, fail to provide either the basis for a national power grid or a force for new secondary manufacturing. At best, they foster dependent industrial development and staple production and, at worst, they provide surplus electricity for continental, rather than national, use.

In Chapter 2, I demonstrate that the expansion of provincial hydro systems, despite substantial attempts by federal and provincial governments, has not been integrated into national or regional power grids. I also show that the free trade agreements and US transmission regulations have contributed to the regional integration of several provincial utilities with US transmission groups or power pools. In Chapters 3 to 7, evidence from five hydro expansions in specific provinces and within specific timeframes further demonstrates recurring development patterns that are part of hydro expansions. These include privatization reversals, dependent industrial development, continuation or intensification of staple extraction, overbuilding of hydro infrastructure and creation of surplus power, power exports, and provincial transmission system integration with US regional transmission groups.

As noted in the Preface, Chapter 3 (on Niagara Falls) focuses on the first three decades of hydro development in the twentieth century and offers a brief history of the pre-1900 period. The more recent period of major hydroelectric-system expansions elsewhere in Canada began in the early 1960s and, except for projects under consideration in Manitoba, Labrador, and Québec, ended in 1994. In Chapters 4 to 7, I concentrate on power projects at Churchill Falls, on the La Grande River flowing into James Bay, and on the Nelson and Peace Rivers (Figure 1.1). Also, where appropriate, I trace the influence of this northward expansion on local industrial transformation up to the mid-1980s (when manufacturers ceased reporting electricity consumption to Statistics Canada) and examine its role in power exports and in the integration of a restructured Canadian power system with its American counterpart up to the late 1990s. Insights from the new Canadian political economists further an understanding of expansions of provincial hydro systems in the context of national policies, regionalism, differential continental integration, and resistance in an era of fragmented state sovereignty. The conceptual framework used with regard to the provincial cases may be applied both diachronically and synchronically. In Chapter 8, the conclusion, I will review findings at both the supra-provincial level and at the provincial level.

2
Avoiding National Power

The federal government has not, thus far, taken an active role in planning for or assisting with the development and use of electric energy on a national basis. Planning of electric power production has been almost entirely provincial, and provinces have tended not to look beyond their own borders. From discussions and investigations undertaken on a preliminary basis, it seems clear that the planning of electrical power development with transmission on a regional or national basis would make possible substantial economies and would permit the early development of major hydro-electric sources in remote areas.[1]

– Walter Dinsdale, Minister, Department of Northern
Affairs and National Resources, 6 December 1961

In this chapter, I outline the clash of interests from the 1960s to the 1990s over the national coordination of Canadian electric power development and its transmission – a clash that began when technological advances made trans-Canada power transmission possible (see Figure 1.1). At the federal level, advocates of forming a national power grid promoted its economic advantages over individual provincial systems. At the provincial level, however, there was a desire to expand the generation of hydro-electricity in order to advance provincial industrial development and to profit from exports to the US. Exports, in turn, required transborder utility integration.

First, I introduce some of the ideas of national power grid advocates and of utility experts, policy makers, political economists, and others. They identify the technical, economic, and environmental benefits of a national system, which they claim would increase the security of electricity supply, reduce the requirement for new power plants, and decrease pollution by replacing fossil-fuel-burning electricity with hydro-based electricity. Others warn that such benefits are easily undermined by the political nature of power development, which requires its benefits to be distributed across jurisdictional boundaries (Brodie) and demands centralized control. Such national grid initiatives also have the potential to be undermined by allowing exports and by the resistance of some provinces to the national benefits that they might produce. Following this debate, I show how the Diefenbaker Cabinet supported the federal-provincial initiative of developing a national power grid by proposing that remote reserves of water power in Labrador, Québec, Manitoba, and British Columbia (see Chapters

3 to 7) be integrated to serve electricity supply and industrial development on a national scale. The Pearson Cabinet lost control over the national electricity policy when (1) it proposed to allow both the early development of northern projects and their interconnection with US utilities in order to transport both 'temporary' and long-term exports and (2) it failed to obtain Québec's agreement either to participate in federal-provincial power-grid initiatives or to allow the transport of electricity from Labrador across its territory.

I then outline the provinces' proposal, during the so-called oil crisis of the 1970s, to build a national grid, as well as their attempts and failures to advance regional grids linking groups of provinces in the East, the Maritimes, and the West. Thereafter, I address the process of continentalizing electricity policy through the free trade agreements – agreements that led to the weakening of the NEB's export regulations, to the adoption of US regulatory approaches requiring the deintegration of Canada's largest public utilities, and to the opening of transmission systems to US power suppliers in several Canadian provinces as part of electricity trade reciprocity. To understand these outcomes and to provide a transition to the Niagara Falls, James Bay, Churchill Falls, and Nelson and Peace Rivers cases that follow, this chapter concludes by briefly examining provincial resistance to federal plans and the independent roles of the provinces in the planning of power systems.

National Power Grids

During the 1950s, 1960s, and 1970s, a number of federal and provincial politicians, civil servants, and utility executives, foreseeing economic advantages for Canadian society as a whole, advocated the integration of provincial electric systems into a national power network.[2] One of the first to make the case for a national power grid in Canada was David Cass-Beggs, a Fabian socialist, professional engineer, and former chairman of three western Canadian hydro utilities.[3] Ahead of its time, his vision went beyond the integration of western provincial networks. He 'is credited with being the first to propose and seriously analyze the benefits of a national power grid in Canada.'[4]

In his paper, 'Economic Feasibility of Trans-Canada Electrical Interconnection,' presented in 1959 at a conference of the Engineering Institute of Canada, Cass-Beggs proposed the installation of a 3,000-mile-long, coast-to-coast national power network to be completed by 1965.[5] He argued that 'large amounts of energy may be moved from the regions in which they naturally occur, or are not yet developed, to those areas increasingly in need of new resources.'[6] Thus, he suggested supplying central Canada with surplus electricity from other regions and building a national transmission line that would be cooperatively owned by provinces and utilities.[7]

Alternatively, he envisaged the transmission line as a 'common carrier that would transmit, for a fee, the energy that would be bought and sold in deals between the utilities.'[8]

Cass-Beggs's Canadian plans for a national power network were similar to those envisaged in the Soviet Union, Great Britain, and Germany.[9] Whereas Cass-Beggs stressed cooperative control for Canada's national transmission system, planners in Germany, for example, proposed that industrial development required centralized control over a unified system. Hughes writes that after the First World War, 'informed persons determined to maintain or establish the momentum of industrialization' in Germany saw a 'unified electricity supply as a national imperative.'[10] That country planned to nationalize its electrical system in 1919 by taking the following steps: (1) 'electricity would be fed into this all-German pool, or network, by power plants throughout Germany'; (2) the country 'would acquire centralized control over all major power plants'; and (3) with this centralized control, the country 'would expand and interconnect the individual plants and also existing or planned regional systems.'[11] Despite the economic benefits such a system would have provided, resistance to it on the part of certain political and economic interests soon surfaced. Later, the same sort of thing would occur in Canada.

In Canada, as in Germany, the scheme to interconnect growing regional systems appeared technically and economically sensible. Since the 1920s in Europe and since the 1960s in Canada, utility executives, network planners, analysts of the electricity industry, and government leaders similarly identified the economic, technical, and environmental benefits of an integrated power system.[12] These benefits were best summarized in 1994 by the Department of Natural Resources, which argued (albeit with reference to international trade) that 'interconnections improve the economics and security of electricity supply and they reduce the level of capacity needed to meet peak loads;'[13] furthermore, 'non-coincident peak loads [in five Canadian time zones, for instance] can allow utilities to share generation and realize economic benefits,' and 'agreements for mutual generation support [mean] that new power plant requirements are decreased.'[14] The department further indicated that 'interconnections also improve the flexibility of electricity supply, making it possible to minimize costs by replacing the highest-cost, and most polluting generation,' such as oil-fired generation or nuclear power, with hydroelectric energy from other areas.[15] Furthermore, 'where many communities [or provinces] can be connected together via an electric grid and facilities can be shared, the capacity reserve requirement can be reduced substantially.'[16]

In 1960, Cass-Beggs pointed out some other benefits specific to the Canadian situation. Failing to take into account regional resistance, he thought that it would be advantageous for Canada if electricity were to flow from east and west towards the centre of the country.[17] He proposed that Ontario first use all its hydro-generating capacity and some electricity from coal-fired (and stand-by thermal) plants but then import surplus electricity from Québec in the East and from Alberta, Saskatchewan, and Manitoba in the West.[18] The reduction in capacity – the provinces would likely construct ahead of need – would yield, he estimated between 1958 and 1965, 'investment and operating savings' of '$430 million,' which might be sufficient to 'meet the initial and annual costs of a coast to coast interconnection' of power plants.[19]

In Ontario, a precedent for the successful interconnection of power plants had already occurred during the early years of the twentieth century in the cooperative integration of Ontario's power system. It began with the gradual reversal of privatization of water-power rights on the Canadian side of Niagara Falls. Private US companies that had obtained these rights had failed by the beginning of the twentieth century to develop transmission lines into Ontario, preferring to tap into the more lucrative market in New York State. When Sir Adam Beck took over the leadership of the Ontario Power Commission in 1904,[20] he appointed one of his commissioners, E.W.B. Snider, to head a commission of inquiry into Ontario's power supply. Snider's report, filed in March 1906, recommended that small southern Ontario towns buy and distribute electricity at cost and build a cooperative transmission system linking the major towns with Niagara Falls plants.[21] In part, as a result of the acceptance of his recommendations, the new power lines eventually reached Berlin (now Kitchener), Ontario, on 11 October 1910.[22]

In this way, Ontario Power commissioners took the first steps towards the unification of much of the electricity network within Ontario. Later, following the experience of the First World War, Prime Minister Mackenzie King expressed the prevailing sentiment against exports to the US, in a period when southern Ontario developed its intraprovincial network while also exporting power. The resulting power shortages made it clear to utility executives, industrialists, and residential customers that, to use Sir Henry Drayton's resonant phrase, 'power exported is power lost.'[23] Arguing that power development would add momentum to industrialization, King insisted that 'power ... shall be utilized *within the Dominion* to stimulate Canadian industry and develop the natural resources' [my emphasis].[24] However, Armstrong informs us that in the 1920s and subsequent decades such public proclamations were also accompanied by both private developers' and governments' proposals to build plants for power export to the US.[25]

By the 1960s, advances in long-distance technology had brought within reach the possibility of the nation-wide interconnection of provincial power systems. Earlier, during the first decades of the twentieth century, transmission technology was capable of transporting electricity from Niagara Falls to southern Ontario communities and was later extended to supply most of the province. After the Second World War, various parts of the country, 'in response to rapid electrification and installation of larger hydro and thermal generating stations,' introduced much higher voltages of transmission lines into commercial operation.[26] For example, in 1965, Hydro-Québec installed the world's first 735-kV-class transmission system extending over 1,100 kilometres from the Churchill Falls generating station in Labrador to Montréal.[27] By the 1960s, Canada had become a world leader in developing techniques for reducing losses in transmission distribution.[28]

The installation of transmission lines across provincial or US state boundaries, like that of highways or railways, however, creates fundamental economic and political difficulties.[29] When applying this insight to national energy transport in Canada, Michel Duquette argues, if nationalizing policies in Canada go too far through 'forced centralization,' then they will be perceived by the periphery as internal colonialism.[30] Therefore, to 'escape' the perceived or real burden of federalism, provinces may become overwhelmingly committed to the logic of foreign markets in their industrial strategy.[31] Focusing on the decision makers involved, Hughes points out that *politicians* become involved when plants are interconnected across the boundaries of different political units (e.g., provinces, states, cities, and counties).[32] Furthermore, utility managers and their consultants become involved in the integration of different loads and plants of different characteristics (i.e., integration of industrial or residential demands in different time zones or the interconnection of hydro, coal, and nuclear plants) when large-scale electrical systems are planned. Although Canadian politicians and planners throughout the twentieth century have understood that a national scheme would require a form of cooperative or even centralized decision-making agency, they nevertheless hoped that, even without unanimity, the economic and technical rationale would prevail. As it turned out, the advocates of the national power grid underestimated the political difficulties involved in obtaining provincial consensus.

Given the technical and economic feasibility of a national power grid, the federal government took several steps to establish the necessary provincial cooperation. During the early 1960s, controversy raged in the House of Commons, where members of Parliament debated whether provinces would share some of their hydro-power resource advantages to establish a national power grid or whether some provinces would prefer developing only politicized provincial power grids. They also debated which river sites could be developed to supply a national grid. Over the years, the federal Cabinet and several federal departments – Northern Affairs and National

Resources; Trade and Commerce; and Energy, Mines and Resources (not to mention the NEB) – became involved in Canadian electricity policy.

The NEB, established under the National Energy Board Act in 1959, was given the responsibility to regulate the interprovincial transmission and international trade of energy, to hold public hearings, to decide on export applications (Table 2.1), to report to the federal Cabinet, and to advise the federal departments concerned with energy policy. I will first show how the Diefenbaker Cabinet took steps to advance the technical and economic advantages of a national system, and then I will show how the Pearson Cabinet lost control over this system. While allowing exports, both prime ministers attempted to get the premiers to support a national electricity policy in order to give momentum to Canadian industrial development. This policy turned out to be self-defeating because several provincial governments and some utilities decided to build plants designed, in part or entirely, to export power to the US. Building plants to supply electricity within an integrated national or regional network in Canada was not a planning priority. Unexpectedly, the desire to export low-cost Canadian electricity to US utilities and their industrial customers actually undermined the strategy of trying to attract US and other non-Canadian industrial firms to new supplies of electricity in Canada.

Federal Initiatives in the 1960s

In order to examine the national power network plan, certain questions need to be raised. What were the major steps taken towards the development of a national power network? What political difficulties arose? What help could the provinces be expected to give to the development of a

Table 2.1

Provincial and Canadian government functions in the development of electricity supply

Function	Provinces	Canada
Planning	Utilities	
	Governments	
Generation design	Utilities	
Transmission	Utilities	
	Transmission Administration	
Distribution	Utilities	
	Power Pool Council	
Export regulation		National Energy Board
Interprovincial trade	Utilities	National Energy Board
regulation	(unwritten rules)	(authority limited)

Source: Canada, Natural Resources Canada, Energy Resources Branch, Energy Sector, *Electric Power in Canada 1996*, Table 1.2, 'Electricity in Canada – Who Does What,' p. 8.

national power network? Would provinces cooperate in interconnecting their generating stations? How would Québec react to federal ideas about interconnecting transmission lines between neighbouring provinces? Would it be possible to develop a national network while also allowing exports from large power projects? What led to the demise of the national power network plans?

The Diefenbaker Cabinet and the National Power Grid

In the early 1960s, Cabinet ministers in John Diefenbaker's Conservative government discussed the concept of a national power grid, invited all premiers to participate, set up a working committee (including most of Canada's electricity elite), and engaged consultants to report on the technical and economic merit of long-distance transmission and to assess the benefits of a national grid as opposed to individual provincial hydro networks.

Walter Dinsdale, Diefenbaker's minister of northern affairs and national resources, was the major advocate of the national power network. He met with David Cass-Beggs, then the president of the Canadian Electrical Association, on 1 December 1960 and informed him that the Conservative government had established a committee two years earlier 'to consider some aspects of long-distance transmission in Canada.'[33] On 17 January 1961, he informed Cass-Beggs in writing that 'Newfoundland, Quebec, Manitoba, British Columbia and the Yukon have large but remote reserves of water power which in all probability will more than satisfy the local industrial development needs. Large blocks of power are therefore available at these sites for transmission to the more industrialized areas of Canada.'[34]

Dinsdale circulated the 'Long Distance Power Transmission' memorandum dated 6 December 1961 among his Cabinet colleagues, and they debated it in Cabinet on 19 December 1961. It proposed that the Federal government take the initiative, through discussions and investigations, in the planning of the electric power development and transmission on a national or regional basis.[35] As his memorandum states, such a plan had not been tried before:

> The federal government has not, thus far, taken an active role in planning for or assisting with the development and use of electric energy on a national basis. Planning of electric power production has been almost entirely provincial, and provinces have tended not to look beyond their own borders. From discussions and investigations undertaken on a preliminary basis, it seems clear that the planning of electrical power development with transmission on a regional or national basis would make possible substantial economies and would permit the early development of major hydro-electric sources in remote areas.[36]

To illustrate the technical and economic merit of his national plan, Dinsdale provided an example of the interprovincial sharing of generation and transmission facilities: 'A small-scale example of this approach is the Atlantic Provinces Power Development Act of 1958. Up to March 1961, the recoverable cost of transmission facilities for Nova Scotia and New Brunswick was about $10.7 million.'[37] He tried to convince his colleagues that 'similar advantages could accrue to other regions of Canada if long-distance transmission facilities could make possible the coordination, first on a regional and later on a national basis, of Canada's reserves of hydroelectric energy and power from thermal sources.'[38] He informed his fellow ministers that 'recent indications are that some provinces might favour a meeting of responsible provincial ministers and power experts to discuss long-distance transmission problems with federal ministers, if the federal government were to take the initiative.'[39] Among other items, he recommended in early 1962 that the government call a meeting of provincial ministers to discuss the long-distance transmission problem, to send an invitation to the premiers, and to employ a suitable technical consultant. Although the technical problems of alternating versus direct current long-distance electricity transport appeared surmountable and the economic benefits promising, political difficulties soon emerged at the provincial level.

Dinsdale had to tell his Cabinet colleagues on 19 December 1961 that he had had an opportunity to talk to Manitoba and some other provinces and found that 'little help could be expected from certain quarters.'[40] Therefore, he suggested it would 'not be a good time to call a meeting'; instead, he thought it better first to hire an engineering consultant to study the economics of the proposal. In response, his Cabinet colleagues raised several points. Some said it would be better to meet with provincial representatives before hiring the consultant; 'otherwise the provinces might resent the move as a form of pressure.' Others thought that the whole idea might be scuttled since 'there were signs that Alberta, Quebec, and British Columbia would not cooperate.' Another colleague at the same Cabinet meeting agreed, reasoning 'that Canada was being more and more fragmented and that something of a national character such as this plan was needed to offset this tendency.' The Cabinet decided to call a meeting of provincial ministers early in 1962 to discuss the problem of long-distance transmission of electrical power in Canada; the premiers would be sent letters of invitation, the hiring of consultants would be postponed until after the meeting, and the proposal would be announced in the Speech from the Throne.

Prime Minister Diefenbaker included the national electricity grid in his 'national development policy.' In his Speech from the Throne on 18 January 1962, the Honourable Roland Michener, governor general of Canada, revealed the key elements of Diefenbaker's national development policy:

'The provision of low cost electric power is one of the most important factors in the economic growth and industrial development of Canada. As a further step in the national development policy, my Prime Minister has invited provincial governments to join with the federal government in early discussions of the steps that might be taken toward the establishment of long-distance power transmission to link provinces and eventually the different regions of Canada.'[41] Diefenbaker stressed in his letter of invitation to the premiers 'that interprovincial connection and arrangements would make possible substantial economies in reserves and in capital requirements.'[42] He tried to convince them that provinces would save large sums of capital because in an interconnected system their capacity reserves (and therefore the number of plants they needed) would be reduced and that new supplies of hydro power distributed over a national network for industrial expansion could provide Canada with an advantage over other countries.

Additional political difficulties in launching the national power network, however, soon emerged. Prime Minister Diefenbaker informed his Cabinet on 16 February 1962 that 'the Premier of Quebec [Jean Lesage] had stated that he would not attend such a meeting' and that 'he had not received a reply from [Joseph Smallwood] the Premier of Newfoundland' on his 'proposal that federal-provincial discussions should be held on steps that might be taken toward the establishment of a national power grid.'[43] In his letter of 24 January 1962 Lesage took the position that 'cette conférence aurait pour object une discussion sur le transport à longue distance de l'énergie électrique entre les provinces. Nous considérons cette question de juridiction provinciale' [the objective of this conference would be a discussion about long distance transportation of electric power between the provinces. We believe this issue to be under provincial jurisdiction].[44] He added in the same letter that he would preside over next summer's premiers conference in Victoria and that transportation of electrical energy was on its agenda. Diefenbaker did not easily give up. He tried to convince Lesage to attend the Federal-Provincial Conference on Long-Distance Transmission of Electricity in Ottawa on 19 March 1962 and wrote further letters on 3 and 22 February 1962, respectively, stating, in one letter: 'les projets d'aménagement interprovinciaux intéressent et préoccupent le Gouvernement fédéral' [inter-provincial development projects are of interest and concern to the federal government].[45] To leave no doubt about Québec's position, Lesage sent off another reply to Diefenbaker, on 2 March 1962, two weeks before the conference, in which he minced no words: 'La province de Québec, tout en étant déterminée à utiliser ses richesses naturelles pour favoriser son développement économique, est bien disposée à faire avec ses provinces soeurs des arrangements d'interêt mutuel mais elle n'entend pas accepter de le faire sous la tutelle du gouvernement fédéral.' [The province of Québec, though determined to use its natural resources to favour its own economic development, welcomes making arrangements of mutual interest with its sister provinces but it does not intend to agree to do so under the tutelage of the federal government.][46] The Cabinet, in an earlier meeting, had decided that the Ottawa meeting would go ahead even without unanimity.

On 15 March 1962, four days before the federal-provincial meeting, Diefenbaker proposed that it consider three principal questions: '(a) the advantages of regional and inter-regional (and in certain circumstances international) links, (b) the main problems involved in long-distance power transmission, and (c) the roles which the Federal and provincial governments might play in planning, research and development in this field.'[47] The Cabinet approved Dinsdale's revised memorandum as a general guide for the conference and recommended that the minister of trade and commerce, representatives from mines and technical surveys, northern affairs and national resources, and other designated officials also be present.[48] Dinsdale also anticipated provincial concerns with surplus power and its export: 'Provincial representatives will almost certainly raise the question of the Federal government's policy on the export of electrical energy. Premier Bennett [of BC] will doubtless raise this issue in relation to the Columbia and Peace Rivers, and Premier Roblin [of Manitoba] in relation to the Nelson. Indeed the development of some of the major hydro power sources may not be possible unless a portion of the added power is exported while surplus.'[49] Prime Minister Diefenbaker, as reported later by the NEB, in his opening remarks at the 19 March 1962 federal-provincial conference explained why he thought the initiative for a national power system was necessary: 'To begin with, he asserted that the long-distance transmission of electrical energy would yield great benefits to the participating provinces. Second, he saw a national electricity grid as the equivalent of the transcontinental railways, the trans-Canada highway, the nation-wide civil aviation system, and the cross-Canada radio and television networks. All of these, he said, 'are links helping bind the country together.'[50] Cabinet records indicate that at this Federal-Provincial Conference in Ottawa several steps were taken to cooperate in the examination of technical, economic, and policy problems pertaining to the national power grid.[51] The meeting decided: (1) to set up a joint working committee to ascertain the nature of the technical and economic studies required; (2) that 'the federal government should pay the total cost of the consultants' fees' because the provinces felt 'they would be put to considerable expense in providing data for the consulting engineers'; (3) that two well-known engineering firms, Montreal Engineering Company and H.G. Acres and Company, be employed jointly for the initial study (to enunciate the problems, propose a method of solution, and draw up terms of reference);

and (4) that a second, more detailed study should follow, indicating a time schedule and estimated costs. As will be shown below, the engineering studies indicated that the economic benefits of interprovincial coordination with regard to electricity supply and of a national power grid (to be completed by 1975) far outweighed those of the individual provincial systems.

Jean Lesage did not attend the meeting in Ottawa, but he chaired the Third Provincial Premiers Conference in Victoria on 6-7 August 1962 the following summer. At this conference, Duff Roblin, the premier of Manitoba, expressed his province's strong support for the national power grid: 'We are definitely interested in the concept of a national grid. We do not visualize it springing into existence from Atlantic to Pacific overnight. We think the first step is to develop and maintain one major link, and, if it works to the satisfaction of all concerned, we could develop it across the country.'[52] Referring to the Ottawa meeting and to the steps taken there concerning how northern hydro projects might be integrated into the national grid, Lesage had the following exchange with Premier W.A.C. Bennett (in the absence of Joseph Smallwood, the premier of Newfoundland):

> Mr. Lesage: Their [the Diefenbaker government's] view is not very clear. I was not very clear. I was not at the conference, of course.
>
> Mr. Bennett [premier of BC]: To be fair, they were seeking advice. They wanted to know about our experience. ...
>
> Mr. Lesage: I was thinking of the development of Hamilton Falls [later Churchill Falls]. I have discussed this with Mr. Smallwood a number of times. He would like the Québec Hydro to go in fifty per cent in the cost of the development of Hamilton Falls. The horsepower we could derive is very interesting, not only for Québec but possibly, also, for Ontario and the Maritimes.[53]

This comment by Lesage, made in the context of the national power grid discussion, is in sharp contrast to Smallwood's and Lesage's actions three years later, as will be shown below. At the conference, the premiers also discussed several other issues related to the national power policy: interprovincial coordination, ownership of the national grid (provincial or federal property), the Ontario electricity market, and nuclear power.[54] Manitoba received extra financial and technical help from the federal government for its transmission line from the Nelson River projects. Premier Roblin stated that his province could benefit from the federal government's national grid development without the federal government necessarily owning or managing it.[55] British Columbia's premier, Bennett, stressed that 'we would not want the federal government to own the national grid within the province,' and Lesage added, on behalf of Québec, that 'we are not thinking in terms of federal control.'[56] When Allan Blakeney of Saskatchewan raised the problem of the development of large hydro projects without the assurance of having Ontario as a base market, Premier John Robarts of Ontario showed little interest and indicated 'that Ontario is investigating nuclear power, and that there are ample resources of uranium in the province.'[57] Ontario opted to take the nuclear path, and Québec anchored the economic program of its Quiet Revolution to the nationalization of privatized hydro power.

According to historian René Durocher, the most spectacular economic accomplishment of Québec's Quiet Revolution[58] was the nationalization of electricity companies and their integration within Hydro-Québec (Commission Hydro-Electrique du Québec).[59] National coordination over hydroelectric resource development would have necessitated a loss of autonomy that was antithetical to the nationalism that permeated Québec policies and reforms. Electricity, in fact, became the symbol for economic liberation of the people of Québec, as the headline on the front page of *La Presse* on 1 October 1962 shows: 'L'électricité: clef qui nous rendra maîtres chez nous.' [Electricity: the key that will make us masters in our own house.][60]

The Pearson Cabinet and the Avoidance of National Power

The following spring, on 22 April 1963, Lester B. Pearson and his Liberal government took power from the Diefenbaker Conservatives, and Mitchell Sharp, the new minister of trade and commerce, took over the initiatives dealing with the national power grid. Soon, interest in long-term exports outweighed initiatives to formulate a strong national power policy. On 26 September 1963, the Cabinet approved Sharp's memorandum on how to state publicly the Pearson government's power policy.

In effect, the government would announce its readiness to approve the export of large blocks of electric power for periods of up to twenty-five years in cases where the NEB so recommended. Provisions would have to be made for the recapture of such exported power over a period of years, in stages commensurate with the need or ability of the Canadian market to absorb it, and upon giving the importing entity enough notice to arrange for replacements. Failure to adopt such a policy would be likely to frustrate such major power developments as that at Hamilton Falls (later Churchill Falls). The NEB estimated that, under such a policy, power exports might run to $70 million annually.[61]

To make this announcement more acceptable to members of the public who did not favour exports of a national resource such as electricity to the US, Pearson's minister of 'NA&NR [Northern Affairs and National Resources] said that the statement proposed by Mr. Sharp in his memorandum would be improved if the savings to Canadian consumers were given more emphasis and if it were made clear that the exports of power to foreign buyers would not be contemplated in terms more favourable than those available to comparable Canadian purchasers.'[62]

Sharp's two-directional policy amendment allowed continentalization initiatives to become the dominant policy. Sharp indicated that the federal government would now pursue two directions in distributing the benefits from Canada's large, low-cost power sources: one required the interconnection of Canadian power systems nationally, the other required their connection continentally. In his power network policy statement to the House of Commons on 8 October 1963, he stated: 'The government has decided to develop and carry forward effective policies embracing two essential concepts; first, encouraging development of large low cost power sources and the distribution of the benefits thereof as widely as possible through interconnection between power systems within Canada; second, encouraging power exports and interconnections between Canadian and United States power systems where such induce early development of Canadian power resources.'[63] Sharp's national power policy promoted the accelerated construction of the largest plants in Canada's history and increased integration of the new northern power with US networks. In this way, the continentalization of energy was allowed to advance. The demise of the national plan resulted not only from the continental bias of the Liberal government, but also from the historic constitutional division of political powers and from a lack of consensus among the provinces.

Once Sharp had given the signal that his government encouraged 'interconnection with the United States' and the 'early development' of power projects, the Pearson government appeared to lose the ability to coordinate the hydroelectric development schemes of various provinces and private developers in the country. In the spring of 1964, six months after he revealed his two-directional, continental-national power policy in the House, Sharp brought disturbing news to the Cabinet. He told his colleagues that Québec had a measure before its legislature 'which would impose provincial controls on the export of electric power from the province,' and he maintained that the Hamilton River (Churchill Falls) development 'should not be frustrated by provincial restrictions on the transmission of power.'[64] Not only would such legislation put a damper on federal export jurisdictions, but it would also set a problematic precedent for any trans-provincial transport of electricity. During the summer of 1964, any semblance of national power policy had evaporated: provinces were simply off in all directions. Sharp informed Prime Minister Pearson and his ministerial colleagues of a disturbing trend in provincial approaches to power development: 'He had seen assessments that Canada would need within the next few years an increase in energy four times the capacity expected to be produced by the Hamilton Falls development [Churchill Falls's capacity was expected to be 5,000 MW, Canada's entire capacity in 1960 was 23,035 MW]. The various provinces and private interests were proceeding in a completely unco-ordinated manner in developing energy facilities.'[65] He was concerned not only about the acceleration of the building of hydro plants in the country, but also about the fact that Joseph Smallwood appeared to be unaware that a federal government export permit was needed to export power from Churchill Falls. He noted that 'before giving such approval the NEB would need to take account of the requirements of the Maritime Provinces, Ontario, Québec and possibly other Canadian provinces.'[66] Nevertheless, he left further preparations for a national policy to the export-oriented NEB and informed the Cabinet that 'the NEB was now gathering information on this problem which it would study and analyze with a view to the elaboration of a national policy.'[67] Prime Minister Lester Pearson emphasized that it was important that the Canadian government 'give guidance and provide coordination to the various parties concerned within the framework of a well conceived long-term policy.'[68]

As it turned out, two years after the election of the Pearson government, coordination on power policy was in decline at the provincial level, and Pearson's minister of trade and commerce discovered that the responsibilities for the power question were divided among federal departments and that provinces had given information to one department on the condition that it not be shown to others. Sharp found on 25 March 1965 that the coordination of various aspects of electricity policy was difficult: 'The NEB had broad terms of reference under its Act but other agencies also had responsibilities that were difficult to reconcile with the responsibilities that the NEB Act implied.'[69] Discussion in Cabinet pointed to the fact that the NEB was not established simply as a judicial board and regulatory body to license exports, but that 'it had other responsibilities [e.g., as stated above, to give national policy advice] by Act of Parliament.' Furthermore, it was pointed out that 'provincial authorities had agreed to provide information to the department [of Northern Affairs and National Resources] on the basis that it would not be given to the NEB.' Cabinet reached consensus that 'overall the coordination was essential for the determination of a national power policy.'[70] The issue of coordinating power policy was passed on to the Cabinet Committee on Trade and Resources.

Pearson's power diplomacy took a disastrous turn at the interprovincial level when he tried to mediate between Premier Smallwood and Premier Lesage. Pearson told his Cabinet colleagues on 6 May 1965 that 'Mr. Smallwood still hoped it would be possible to arrange for the transmission of Hamilton Falls power through Quebec on an acceptable basis.'[71] Pearson had discussed Hamilton Falls with Lesage on 3 May 1965 and with Smallwood the next day. Smallwood told the press that Brinco (British Newfoundland Corporation)[72] would not sell Hamilton Falls power to Québec

– a statement that enraged Premier Lesage, who assumed that the prime minister had something to do with it. Lesage responded angrily to the press that day, stating that Québec would buy all of the electricity from Churchill Falls or none all and would not allow others to build a transmission line across its territory or to transport this electricity to other provinces or the United States. *Le Devoir* quoted him on its front page on 5 May 1965:

> La condition première et absolue d'une entente entre Brinco et l'Hydro-Québec sur l'exploitation des chutes Churchill (anciennement Hamilton) est que toute l'énergie qui rentrera dans le Québec devienne la propriété de l'Hydro-Québec.
>
> Cette condition ... a toujours été la même et nous négocierons jamais sur une autre base. Jamais nous ne permettrons, à quelque condition que ce soit, de laisser construire par d'autres une line de transmission sur le territoire du Québec ou de laisser transporter par d'autres l'énergie produite aux chutes Churchill quelle que soit la destination de cette énergie, les Etats-Unis ou les autres provinces.
>
> [In order to reach an agreement between Brinco and Hydro-Québec on the exploitation of Churchill Falls (formerly Hamilton Falls) the primary and absolute condition is that all energy that will enter Québec becomes property of Hydro-Québec.
>
> That condition ... has always been the same, and we will never negotiate from another base. We will never permit, under any condition, others to build a transmission line on Québec territory, or let others transport the energy produced at Churchill Falls whatever the destination of that energy, whether it be the United States or the other provinces.]

This position made the development of a national power grid next to impossible. The following day in Ottawa, the Cabinet made the point that in 'light of recent events, it was important that negotiations be carried on directly between the two power corporations concerned [Brinco and Hydro-Québec], without interference from the provincial governments.'[73] These direct negotiations eventually led to the Churchill Falls power contract in 1969.[74]

By the time the engineering consultants presented their carefully prepared report on the technical and economic advantages of a national power system, the Liberal government's power policy had all but disintegrated. In 1964, the Federal-Provincial Working Committee on Long Distance Transmission included representatives from Canada's electric energy authorities, who represented both provincial and federal interests.[75] In addition, the Committee on Long Distance Transmission consisted of senior engineers and planners from two federal energy offices, and nearly all provincial hydro utilities participated in the studies of the future national energy grid.[76] Hydro-Québec was the only utility in Canada whose officials did not take part in the Federal-Provincial Working Committee on Long Distance Transmission.[77] The utility's absence, combined with Lesage's position on the transportation of electricity across Québec, contributed to the problems of developing a national power grid.

Nevertheless, at the same time, even though Diefenbaker's Conservative government had lost to Lester Pearson's Liberal Party in April 1963, the committee, which had been established by the former, continued to plan Stage I of the national power grid with H.G. Acres and Company. And in 1964 it went on to plan Stage II with Ingledow and Associates – a firm of engineering consultants based in Vancouver and hired to report on the technical and economic feasibility of such a venture. As the NEB reported later, 'Stage I was carried out by H.G. Acres and Company. It examined the merits of a coast-to-coast high voltage transmission line, which would proceed along a generally southern route from Halifax to Vancouver, in close proximity to Canada's major load centres. Variants of this main route indicated how the hydroelectric resources in Labrador and the Peace River could be brought within the scope of the national plan.'[78]

Based upon Acres's conclusions that 'such a project would indeed facilitate the development of Canada's remote hydroelectric resources, and that it would yield substantial benefits to the national and provincial economies,' the federal and provincial ministers accepted the recommendation that the assessment move to the second stage.[79] Similarly, the Ingledow plan called for the development of hydroelectric sites (discussed extensively in chapters below) in Québec, British Columbia, and Manitoba and the deferral of construction of coal-fired sites generating stations on the Prairies and in the Maritimes.[80]

In Stage II of the national power grid study, Ingledow and Associates conducted a comparative analysis of the benefits of two development plans. One plan described the development of hydro systems according to individual provincial interests, and the other proposed a provincially interconnected national power network from Vancouver, British Columbia, to Corner Brook, Newfoundland (see Figure 1.1).[81] In reports presented in 1966 and 1967, respectively, Ingledow advised the working committee that a national power network was not only possible, but also technically more efficient and economically more advantageous for Canada as a whole than were separate provincial systems. The Ingledow group concluded that 'the greatest economy is to be derived through the interconnection and coordination of all systems on a national basis.'[82] The report estimated that a national power network could save $644 million within a thirty-year period (1970-99) if implemented by the mid-1970s.[83] The study also pointed to the need for some form of central authority to coordinate the planning, development, and operations of the national network.[84] Instead, in what was to become a pattern in the next decade as well, after a variety of engineering firms with extensive experience in hydroelectric development found the national power network to be technically feasible and economically beneficial, political decisions undermined the considerable efforts to establish such a network.

Despite the recommendations of engineers to proceed with a national network, the Federal-Provincial Working Committee decided against a major new national grid initiative, favouring regional ties rather than a trans-Canada electrical network.[85] The committee gave several reasons for this: (1) much uncertainty was associated with the assumptions upon which this study was based; (2) the projects proposed were capital-intensive and financially risky; and (3) the prospect of major new north-south interties (power lines) between Canada and the United States, rather than the east-west interties between the provinces, appeared to be a more attractive option to some Canadian utilities.[86] Moreover, by the time the report was finished and the committee had stated its preference for regional power grids, two major hydroelectric projects at Nelson River (Manitoba) and Churchill River (Labrador) had already moved ahead as independent initiatives and were, in part, to be developed for export to the US.

The possibility of realizing the technical and economic advantages of a national system in the 1960s appears to have been totally undermined both by federal and provincial politics as well as by private and public utility interests. Provinces followed separate development strategies, and this militated against the formation of a national power grid: Ontario pursued the nuclear power option, British Columbia expected its Columbia River benefits to finance its Peace River development[87] and dreamed of exports to California from the latter, Québec planned to annex Churchill Falls and block transmission access across its territory, and Manitoba hoped to export power to the US from the Nelson River plants over federally financed power lines. Each province appeared to follow Sharp's policy of the accelerated pre-building of projects for 'temporary' exports and to believe that new supplies of electricity would strengthen and diversify their weak industrial bases. Rather than developing a coordinated national power network, the provinces (though with some export arrangements with neighbouring Canadian utilities) opted for separate provincial network development so as to suit their local industrial transformation and export strategies.

Provincial Initiatives in the 1970s
In the 1970s, when an attempt to integrate provincial power systems again became politically important because of expected oil shortages, the provinces, rather than the federal government, took the initiative to develop

a national electricity network. In 1972, after two major engineering studies had been supportive of a national power supply, David Cass-Beggs, then chair of BC Hydro and a critic of the reduced benefits that 'isolated electrical regions with token ties between them' would bring, supported relaunching the effort to build a 'true national grid' – an idea soon supported in March 1973 by yet another engineering firm, Leighton and Kidd, in its brief to the federal government. In February 1974, Energy Minister Donald Macdonald (who, as head of the Macdonald Commission in the 1970s, provided some of the concepts for an economic union with the US) suggested that 'a national power grid is now much closer to realization in Canada because of the impetus of the energy crisis.'[88] The federal government of Pierre Elliott Trudeau, however, appeared lukewarm towards a national power network plan. The federal government still held the position that 'any national grid initiative would have to come from the provinces.'[89] Its support was confined to sharing 50 percent of the cost of interconnection studies; and in cases where provincial or utility resources were inadequate, it provided loans of up to 50 percent of the costs of regional interties – meaning, in this instance, both inter- and intraprovincial transmission.

On 23 and 24 January 1974, at the First Ministers Conference on Energy (a response to the oil crisis) and in later follow-up correspondence, all provinces supported the renewed initiative to develop a national grid.[90] In the summer of 1974, Premier David Barrett of British Columbia proposed 'a linking of Canada's eastern and western transmission systems by a 750 km line from the Nelson River to James Bay in Quebec' and suggested that a similar link between the Peace and Nelson Rivers might be feasible.[91] Québec's minister of natural resources, however, doubted that a national grid would be economically feasible and technically reliable, and he argued that 'interprovincial and international interconnections' with neighbouring provinces and the US 'had already been shown to yield significant economic advantages.'[92]

On 12 April 1976, an advisory body composed of provincial government officials, the Interprovincial Advisory Council on Energy (IPACE), initiated a new study of a national power grid; in this case, the federal government was granted only 'observer's status.'[93] This group proposed a massive new east-west transmission system extending from Selkirk, British Columbia, to Eel River, New Brunswick, but 'did not propose to strengthen transmission lines to the United States.'[94] The economic benefits of the system proposed in the council's 1978 study, although sensitive to future fuel prices, were estimated to derive from replacing oil-fired generation with less costly hydro generation, the reduction of generating reserve, the diverse use of generating equipment across different Canadian time zones, the enhanced security of supply, and a flexible response to changing energy policy. The council's proposed conceptual plan also raised questions as to whether enhanced north-south electricity trade would be better for Canada than east-west trade, whether there was a need for an organization to administer a 'system of strengthened interprovincial connections,' and – an issue that was at the centre of the Victoria Conference sixteen years earlier – 'which level of government would have jurisdiction over a national grid.'[95]

Because Canada's Constitution grants the provinces the jurisdiction to plan, construct, and operate hydroelectric systems, some provinces were concerned whether an interprovincial electricity network would be safe from federal government interference. They consulted Professor Leo Barry of the Dalhousie Law School, who argued in 1978 that there 'is no way in which the provincial governments can obtain absolute assurance that the federal government will never become involved in the development and administration of an interprovincial electricity network.'[96] The proprietary question was also raised in the Network Study Group's report, which wondered whether provinces should own only those interprovincial transmission lines that were needed for intraprovincial development. The report proposed a voluntary organization to promote interconnection and pooling (sharing of electrical facilities) with the idea that this would eventually evolve into a more structured form as interconnections were strengthened. In a weakened approach to interprovincial coordination, the IPACE Council would merely review provincial plans of interconnection and pool agreements to discern economic benefits; each province, however, would continue to plan its own system.

Like engineering reports and previous studies based on information from utilities across the country, the IPACE study concluded that arrangements for a national network of 'strengthened interprovincial interconnections can be made provided that all parties have a desire to find a mutually satisfactory solution and are willing to extend their cooperation to that end.'[97] On 12 December 1978, in an announcement that was similar to the 1967 avoidance statement by the Federal-Provincial Working Committee with regard to the first national grid proposal, the Interprovincial Advisory Council on Energy drew back from endorsing the second set of national power network plans by announcing that the belief in net benefit of strengthened interconnections was premature, that further studies were under way in the western and Atlantic provinces, and that the council would follow these developments.[98]

Here we see a pattern emerging. First, carefully collected information is evaluated by experts in reports that proclaim the technical and economic advantage of a national power grid; then these reports are followed by avoiding proceeding with the grid's installation. Part of this pattern is due to the division of political powers over electricity development in the Constitution (as will be discussed below), and part is due to the unwillingness of provinces to delegate at least some authority to an extra-provincial body. Other issues would need to be resolved by an extra-provincial electricity agency in order for a national grid to become a reality: provincial veto rights over national grid projects, unresolved jurisdictional ambiguity as to interprovincial trade responsibility, the proposed sharing of costs and benefits by provinces, the coordination of international trade, and the type and location of new generating facilities.[99] Twice the federal government has supported the initiation of a national power system yet has postponed the development of a national grid. Subsequently, three initiatives to develop regional grids were undertaken in the 1970s and 1980s: an eastern grid, a Maritime grid, and a western grid. Have these initiatives succeeded in developing regional electricity grids?

Regional Grids

Steps towards the development of the eastern grid were taken in September 1976, with the decision by Québec and the Atlantic provinces to study the idea of a regional grid. One of the short-term options in the 1977 to 1985 period was to route Hydro-Québec's considerable surpluses through the existing transmission system and to replace oil-fired generating stations in Atlantic Canada.[100] As a more long-term option, Newfoundland proposed to develop its Gull Island site (located downstream from the Churchill Falls hydro plant on the Churchill River in Labrador) and to supply the Atlantic provinces and the national power grid with electricity. The March 1976 report of the Committee on Interconnection between Québec and the Atlantic Provinces found that there would be economic benefits to this arrangement because of its lower reserve requirements (together with improved stability and reliability) as well as environmental benefits due to the reduction of air pollution through the phasing out of thermal plants. Still, despite the fact that the study pointed to some obvious benefits available at little cost to the participating provinces, no sales of the type suggested were ever made by Québec to the Maritime provinces. Interests had shifted. Québec became more interested in selling its surplus power to the US, and Newfoundland (the strongest advocate for a national power grid) attempted to reopen the Churchill Falls contract with Hydro-Québec, bringing to an end initiatives for this regional arrangement.

From 1976 to 1979, three Maritime provinces and the federal government, taking steps towards the development of another regional grid, held discussions to form the second regional interprovincial integration initiative aimed at the formation of the Maritime Energy Corporation (MEC).[101] This corporation 'was to be a regional generation and transmission utility, jointly owned by New Brunswick, Nova Scotia, Prince Edward Island and the federal government.'[102] Again differences emerged among the federal government, the provinces, and the utilities, and this regional cooperation initiative failed.[103] The NEB wrote:

> The failure of the MEC negotiations points out that for the success of a regional cooperative initiative, political consensus is as important as the appeal of economic benefits and approval of the industry. From early on the MEC negotiators failed to ensure that the provincial governments were informed and supportive of what they were doing. Too often the thinking of federal and utility officials outpaced that of the provincial governments; the federal and utility negotiators would come to an agreement on an issue, only to have negotiations stall when the provincial governments announced that it wasn't quite what they had in mind. There were even significant differences of opinion between provincial governments and provincial utilities, even *provincially-owned* utilities, which developed from time to time. It is not surprising that a regional consensus was not reached when often there was no consensus within a province.[104]

In April 1978, at the Western Premiers Conference in Yorkton, Saskatchewan, plans for a western power grid – connecting the power systems of British Columbia, Alberta, Saskatchewan, and Manitoba – were announced; this grid was expected to have significant economic advantages.[105] The technical review team established to study this grid concluded in February 1979 'that substantial benefits would flow from the interconnection of four provinces.'[106]

As with other regional studies, divergent political and private interests could not agree on the benefits of this western grid. Utilities were sceptical and British Columbia – now under the Social Credit government of William (Bill) Bennett – argued on 9 August 1979 that the report 'presented insufficient grounds to participate.'[107] After a second report – a review of the first western grid report commissioned by Alberta – confirmed that the regional interconnection of electrical systems was a good idea, Alberta, Saskatchewan, and Manitoba signed the Western Electric Power Grid Study Agreement on 22 March 1980. Based on technical information supplied by provincial utilities, the Western Electric Power Grid Study estimated that the regional grid system would yield a net benefit of $150 million.[108] Furthermore, the study found that the construction of the Limestone hydro project in Manitoba would have less harmful effects on the western environment than would the proposed coal-fired thermal plants in Alberta and Saskatchewan. While building the hydro plant in Manitoba would increase investment and employment in that province, in the other two provinces it would reduce investment and lessen employment in this economic sector and reduce investment, and most of the electricity sold would flow to Alberta.[109]

During the negotiations of the next steps towards the realization of the western grid, difficulties emerged: while Manitoba and Alberta announced on 22 October 1981 that provincial energy ministers had reached an accord, Saskatchewan claimed they had not.[110] While attempting to finalize the wording of the agreement, 'each party wanted the others to bear more risk in case of net loss.'[111] Negotiations stalled. Alberta and Saskatchewan found the prospect of lower employment and deferred investment at a time of economic downturn unattractive.[112] The western power grid was postponed. Subsequently, Alberta became interested in developing Slave River hydro, and Manitoba signed an export contract with the Northern States Power Corporation of Minneapolis in 1984, with electricity to be generated at its Limestone project on the Nelson River.[113] Thus, the western portion of an eventual national power grid did not proceed. In the 1990s, in the absence of a Canadian consensus and with the free trade agreements in effect, US electricity trade policy has started to have its effect.

When taking steps towards federally initiated or provincially initiated plans to build a national power grid, planners and politicians failed; similar attempts by more than two provinces to form western, eastern, or Maritime regional power grids were equally unsuccessful. Only when two provinces found electricity trade profitable (not always on equal terms, as the Churchill Falls case shows), did they establish provincial interconnections. These interconnections between provinces, however, remain weaker than those largely dedicated to export between provinces and US states. In 1996, for instance, '37 major provincial inter-connections' were capable of transferring 'about 10,245 MW' (with the Québec-Labrador interconnection accounting for about half of this capacity), whereas over 100 international lines, of which 37 are bulk power interties, had a total transfer capability of 18,900 MW.[114] An examination of the east-west trade of electricity in Canada reveals that provinces and utilities were 'willing to participate in profitable interprovincial cooperative projects,' but interprovincial agreement was difficult to achieve (1) when the sharing of risks and benefits was not clear; (2) when more than two provinces were involved in negotiations; (3) when new decision-making institutions (other than the provinces) needed to be created; and (4) when wheeling services (transporting electricity for a fee) were required.[115]

Which level of government, then, should legislate policies regarding the trans-provincial transportation of electricity in the 1990s? The NEB's 'Review of Inter-Utility Trade in Electricity' reported in 1994 that, according to Hydro-Québec, 'unless the federal government made use of the exceptional powers conferred upon it by *Section 91* of the *Constitution Act*, it would be unlawful to compel provinces to sacrifice their individual interests to a higher Canadian interest.'[116] Based on information gathered in its electricity review survey of provincial governments' views about electricity trade in the early 1990s, the NEB conjectured that 'B.C., Manitoba, Saskatchewan, Ontario, Québec and New Brunswick would oppose any formal involvement by the federal government in this area [of wheeling, e.g., trans-provincial transport of electricity for a fee]; Nova Scotia, P.E.I. [Prince Edward Island] and Newfoundland would support federal involvement; and Alberta's position is unclear.'[117] The Atlantic provinces, which have no direct access to Ontario or US markets via Québec, support formal involvement by the federal government, while provinces west of Québec do not (although some provinces may involve the federal government in order to settle disputes).

For a number of years in the early 1980s, the Trudeau government's National Energy Policy (NEP) (mostly concerned with the oil and gas sector) slowed the drift into energy continentalism. However, soon domestic political forces accelerated integration of the electricity sector with that of the US.[118] For instance, in 1984, the Conservative government of Prime Minister Brian Mulroney dismantled Trudeau's NEP and permitted the transformation of energy into a continental commodity.[119] In 1985 in Québec City, Mulroney and President Reagan decided to open the US-Canada energy markets to cross-border trade.[120] In this way, Canada increasingly came under the influence of US electricity policy. Since more than 90 percent of the electricity generated in Canada between 1985 and 1996 was used by industrial, domestic, and commercial customers in Canada, and less than 10 percent was exported to the US (fluctuating from a high of 9.4 percent in 1987 to a low of 3.5 percent in 1990 and rebounding to 7.7 percent in 1996), the emphasis on continentalizing the Canadian electricity market is disproportionate.[121] Thus the next move – including electricity in a free trade agreement with a foreign country (whose customers buy only a small portion of the electricity generated in Canada) before developing an interprovincial electricity trade policy – demonstrates the weakness of national forces and the strength of continental forces in Canada.[122]

In 1988, with the Canada-United States Free Trade Agreement (FTA) about to take effect at the beginning of the new year, the problem of finding a strategy for Canada's interprovincial electricity trade took on a renewed urgency. Because the NEB has some responsibility for interprovincial trade and for advising the minister of the Department of Natural Resources Canada (formerly the Department of Energy, Mines and Resources) on electricity policy, on 19 September 1988, Marcel Masse, the minister of energy, mines and resources, asked the chairman of the NEB, Mr. Roland Priddle, what measures could be taken to 'eliminate barriers of interprovincial trade.'[123] In particular, he requested advice on the transmission and generation of electricity so that the federal government could provide leadership in two areas: (1) in encouraging greater cooperation between utilities in systems planning and development; and (2) in clarifying provincial rights to transport electricity through neighbouring provinces to more distant markets.[124] As already stated, utilities generating electricity in Labrador, Nova Scotia, and Prince Edward Island have no direct access to Ontario consumers, and so their electricity would have to be carried to that province on lines that pass through Québec, whose government, until recently, has been opposed to such transmission.

The focus of the NEB, however, did not remain national. The terms of reference in this examination of trans-provincial electricity transport, also called the 'Electricity Review,' were soon expanded to include, in effect, the elimination of provincial transmission barriers for electricity – more specifically, before long US suppliers were regarded as recipients of 'greater access to interprovincial and international power lines.'[125] As it turns out, the concern to eliminate barriers to interprovincial electricity trade had less to do with creating a national Canadian electricity market than it did with opening the Canadian market for electricity suppliers from the US – in other words, changing Canada from an 'electricity hinterland' of the US to a country engaged in transborder electricity trade reciprocity.

Electricity and the Trade Agreements

Given that electricity exports constitute such a small fraction of the electricity generated in Canada, why was electricity included in the provisions of the Canada-US Free Trade Agreement, which came into effect on 1 January 1989? One report by the Department of Energy, Mines and Resources reasoned that (1) since electricity is not covered in international trade agreements, the US government might listen to its protectionist midwestern coal interest groups, who were lobbying against Canadian electricity imports, and impose restrictions on electricity imports;[126] (2) the US and Canada would benefit when Canadian electricity surpluses were used to displace more expensive electricity generated in coal- and oil-fired facilities;[127] and (3) the agreement 'should assist electricity export efforts for a number of provinces, Quebec and British Columbia in particular.'[128] In addition, in part so that Canadian producers could gain access to the American market, the FTA allowed US electricity importers to accrue access rights to Canadian generating capacity. Whatever reasons were given for including electricity in the FTA, supporters assumed that electricity would flow largely from north to south rather than from south to north.

At the end of the first year of the FTA, on 11 December 1989, the chairman of the NEB, Mr. Robert Priddle, in a speech to his US counterparts, highlighted the benefits of the FTA by calling it 'a broad agreement to assure the freest possible bilateral trade in energy [oil, gas, and electricity], including non-discriminatory access for the US to Canadian energy supplies and secure market access for Canadian energy exports.'[129] He told them that although 'the Board must satisfy itself that, among other things, the quantities of energy exported do not exceed the surplus remaining after making due allowance for reasonable foreseeable Canadian requirements,'[130] the FTA had resulted in a new electricity export policy that reduced electricity export regulation.[131] For instance, he said, while stressing that Canadian utilities have fair access to the electricity surplus about to be exported, 'both the surplus and price tests will be effectively eliminated' and replaced by 'essentially market-based' export regulations.[132] In addition, the requirements for hearings by the NEB were drastically reduced. Since the 1960s, as I shall demonstrate in some of the following chapters, Aboriginal peoples, farmers, environmentalists, and economists, among others who opposed building hydro projects for electricity export, had at least been able to make their voices heard in public export hearings. Since the 1990s, however, according to the new NEB policy (of 6 September 1988, legislated as an amendment to the NEB Act on 1 June 1990), only if the board recommends and the Cabinet agrees that a proposed export of electricity or construction of an international power line requires such scrutiny will hearings be held; otherwise, 'routine proposals,' Roland Priddle said, 'will be granted permits without public hearings.'[133] He expected that the board would reduce its regard for the nationality of corporations trading in energy matters and informed his American audience that the NEB had become increasingly part of the North American trade regulation business.[134] He noted that 'Article 905 provides for direct consultation between the respective energy departments about regulatory actions of the NEB or FERC [US Federal Energy Regulatory Commission] which are considered discriminatory' against energy traders.[135] He foresaw that, like the Canadian oil and gas pipelines, Canada's electrical transmission system would become part of the North American transmission system. The chair of the NEB also mentioned that the provision whereby US importers would accrue proportional access rights to their share of the Canadian electricity supply did not have to be evoked in the first year of the agreement.[136]

The second major trade agreement, which includes energy in general and electricity in particular, is the North American Free Trade Agreement (NAFTA). In effect since 1994, its signatories are Canada, the United States, and Mexico. It differs most significantly from the FTA, according to André Plourde in his analysis for the C.D. Howe Institute, 'Energy and the NAFTA,' in its provision for the insurance of greater respect for 'subfederal regulatory entities' (e.g., US state or Canadian provincial regulatory commissions) and 'compliance with the national treatment provisions' regarding the organizations trading in electricity; otherwise, if not treated like domestic firms, foreign firms could claim compensation.[137] Another significant addition to NAFTA is the provision that 'the Parties [should] confirm their full respect for their Constitutions.'[138] In the Canadian case this could provide recognition that ownership of energy resources is vested in the Crown and, more broadly, that the 'distribution of powers and responsibilities among federal and provincial jurisdictions' must be respected.[139]

Although most provisions in NAFTA extend to all three countries, the manner in which Mexico included energy and electricity in NAFTA differs significantly from the way in which Canada did so in the FTA. Whereas in the FTA Canada granted importers proportional rights of access to Canadian electricity should there be a need to curb exports, Mexico exempted itself from such a clause.[140] Whereas Canada extended national treatment to foreign energy corporations in the FTA, Mexico restricted foreign-controlled electrical generation to a corporation's 'own-use, cogeneration, and independent power production, so long as the electricity is not intended for public use.'[141] Unlike Canada's NEB, which primarily regulates the sales of surplus power, Mexico's Comisión Federal de Electricidad (CFE) retains the authority to buy all surplus power.[142] Furthermore, whereas Canada's move towards deintegration of its power systems make it vulnerable to private takeovers, the Mexican state reserves to itself the rights in 'the supply of electricity as a public service ... including ... the generation, transmission, transformation, distribution and sale of electricity.'[143] However, Mexico allowed Canada and the US to bid for services and construction contracts in its energy sector.[144] Yet, unlike what Canada did in the FTA, in NAFTA Mexico disallowed assured proportional access to US electricity importers, insisted that the US respect its Constitution, and restricted foreign ownership and restructuring in its energy sector.

Subordination to US Electricity Policy

The free trade agreements also promoted neo-liberal restructuring (including deregulation, deintegration of utilities, minimal government intervention, and opening investment opportunities for new suppliers) and continental integration in the energy and electricity sector at an administrative level, thereby making it more difficult for future governments to backtrack.[145] While the Canadian government has avoided or even weakened its national approach to electricity policy, the US energy policy is starting to fill the Canadian void. I shall illustrate the implications of this US policy for Canada's provincial utilities, as they continue to export electricity to the US.

In the US, Congress passes legislation dealing with energy policies. Such policies give powers to the Federal Energy Regulatory Commission (FERC) to influence the generation and regulate the transmission of electricity.[146] In the 1980s, according to Canada's NEB, FERC 'was proactive in trying to introduce more competition into the US electricity supply industry'[147] and in allowing industrial customers access to electricity generated by firms that were unencumbered by debts for nuclear plants or other 'stranded assets.' In October 1992 the US Congress passed the Energy Policy Act to provide FERC with the 'power to order transmission-owning utilities to provide access and wheeling to others.'[148] This act was a response to the problems that FERC experienced at the state level – a reaction to the implementation of its neo-liberal strategy to introduce choice and competition into the US utility supply industry and remove barriers to power grids for fresh supplies of electricity from new generating facilities.[149]

FERC issued Final Rules, to which some Canadian utilities that are members of US regional transmission groups began to conform, such as Order No. 888-A (3 March 1997) entitled 'Promoting Wholesale Competition Though Open Access: Non-discriminatory Transmission Service by Public Utilities,' (a clarification of rules in Order Nos. 888 and 889) which states: 'At the heart of these rules is a requirement that prohibits owners and operators of monopoly transmission facilities from denying transmission access, or offering only inferior access, to other power suppliers in order to favor the monopolists' own generation and increase monopoly profits – at the expense of the [US] nation's electricity consumers and the economy as a whole.'[150] FERC predicted that, in the US, 'unbundled electric transmission service will be the centerpiece of a freely traded [sic] commodity market in electricity in which wholesale customers can shop for competitively-priced power.'[151] Washington-based FERC accepts that the functional unbundling (administrative separation) of a utility's business functions (e.g., generation, transmission, and distribution of electricity), 'not corporate divestiture,' is 'sufficient to remedy undue discrimination' in access to transmission facilities.[152] At least for now, privatization of utility assets is not required, but FERC argues that the strict administrative separation (unbundling) 'of wholesale generation and transmission services is necessary to implement non-discriminatory open access transmission.'[153] As a result, in the mid-1990s, in order for BC Hydro to become a provider of wholesale transmission services to local and US power companies selling electricity in British Columbia, and to obtain a licence from FERC to sell British Columbia's surplus electricity to the US market, BC Hydro restructured its operations into three entities: BC Hydro Transmission and Distribution (administering the power grid), BC Hydro Power Supply (supplying power from generating facilities), and the BC Power Export Corporation (Powerex).[154] In other provinces, major public utilities restructured their operations in similar ways in order to provide transmission services to US utilities and to sell electricity from James Bay and the Nelson and Churchill Rivers to the US market.

Since it had neither the technical capability nor the desire to get involved in taking on the administrative burden of regulating transmission access and wheeling, FERC introduced rules under which self-regulating Regional Transmission Groups (RTGs), whose membership would be open to both utilities and non-utility owners of generating facilities, could be established. For instance, the members of the Western Systems Coordinating Council are involved in establishing one or more such groups in accordance with FERC's rules. Among the rules is one pertaining to 'reciprocity.' Under this rule, an RTG member (utility or non-utility) would have the right to wheel power over the transmission lines of other members, but, in return, under the 'reciprocity' rule it 'would have to afford other members the same right over its system.'[155] Other rules require that RTGs file information about transmission pricing and 'good faith' wheeling agreements[156] with FERC, which acts as the tribunal of last resort should transmission group members be unable to resolve their transmission access disputes.

This restructuring of the US utility sector has implications for Canadian provincial utilities that continue to export electricity to the US. Unlike FERC, Canada's NEB does not enforce, and does not appear to have the power to enforce, the reciprocity of access to provincial transmission systems in Canada. As I have already demonstrated, in the 1990s, with the demise of the initiatives for a national power grid, national power network policies have been avoided and the trans-provincial movement of electricity has remained a problem. The NEB observed in 1994, in its *Review of Inter-Utility Trade in Electricity*, that the federal government has not yet established any conditions under which the trans-provincial movement of electricity could take place: 'federal jurisdiction,' according to the NEB, 'under existing legislation is limited to the export of electricity, and the construction and operation of international and designated interprovincial power lines. Existing legislation does not permit the federal government to regulate interprovincial electricity trade, or transmission access and wheeling.'[157] Interpretations by other participants influential in formulating federal electricity policy indicate that the NEB may be overstating the powerlessness of the federal government. In 1996, Natural Resources Canada (the department largely responsible for federal energy policy) and the Canadian Electricity Association took the position that the provincial and federal governments hold concurrent powers over interprovincial electricity sales; however, should conflicts over interprovincial electricity trade arise, then federal regulatory paramountcy prevails (an issue discussed in more detail below).[158] Nevertheless, in the late 1990s, Canada's interprovincial electricity transport regulations appear to be much weaker than the US's interstate electricity transportation regulations. Canada had not formulated a strong interprovincial electricity trade policy by the time the US regulated the opening of its transmission lines to a more diverse group of power suppliers.

Until the 1980s, Canadian power suppliers, such as Hydro-Québec and BC Hydro, were able to negotiate long-term, large-capacity contracts with

US power companies; in the 1990s, however, such contracts have been difficult to arrange, and provincial utilities have had to accept that they can participate primarily in 'competitive bidding for relatively small blocks of power sales of limited duration' in the US market.[159] The NEB's review study indicates that this less predictable access to the US electricity market has led the board to grant blanket export permits that allow utilities to bid for certain short-term (up to three years) power exports with reduced regulation.[160] For example, 'in the case of the Export Permits granted by the Board in 1992 to BC Hydro/POWEREX [British Columbia Power Exchange Corporation] and to Manitoba Hydro, the exporters have the freedom to engage in firm and interruptible sales transactions that have durations of up to three years (except that Hydro-Québec's permit allows for five years) without prior Board approval.'[161] To improve their US market access, some utilities may consider becoming members of US regional transmission groups. For example, BC Hydro and TransAlta Utilities are members of the Western System Coordinating Council and are participating in coordination and transmission planning in western US systems. Such membership, however, might be made conditional on the Canadian utilities being in compliance with FERC guidelines and on participating in transnational regional planning. Not only British Columbia, but also Ontario, Québec, and Manitoba have strengthened their transnational, regional infrastructure and opened up several continental axes that have become interconnected with the American regional markets.[162]

Federal-Provincial Division of Authority
The division of powers under Canadian federalism, whereby provinces control the development of natural resources and the federal government controls their export, has reduced the possibility of formulating national electricity policies. The federal Department of Natural Resources Canada and Section 92A of the Constitution Act, 1982, assert that trade in electricity and the installation of international transmission lines is subject to the prevalence of federal jurisdiction (with concurrent federal and provincial powers over interprovincial trade and exclusive federal powers over international trade), whereas the planning, development, and distribution of hydroelectric resources within provinces is the responsibility of each province (Table 1.1).[163] However, Christopher Armstrong, in his historical examination of Ontario's conflicts before the 1940s over power rights and the Constitution, found that the federal government challenged the jurisdiction of the provinces over development of water power. He argues that the constitutional ambiguity concerning which level of government has user rights to flowing water in rivers that are suitable for navigation, canals, and water-power developments led premiers of Ontario and Québec in the 1920s and early 1930s to assert their provincial rights against federal

claims over water-power rights when these claims were associated with canal construction.[164] Michel Duquette argues 'that the constitution unambiguously grants to the provinces' rights over developments such as James Bay in the 1970s.[165] Yet the federal government, according to Duquette, has used interventionist tactics to control provincial export quotas and electricity production itself, especially when Québec's hydro developments affected international and interprovincial trade and gave rise to environmental issues.[166] His interpretation of unambiguous provincial jurisdiction over electricity trade and transport conflicts with that of the federal institutions that contribute to the formulation of Canadian electricity policy.

The Energy Resources Branch of the department of Natural Resources Canada reported in 1996 that 'under the Canadian constitution, electricity is primarily within the jurisdiction of the provinces' and that, 'as a result, Canada's electrical industry is organized along provincial lines' (see Table 2.1).[167] The branch takes from the Constitution, Section 92A (1) (c), the assignment to the provinces of explicit responsibility 'for sites and facilities in the province for generation and production of electrical energy.'[168] On matters of transmission, however, the Canadian Parliament has 'legislative authority over transportation undertakings that cross interprovincial boundaries or the international boundary.'[169] Furthermore, in matters of electricity trade, the federal and provincial governments have 'concurrent powers with regard to regulation of interprovincial electricity sales'; however, 'if a conflict between federal and provincial laws over interprovincial trade should arise, the federal law will prevail.'[170] In practice, the Canadian electricity trade is largely determined by provincial utilities entering into purchase and sales agreements with other utilities, and the role of the Canadian government in achieving electricity policy objectives has been to seek 'provincial policy consensus and cooperation.'[171] Such a consensus has not been achieved over the proper means of integrating major power developments in order to supply national or regional power grids.

Provincial Development Strategies

Until 1961, as Walter Dinsdale stated, the planning of electric power production had been almost entirely provincially based (at least in the previous three decades). The federal government had not 'taken an active role in planning for or assisting with the development and use of electricity on a national basis.'[172] Dinsdale thought that a nationally or regionally integrated transmission system would bring substantial economic benefits and would allow for the early development of major remote hydro projects and the flow of surplus electricity to the more industrialized areas of Canada. Once constructed, however, distant northern generating stations in Labrador, Québec, Manitoba, and British Columbia did not supply electricity to Canadians on a regional or national basis.

Why was the development of major hydroelectric sources in remote areas not integrated to serve Canadians via regional or national transmission systems? At the federal level, there were three prominent reasons for this: (1) provincial resistance to federal plans; (2) the desire to allow the early development of power projects and international transmission; and (3) the trading preference for viewing electricity as a continental rather than as a national commodity. Diefenbaker's 1961 initiatives to plan transmission on a national basis failed in part because provinces resisted the prospect of having an extra-provincial authority involved in deciding or planning power production and in building a trans-continental power grid. In the Pearson government, Mitchell Sharp encouraged the early development of remote hydro projects together with their international interconnection to US systems and (as widely as possible) to interprovincial transmission. Sharp's policy of allowing exports de-emphasized transmission on a national or regional basis and allowed the NEB to approve long-term exports to the US. In addition, the possibility of coordinating power development nationally was further weakened by the free trade agreements, which advanced provincial power planning for export and enabled electricity to become a bilaterally traded commodity.

In most decades since the 1960s, domestic political forces in several provinces encouraged resistance to federal involvement in energy policy and favoured the development of electricity as a continental commodity and as a factor in industrial diversification. Some provinces, such as those in the Atlantic region of Canada, that have to transmit electricity across a neighbouring province if they want to reach Ontario or the US market, preferred the imposition of federal authority over transmission access; other provinces, however, such as those that had direct access to the US market or to neighbouring provinces requiring electric supply, did not favour federal involvement. British Columbia, Alberta, Manitoba, Ontario, and Québec generally resisted extra-provincial authority over wheeling, and yet these same provinces have increasingly accepted the wheeling reciprocity and regulatory requirements of a foreign institution, namely, FERC. Ownership or control over a national power network (whether by the federal government or by the provinces) remains contentious. Only when provinces find electricity trade profitable do they tend to establish interconnections with neighbouring provinces: the Churchill Falls connection with Québec, albeit on unequal terms, serves as an example. Divisive political conflict around issues of 'where' and 'how' to distribute benefits across geographical space, as Janine Brodie discusses, continues to be evident.[173] For instance, in 1981 the Western Power Grid initiative failed, in part, because British Columbia was not interested, as most employment and investment benefits would have flowed to Manitoba and most of the electricity to Alberta. Arising at a time of economic downturn, Alberta and Saskatchewan felt that this circumstance would have meant deferred investments and less employment for them. The regional politics of expanding the hydroelectric system in Canada since the 1960s shows that, instead of coordinating power development in the national interest, provinces want to retain their independent roles in the planning of provincial power systems and to use new supplies of power for industrial diversification and export revenues. Much as Ontario decided to develop nuclear power for its modernization goals and export the surplus to the US, Québec made Hydro-Québec a cornerstone of its economic development during the Quiet Revolution and, later, dedicated its capacities directly towards export; British Columbia planned to bring industrial development to its North and to export to 'power-hungry' California; and Manitoba advanced its Nelson River project for industrial development in the province and for exports to the mid-continent. Since the national power network and regional networks spanning more than two provinces were not developed in Canada, in part because of provincial development interests, this analysis of major hydro expansions will now proceed at the provincial level.

In the mid-1990s, when publicly owned electric utilities were broken up into separate functions such as generation, transmission, and distribution (in villages, towns, and cities), the question of why and under what circumstances utilities became public and integrated was rarely asked. Before 1906 at Niagara Falls, the province of Ontario left the supply of electricity to the private sector. Then, mostly small-town manufacturers agitated and achieved public power at cost because private utilities did not supply them with Niagara power. As Chapter 3 shows, some of the same development patterns associated with expanding the hydroelectric system – privatization reversal, technological dependence, export problems, and unpredictable industrial consumers – recurred in other provinces.

3

Niagara Power Repatriation (Ontario)

> For more than fifty years every observant tourist who found
> his [or her] way to Niagara Falls has been impressed by the
> vigorous industrial development on the American side and the
> complete absence of any such development on the Canadian side
> of the Falls.[1]
>
> – James Mavor, *Niagara in Politics: A Critical Account of the
> Ontario Hydro-Electric Commission*

Introduction

During the first two decades of the twentieth century, Ontarians argued over whether hydroelectricity developed under *private* ownership or under *public* ownership would best advance industrial growth at Niagara Falls and in the province. Supporters of the private enterprise view maintained that, if private capital was allowed to make a profit from hydro development without government constraints, then 'vigorous industrial development,' which James Mavor observed in Niagara Falls, New York, would be repeated on the Ontario side. On the American side, private developers owning water rights had initiated development in the 1870s by building diversion canals from above the Falls. They sold the use of flowing water (hydraulic power) so that the waterwheels (turbines) of 'industrial tenants' could drive the machinery in grist and pulp mills on canal lots and cliff sites.[2] Starting in 1881, private power companies generated and sold power to electro-processing industries (chemical, abrasive, and electroplating), paper mills, and food-processing companies.

Public power advocates, on the other hand, predicted that once Niagara power was unburdened by private profit it would 'attract to the district the enterprise of others' and would be an 'incalculable stimulus to the productive and competitive efficiency' of manufacturers in Ontario.[3] Reliable public power, they argued, could be brought to areas that had so far experienced energy shortages due to dry hydro sites during the summer and US coal supply problems in the winter. Power, they insisted, was essential for the survival of manufacturing in the province. In the words of the founder of Ontario Hydro, Sir Adam Beck, 'an abundant supply of motive power is to the manufacturing arts what blood is to the human body.'[4] Thinking on a larger scale, E.W.B. Snider, a miller, farm implement manufacturer, lumberman, and former provincial politician, believed that electricity 'would easily enable our manufacturers to make Ontario the workshop of the Dominion and to control the home market for manufactured goods in Canada.'[5]

Mavor, holding to his 'hydro by private enterprise' view, overlooked two potential developments: (1) that the failure of private power companies to provide adequate energy for industry can become a source of special interest pressures to start public hydro ventures and (2) that such a failure can stem from private hydro companies not Canadianizing their operations. Whether supporters of public power or not, both camps overlooked the stifling influence that foreign direct investment and the self-interest of private hydro plant owners can have on industrial development.

Reversing the Privatization of Niagara Falls

Initially, Ontario politicians believed that the development of energy resources by the private sector would naturally result in industrial growth. Holding such optimistic views, government officials in Ontario first privatized the water-power rights to Niagara Falls but soon after needed to repossess them. Why this reversal? I will argue that it was necessary because the private owners, especially US investors, failed to supply electricity from their Niagara power plants to the small manufacturers in southwestern Ontario. The latter were eager to modernize their cumbersome steam-driven machinery by replacing it with technologically superior and more profitable electrical machinery. Fearful of becoming locked into industrial backwardness, small-town manufacturers fought for public power and thereby initiated Ontario's repossession of the water rights to the Horseshoe Falls (Table 3.1, see contrast to Nelles's conception in footnote).[6]

Although the power rights to Niagara Falls are apportioned between Canada and the US according to the international boundary, Ontario initially allowed US investors to monopolize power franchises at the Canadian Falls. In 1887, the Queen Victoria Falls Park Commission was established with a mandate to buy the land in the vicinity of the Falls and, shortly afterwards, it entered the hydroelectric business.[7] To help launch the first hydroelectric enterprise, the Queen Victoria Niagara Falls Park Commission granted exclusive power rights not to a developer, but to a speculator. As H.V. Nelles records, in 1887 the commissioners sold exclusive water-power rights to the Canadian Falls to Colonel A.D. Shaw of Watertown, New York, for an advance of $10,000.[8] Nelles distinguishes between ownership of land and users' rights to water that flows over it: 'the water, simply by virtue of passing over private property, was not itself private property; it could be used only in passage' by those who held rights but otherwise remained in the public domain.[9] In addition he affirms that 'by retaining title to waterpower in the hands of the crown and by leasing waterpower privileges instead of selling them outright, the state could demand both a revenue from the industry and prompt performance of construction agree-ments.'[10] The Parks Commission needed such revenue to buy the property next to the Falls and convert it from a gaudy tourist area into a riverfront park that would also accommodate a few stately powerhouses. Without having built such facilities, Shaw sold his monopoly franchise to the US-owned Niagara Falls Power Company. Then, in 1892, this firm incorporated its Canadian water-power rights as the Canadian Niagara Power Company.[11] Shaw, in turn, became its nominal president.[12] From then on, this firm 'had the first choice of location for power development works within the Park' and was expected to be 'the first power company to produce power on the Canadian side of the Falls.'[13] It received the right to draw water from Niagara Falls for generating power and to transmit and distribute electricity for sale outside the riverfront park for 100 years.[14]

Table 3.1

Chronology of the privatization and its reversal at the Canadian Niagara Falls

1887	Ontario sells A.D. Shaw the monopoly rights. He sells them to the US-owned Niagara Falls Power Co.
1896	Niagara Falls Power Co. transmits industrial power to Buffalo from its US-based plants at the Falls.
1897	Supreme Court of Ontario reviews Niagara Falls Power Co.'s failure to construct a Canadian plant.
1899-1903	Niagara Falls Power Co. retains only one-third share of falls. Remaining two-thirds is sold to US-owned Ontario Power Co. and Toronto-owned Electrical Power Co.
1901	Niagara Falls Power Co. starts construction of its Canadian plant.
1903	The Hydro-Electric Power Commission of Ontario (Ontario Power Commission), which advocates public power development, is formed.
1906	Niagara Falls Power Co. exports power to Buffalo industries from its Canadian plant.
1907	Ontario Power Co. exports power to its US industrial customers from its Canadian plant.
1908	Toronto's Electrical Power Co. transports power to Toronto and sells to the US market.
1910	Two-thirds of the electricity generated at the Canadian Falls is exported to the US.
1910	Ontario Power Commission delivers first Niagara Falls power over the public transmission line to Berlin, ON.

Although the Ontario government had privatized the water rights in 1887 subject to timely power development, Shaw and Canadian Niagara held up the building of the needed power plants from 1887 to 1901 in anticipation of higher profits from future electricity exports to New York State. The stalling tactics of the American speculators contributed to southern Ontario's falling behind in capturing early industrial benefits. For instance, until 1886, in the absence of long-distance transmission technology, industries, especially energy-intensive industries and electro-processing industries (those producing abrasives, plating silver, or processing chemicals), found it necessary to locate their operations close to power plants. The Niagara Falls Power Company, well aware of this need, had attracted more than twenty 'industrial tenants' who bought short-distance power in the town of Niagara Falls, NY,[15] while stalling electricity generation and industrial development on the Canadian side of the Falls.[16]

With the invention and installation of transmission lines by 1896, electricity could be brought to industry rather than industry having to locate near generating plants.[17] Transmission technology allowed American owners of the Niagara Falls power monopoly either to supply industrialists further afield in Buffalo and Syracuse or to initiate delivery to Ontario manufacturers in London, Guelph, and Berlin. Their choice became evident on 10 November 1896 when the Niagara Falls Power Company's twenty-mile-long transmission line reached its Buffalo industrial market.[18] Meanwhile, on the Niagara peninsula, scepticism about the US power company grew: 'The spectacular growth sparked by hydro-electric development on the American side of the Falls exasperated the residents of the Niagara peninsula who had long since grown suspicious of the endless excuses advanced by the Canadian Niagara Power Co. for the total lack of progress on its monopoly concession within the park.'[19]

It was then that Ontario's Liberal premier, A.S. Hardy, 'asked the Supreme Court of Ontario to rule whether the total absence of construction prescribed by the agreement at the 1897 deadline constituted a breach of contract.'[20] The court found the terms of the original agreement could not be cancelled until 1899.[21] Before the final showdown, the government and the US utility company found a compromise when 'in July of 1899 the company relinquished its monopoly on the Canadian side of the Falls.'[22] Yet, despite its dismal record, Canadian Niagara retained power rights to one-third of the power rights to Canadian Falls (100,000 hp, or 75 MW), the other two-thirds being available to other private utilities.[23]

By 1903, the Queen Victoria Niagara Falls Commission had granted all available power franchises. Foreign ownership of such power rights was well hidden behind corporate names. While the Canadian-owned Electrical Development Company (EDC, or Toronto Power) referred in its name neither to nation nor province, the two US-owned subsidiaries, the Canadian

Niagara Power Company and the Ontario Power Company, had added 'Canadian' and 'Ontario' to their names (see Figure 3.1) – adjectives which in fact disguised their US ownership.

Canadian Niagara Power Company

After having its monopoly reduced to one-third of the water power at the Canadian Falls, the Canadian Niagara Power Company became the first utility to build a power plant on the Ontario side. Harold Buck, electrical director of its American parent firm (the Niagara Falls Power Company), simply conceived the Ontario plant as a system extension of its two New York State plants.[24] Construction began in 1901, and the first power was transferred from branch to parent by 1905;[25] that is, the parent utility directed its Canadian corporate offspring to export nearly all the power

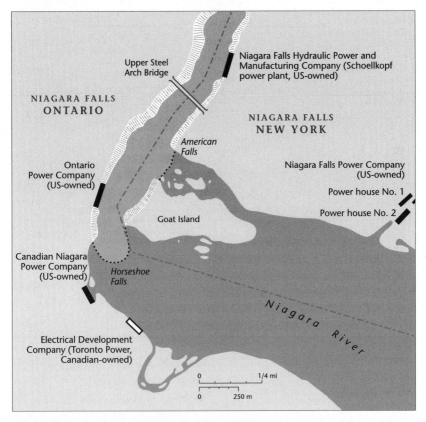

Figure 3.1 Power plants on the Niagara River, 1900-10

Sources: George W. Davenport, *The Niagara Falls Electrical Handbook* (Syracuse: Mason Press, 1904); adapted from 'Powerplants on the Niagara River' (by Geoffrey Matthews) in Pierre Berton, *Niagara: A History of the Falls* (Toronto: McClelland & Stewart, 1992) (used by permission).

from its new Canadian plant back to the US parent's market (see endnote 26 regarding Canadianization).[26]

By 1906, some of this exported energy supplied the short-distance industrial market in Niagara Falls, NY, but the bulk of it went to its long-distance industrial market in Buffalo.[27] The Niagara Falls Power Company dominated both these industrial markets. One was located less than two miles from the company's plants in Niagara Falls, NY, and consisted of twenty-two industries; the other was between fifteen and thirty-five miles from the same Niagara plants in Tonawanda, Lockport, Olcott, and Buffalo, and included more than sixty industrial customers.[28] With a lucrative US industrial market, this American utility shunned the riskier and less profitable Canadian manufacturing market.

Ontario Power Company

In a similar way, the Ontario Power Company, the other US power company owned by a syndicate of Buffalo industrialists, also vertically integrated its structure of supply, transmission, and US customer distribution. In 1900, the Ontario Power Company had received the second franchise from the Queen Victoria Falls Park Commission with the right eventually to develop a 180,000 hp (134 MW) facility at Niagara Falls.[29] Again, its major US customer was its corporate parent, the Niagara Lockport and Ontario Power Company. To assure continued internal power transfers, the corporate parent signed long-term contracts with its offspring in Canada: 'The initial contract between the subsidiary and parent companies was dated 16 July 1904, and called for the delivery of 60,000 horsepower [45 MW] on or before 1 January 1907. The contract was to remain in force until 1 April 1950, with certain provisions for renewal.'[30] As Blake R. Belfield found, 'Niagara Lockport's strategy was to build first a trunk line between Lockport and Syracuse, NY, and then install branch lines from the trunk to smaller urban centers in the region – rather like a general railroad track strategy.'[31] Both US utilities connected their Canadian power plants to their US transmission systems, treated their Canadian subsidiaries as electricity suppliers, and showed little interest in small southern Ontario manufacturers.

Electrical Development Company

For their part, Canadian utility executives showed no more loyalty than did their American counterparts to Canada's nascent industrialists' energy needs. The only Canadian company developing hydro power at Niagara Falls similarly failed to supply industrial electricity to southwestern Ontario towns. On 29 January 1903, the Queen Victoria Niagara Falls Park Commissioners had granted the Electrical Development Company the right to generate up to 125,000 hp (93 MW) of electricity.[32] Its owners, formerly obsessed with railway projects, had formed the Mackenzie syndicate, which was made up of William Mackenzie, a railroad man; Frederic Nicholls, an electrical engineer and head of Canadian General Electric; and Henry Mill Pellatt, a general financier-entrepreneur.[33] They 'represented what was in fact the foundation of a private utility monopoly in Toronto: the head of the traction [electric street car] business, the head of the electric light business [the Toronto Electric Light Company], and the head of the major Canadian electrical manufacturer.'[34] However, they built their plant at Niagara too late to replicate the industrial growth of their US competitors.

Copying US efforts in creating local industrial parks and promoting its own market, 'the Mackenzie Syndicate purchased a huge plot of land (530 acres) in the vicinity of its generating plant at Niagara Falls. It [was] expected that this land would be taken up by manufacturers using electro-chemical processes, or [by] other large power-using businesses.'[35] That strategy failed because, years before, the US-owned Niagara Company had 'established a grip' upon the industrial market in Niagara Falls, Ontario.[36] To strengthen their corporate integration, the owners of the Electrical Development Company established two transmission subsidiaries: the Toronto and Niagara Power Company (to serve 'Toronto and the intermediate territories') and, in 1906, the Niagara Falls Electrical Transmission Company (to compete with Niagara Lockport in the safe, established markets of upstate New York).[37] By 1908, they transmitted electricity from their Niagara plant to their Toronto Power Company (in turn controlled by the Toronto Railway Company) and, thereby, vertically integrated supply, transmission, and distribution.[38]

This turn of events is not what Ontario government officials had hoped for; they had expected private owners to use Niagara electricity to help implant new industries and modernize the emergent small manufacturers in southern Ontario. Privatizing water-power licences had failed as an indigenous industrial development strategy: private owners were simply uninterested in, even opposed to, the wishes of smaller Ontario manufacturers. Private utility owners were preoccupied with the more lucrative industrial markets in Toronto, Buffalo, and Syracuse.

Reversal of Privatization
Ontario's small manufacturers wanted no additional Niagara power exported to Buffalo; they wanted it transmitted to Berlin (now Kitchener). During the first years of the twentieth century, electricity generated on the Canadian side of the Falls was Canadian only by virtue of geography; as Nelles points out, commercially and practically it belonged to American manufacturers.[39] By 1910, 64 percent of the power generation was committed for export.[40] Ontario's small businesspeople became aware of that

after the fact: 'It was not until Ontario businessmen took envious notice of the industrial revolution brought by cheap electricity across the Niagara River in the state of New York that they discovered that their Niagara waterpower had been gobbled up by Americans.'[41]

By 1903, the threat of industrial stagnation had become very real for southwestern Ontario. Small manufacturers started a public power movement whose aims were clear: they did not want regulated power for profit, they wanted power at cost; and they did not want a private urban power monopoly running the hinterland, they wanted their own cooperative utility supplying their communities and industries. To realize their goals, they called for a public power company to coordinate both community and corporate goals. Members of the public power movement envisaged the following steps: negotiating contracts with the private power suppliers, raising capital for public transmission lines and plants, and reversing the privatization of Niagara Falls.

Small-town manufacturers met not only among themselves, but also with key politicians and the public as early as 1902, with the aim of forming the Ontario Power Commission as their progressive utility. For them, bringing progress to towns and factories meant receiving electrical power from public power lines rather than from their own coal-fired steam plants, and running electrically powered equipment rather than steam-powered machinery. A key speaker for the public power movement was E.W.B. Snider. At the Waterloo Board of Trade meeting on 11 February 1902, he suggested that if the members of the board of trade banded together to create an attractive, cooperative market for Niagara power, then the community of Waterloo 'could offer cheap power to manufacturers that would greatly assist [the community's] further progress.' The cooperative utility would help bring a power line from the plants at Niagara Falls to prevent the scattered towns of southwestern Ontario from being left behind.[42] One year later, on 17 February 1903, at a meeting in the Berlin YMCA, Snider recommended to sixty-seven delegates from the main towns and cities in the region 'that the municipalities should build a transmission system only, purchasing their power from one of the existing generating plants at Niagara,' which at eight dollars per horsepower plus transmission, would cost fifteen dollars per horsepower per year at municipal boundaries.[43] During the same month, he led a group of twelve men to meet Premier G.W. Ross and 'begged the province either to distribute hydroelectricity itself or permit the municipalities to do it themselves'; he objected to the formation of a monopoly of Niagara water power and warned that a lack of cheap power would deindustrialize southwestern Ontario.[44] Ross found their arguments persuasive, and the subsequent provincial Act to Provide for the Construction of Municipal Power Works and the Transmission, Distribution and Supply of Electrical and Other

Power and Energy permitted the 'creation of the Ontario Power Commission.'[45] On 12 August 1903, at a general meeting of the interested municipalities and manufacturers in the Toronto City Hall, Snider, P.W. Ellis (a Toronto wholesale jeweler), W.F. Cockshutt (a farm implement manufacturer from Brantford), and Adam Beck (factory owner, mayor of London, and Conservative MPP) were selected commissioners.[46] Beck took over the leadership of the Ontario Power Commission in 1904.

Faced with continuing opposition from the Toronto Syndicate and its Electrical Development Company, Snider headed a commission of inquiry and filed his report in March 1906. The Snider Commission recommended that, since small southwestern Ontario towns could not afford to build a generating station of their own at Niagara, they should propose buying and distributing electricity 'at cost' by means of a municipal cooperative that would 'build and operate the transmission system linking the major towns with Niagara.'[47] The Power Act allowed the commission to proceed with its own transmission network. It first signed a supply contract with the Ontario Power Company on 21 March 1908, then a transmission contract on 13 August 1908 with municipalities in the southwest.[48] On 11 October 1910, the first 'switching on' ceremony took place with great fanfare in Berlin.[49]

In this way, Ontario Power commissioners helped bridge the infrastructural gap and thereby belatedly remedied the failure of private enterprise to supply industrial energy to southern Ontario's producers. This new supply of hydroelectric power allowed manufacturers with an interest in developing Ontario to use more sophisticated machinery in their small factories. Thus, the commission had removed one block to Ontario's application of new technology, but another soon arose in the failure of Canadian business interests to capitalize on their chances to develop an original hydroelectric technology sector. Since the owners of Ontario's generating stations were mostly absentee-owners, they relied heavily on imported turbine and generating equipment.[50] This kind of dependency, so modernization theory argues, constitutes an initial 'diffusion' of technology, which will eventually lead to more self-reliant technological development (similar to that of industrial centres). In interpreting the historical account of the technology transfer in Ontario, Armstrong and Nelles employed the concept of 'diffusion' and found that Canada's emerging electrical utilities borrowed 'business structures' and 'technology' from the US.[51] However, unlike industrial centres such as England and the US, which developed a degree of technological autonomy in producing their generating equipment, Canada remained technologically dependent. In contrast to the invention of the steam engine, which inspired indigenous industrialists to foster the invention of original machinery during Britain's Industrial Revolution, the arrival of electricity-generating equipment at Niagara Falls led

power company executives to rely primarily on the imitation and importation of technology. The non-innovative technology transfer at Niagara
Falls set a historic precedent for future hydro ventures across Canada –
ventures that, typically, have relied on foreign corporations for the initial
technology choices in major Canadian plants.

Imitative and Imported Technology

Since the 1880s governments and municipalities in Canada have invited
established firms from the US to set up their branch firms in Ontario or to
take over existing Canadian ones.[52] As Tom Naylor indicates, many foreign
companies, such as the Westinghouse Company, took advantage of bonus
systems (including tax reductions, grants, stock purchases, and cheap or
free industrial sites) offered by municipalities.[53] For example, in 1895
Westinghouse began exploring Ontario for a site, ultimately settling in
Hamilton.[54] Through this process many Canadians became managers or
owners of foreign branch plants. For instance, Senator Frederic Nicholls, a
member of the Toronto syndicate, advanced the merger of Edison Electric's
two Canadian branches into Canadian General Electric in 1892 and
received a licence to copy General Electric generating equipment.[55] From
1895 onwards, Nicholls owned a controlling share of the company, but
upon his death in 1923, American General Electric purchased and gained
control of this Canadian company.[56] Whether owned by American or
Canadian interests, private or public, all utilities imitated or imported
designs and equipment from the US and Europe, thereby entrenching
technological dependency on foreign firms as early as 1904.

The US-owned Niagara Falls Power Company had already installed two
power plants on the American side of the Falls before it built its third
plant, about a half mile above Horseshoe Falls, in Queen Victoria Falls Park
in Ontario (see Canadian Niagara Power, Figure 3.1). With minor modifications, the firm adopted the hydraulic features of its second US plant
(Power House No. 2, completed in 1904) for its Canadian powerhouse
(Figure 3.1). The eleven generators were manufactured by General Electric
and the turbines were imported from Escher Wyss of Zurich.[57] Harold Buck,
electrical director of the US firm, took charge of the design and arrangement of electrical equipment for the Canadian project.[58] Canadian Niagara's
designs were freely borrowed by the Ontario Power Company, the other
US-owned power utility.[59] Its Canadian powerhouse was not above the
Falls but downstream near river level (Figure 3.1). Ontario Power installed
generators supplied by General Electric, Canadian General Electric, and
Western Electric; it imported turbines from J.M. Voith, a German company, and Welman Seaver Morgan, a US company (Appendix 1).[60] Thus,
unlike British industrialists during the Industrial Revolution, Canadians
did not make the initial choices over hydro technology – Americans did.

Executives of the Canadian-owned Electrical Development Company (Figure 3.1) were also disinclined to be innovative. The equipment design, in its new power plant located on top of the riverbank upstream from the Canadian Horseshoe Falls, was also copied. The eleven generators by General Electric and the turbines by I.P. Morris of Philadelphia were unoriginal and like those of the nearby Canadian Niagara plant.[61] This copying, of course, was not by chance. Belfield notes that 'this bold imitation makes sense if we bear in mind the common manufacturer involved, and the fact that the head for Canadian General Electric, Frederic Nicholls, was also one of the incorporators' of the Toronto-based Electrical Development Company.[62]

The reliance on equipment from Westinghouse and General Electric predominated not only in privately owned utilities but also in the publicly owned Power Commission. Despite the objections of Adam Beck, chairman of the commission, the premier of Ontario, James Whitney, insisted that Westinghouse and the Canadian General Electric Company share the transformer and equipment contracts for the power line from Niagara Falls to Toronto.[63] Once these two US branch plants took root, they dominated core hydroelectric-component manufacturing in Canada in the twentieth century.[64] Interprovincially, by 1904 the influence of the US-owned Niagara Company on technological designs of prime moving equipment extended also to the Shawinigan project in Québec.[65]

Thus, Canadian-owned hydro development was derivative, to say the least. Canadians became dependent on imported technology and US electrical equipment firms; decisions about the transfer of technology were made outside the country or by Canadian managers in US-owned branch plants. Within such branch plants, substitutes for former imports were produced without export rights for management in Canada. Instead of importing a generator from the United States, a copy of such a generator produced in a US-owned branch plant in Canada would be substituted. Usually, export decisions were allowed only from head offices, not from a host country's branch plant. But even though dependently developed, a technological transition had nevertheless been achieved from coal-fired steam equipment inside factories to electrified machinery powered by hydro plants outside the factory. The technical development led to expectations that this new form of energy supply would result in new manufacturing.

Steam, Electricity, and Industry

The Ontario Power Commission argued in 1906 that cheap power means water power, which can be produced at a fraction of the cost of steam power, and that such 'cheap power is the most important factor in the industrial development of a country or province.'[66] However, it is not the most important factor (as will be discussed in subsequent chapters). Cheap power arrived in southern Ontario after several paths of industrialization had already emerged, a pattern that differed from Europe's earlier domestic transition from small manufacturing to the factory system with the discovery of steam power. In Ontario, for example, settler manufacturers, companies extracting natural resources, and implanted US branch firms followed their own trajectories. This pattern of development is evident in surveys that determined the energy need of power customers in small-town Ontario, north and south of Berlin (the first inland Ontario community to receive power from the Falls in 1910). The Ontario Power Commission conducted two surveys of its potential industrial customers, one in 1911 in communities north of Berlin (including Palmerston and Harriston) that were characterized by settler craftshops and small manufactures; another in 1919-20 in towns south of Kitchener (Berlin's new postwar name), including Brantford and London, that revealed a mix of domestic manufacturers and US branch-plant industries. Beyond these regions, in Ontario's North industrial energy was needed mostly for primary processing of extracted natural resources.

As early as 1901, Ontario's small manufacturers had felt severely constrained by the slow transition – it took about fourteen years – from steam production to electric production. Branch plants of US power companies contributed to holding up the use of new technology in inland Ontario communities by exporting Niagara power to Buffalo instead of transmitting it to Berlin. The restricted access to electricity, combined with the pressures of takeovers, competition of branch plants, and lack of bank financing, placed small towns at a real disadvantage.

Since they were still working with steam machinery, small-town manufacturers were falling technologically behind their US counterparts. In 1906, the Ontario Power Commission reported the inefficiencies of steam and the advantages of electricity. It noted 'that the introduction of electric power reduced the average power needed in factories running at varying loads to one-third of what was required when steam was used. This reduction, much greater than was anticipated, was found to be due in part to the fact that in the great majority of such factories, the power used in turning the shafting and countershafting absorbed the largest part of the energy.'[67] Had they been supplied with electricity earlier, one can infer from the commission's report, manufacturers could have produced the same output with two-thirds less energy, saved floor space because of the smaller size of electrical machinery, and operated individual sections of machinery for those working overtime rather than having to run all their equipment simultaneously.

The first power survey of towns and industries, conducted three months prior to 29 August 1911 and prepared for Adam Beck by the chief engineer of the Ontario Power Commission, indicated that most manufacturers still burned coal, wood, or refuse to generate steam power (Appendix 2).[68] The experience of small manufacturers in the communities of Harriston and

Palmerston are typical. Harriston (population 2,000 in 1911) had several firms producing finished goods: Davis Packing Company, J. Merrigan (casket makers), Gunn Coal Storage, G. Walkey (veneer plant and grist mill), Dryden Grist Mill, George Gray (planer mill), Harriston Stove Works, and the Harriston Furnishing Company. An excerpt from the 1911 survey describing the latter provides a typical example of small manufacturers in transition: '*Harriston Furnishing Company* – Have an 80 H.P. Engine Jerome-Wheelock; burn refuse and coal, average one ton per day at present; require steam for kiln and heating purposes. Could use three or four motors totaling 75 H.P. average 45 H.P. Many of the plants in the town [of Harriston] require steam f[or] their process and a[n] [electric] motor drive would permit a more economical steam system being installed and a cheaper result of factory output being obtained.'[69]

Even in towns that had their own power plants, electricity was not always affordable for small manufacturers. Of Palmerston (population 2,500), a town with a planing mill, grist mill, and malting company, the 1911 power survey states: 'The various industries of the Town are very desirous of having electric power; the Town were [sic] apparently unable to supply it from their plant, one reason given being – Cost too great. It appeared to me quite a nice industrial load could be got up in this Town if power were available. Present demands are placed at 250 H.P.'[70]

A diverse group of existing and new industrial customers from London, Guelph, Stratford, Woodstock, and Brantford (typically a mix of smaller Canadian entrepreneurs and US branch plants) continued to apply to the Ontario Hydro Commission for cheap electricity, as its survey from 1919 and 1920 reveals (Appendix 3).[71] In London, some of the largest industrial customers were in the transport vehicle sector: the Ford Motor Company, the Republic Truck Company, and the Stanley Railway Company. Other industrial applicants included the Holeproof Hosiery Company, Carling Brewing and Malting Company, the Nut Crust Company (breadmakers), and London Shoddy Mills. In Guelph, rubber, machinery, and carpet producing companies (e.g., F.E. Partridge Rubber Company, the Sterling Rubber Company, Malleable Iron Company, and Guelph Carpet and Worsted Spinning) requested more electricity. In Stratford, the Canadian Edison Appliance Company and other industrial customers indicated that increased demand called for cheaper, more abundant, power. Woodstock, with its wagon, furniture, and piano companies, also had additional industrial power needs. Brantford's manufacturers of motors, engines, ovens, and kitchens required extra power (Robins and Meyers, Watrous Engine Works, Brantford Oven and Rack Company, and Hamm and Nott).[72] Cheap public power was in demand by groups following different development paths.

In the town of Niagara Falls, Ontario, a small group of mostly US firms had started using electro-processing in silver-plating and other energy-intensive processes in the production of abrasives and chemicals as early as 1895, when the Ontario Silverplate Company (later the McGlasham-Clarke Company, 1905) became the first firm to begin production. A competitor, New York-based Oneida, established its silver-plating branch, Oneida (Canada), in 1916.[73] Cyanamid of Canada, which manufactured calcium cyanamid, a synthetic fertilizer, came to Niagara in 1907.[74] Its two plants were among the largest users of electricity in Ontario. In 1916, 'the Carborundum Company of Niagara Falls, New York, decided to start a Canadian operation in the former Township of Stamford, now part of Niagara Falls.'[75] Predecessors of Nabisco Foods, including the National Biscuit Company (1928), electrically shredded and baked 'Triscuit' as early as 1904.[76] Albeit up to twenty years after the growth of industry on the American side of the Niagara River (in 1881), such developments indicate some belated branch-plant growth – but not of new Canadian-owned industries.

Contrary to the notion that hydroelectricity, once available, diversifies product manufacturing of Canadian natural resources and leads to greater Canadian economic independence from the United States, in northern Ontario, modern hydroelectric power was used for the traditional processing and exporting of pulp and minerals (staples). It was T.C. Keefer, a professional engineer, who thought he foresaw in 1900 a new energy-invigorated industrialism for Ontario. Giving the value-added manufacture of spruce as an example, he argued: 'Heretofore we have cut our spruce into deals and exported it to Europe, and more recently into pulp wood and exported that to the United States; but manufactured by our water power into paper, the raw material would yield this country ten times the value it is now exported for.'[77] Keefer believed that hydro power would enhance Canada's industrial autonomy: 'In the future Canada's own "white coal" of falling water would deliver the Dominion from its "hewer of wood" servitude to American industry and its bondage to American coal.'[78] He expected that some Canadian entrepreneurs would seize the opportunity to manufacture more sophisticated goods and, thus, liberate Canada from continental subservience. However, such concentration on export-oriented commodity extraction continued in northern Ontario, where exploitable resources were still available for invited investors. Nelles summarizes the acceleration of resource and hydro-power extraction: 'New [northern] Ontario more than lived up to its abundant promise during the first decades of the twentieth century. As had been predicted, hydro-electric plants and pulp mills appeared along the northern rivers; iron mines, smelters and steel mills were opened; and the miners of the province, in a wild, exhilarating succession of rushes, uncovered a series of fabulous gold and silver bonanzas.'[79] Some government officials were uneasy about the foreign exploitation and export of such natural resources. As Nelles points out, 'to the director of the Bureau of Mines it seemed unforgivable "that where nature has been so bountiful the citizen folds his arms and the enterprising foreigner is invited to step in to win and carry away the treasure."'[80]

So far, I have demonstrated how private power companies built their Canadian plants for electricity export to US industries. With the public power movement gaining strength, such continentalism in energy integration was temporarily stymied. Southern Ontario manufacturers resisted electricity exports because they needed electricity in order to replace their steam-powered technology with electrical machinery. In this way, they initiated the reversal of US ownership of power utilities in Ontario. Resistance to US ownership in the hydro sector, however, did not extend to other industrial sectors, and public power continued to serve branch-plant expansions and foreign-directed resource processing. US-owned power utilities and their Canadian counterparts developed the first imitative hydro-technology industries in Ontario. Moreover, the uncertainties of foreign-directed economic development caused major coordination problems for the Ontario Power Commission. It became evident that the changing electricity needs of foreign industry were difficult to anticipate.

Predicting Power Use for Unpredictable Firms

The government's attempt to provide electricity for a number of industrialization processes split by private (foreign) capital formation contributed to severe planning difficulties, including excess construction of the hydroelectric infrastructure. That was the case when the Ontario Power Commission expanded its infrastructure to serve small manufacturers, US branch plants, and the export-oriented natural resource-based industries in Ontario. The publicly owned Ontario Power Commission was faced with demands from manufacturers that electricity be turned on for branch plants and turned off for consumption in small towns.

In order to supply municipalities and industries, which increasingly demanded benefits from the cheaper public 'power at cost' rather than the more expensive private 'power at profit,' the power commission initially signed private supply contracts and subsequently bought power plants and power companies. The commission's first purchase in 1914 was the Big Chute plant (4 MW), built in 1909, on the Severn River.[81] By 1917, it had integrated into its system the Ontario Power Company's plant at Niagara Falls, Ontario, built in 1905 to supply industries in Buffalo, NY.[82] By 1920, in the Thunder Bay service area, the commission had added the Cameron Falls development on the Nipigon River to serve pulp and paper companies and to supply the twin cities of Port Arthur and Fort William (now Thunder Bay). In 1921, the public utility officially inaugurated its

Queenston-Chippawa plant, at the time hailed as the largest in the world.[83] And by 1922, the commission had negotiated to purchase the Toronto syndicate's Toronto Power Company, which included the Electrical Power Company's plant at Niagara Falls. It had taken thirty-five years to reverse the privatization of hydro development at the Falls.

Before long, the power commission faced accusations of having built surplus capacity and of having overestimated industrial demand. The 1925 report of a hydroelectric inquiry commission, known as the Gregory Commission, revealed that by October 1921, Chief Engineer Gaby's 'estimates of a demand for from 25,000 [19 MW] to 30,000 h.p. [22 MW] were far from reached.'[84] On the one hand, the Ontario Power Commissioners tried to meet resource company requests; on the other hand, their industrial power consumption estimates were unreliable and made planning plant capacity problematic. In one instance, when the commission's chair was accused of overbuilding the system, he deflected blame onto the Ontario government's failure to make a 'binding contract with the Carrick interest,' also known as the 'the old Tory Timber Ring,' which included the former mayor of Port Arthur.[85] When the Carrick interests had not honoured their request for electricity, the power commission had generated a surplus capacity of '10,000 to 15,000 horsepower [8 to 11 MW].'[86]

Large transnational corporations requested similarly inflated energy demands from the commission, as in the case of the Goodyear Tire and Rubber Company of Canada. Goodyear's manager, C.H. Carlisle, was one of ten representatives of the Canadian Manufacturers' Association who, together with four representatives of the Hydro-Electric Commission, attended a meeting in Toronto on Tuesday, 4 May 1920. The key advocates for the interests of large industrial power consumers were Carlisle and J.G. Perrin, manager of the Willys-Oberland Company. Sir Adam Beck, chair of the Hydro-Electric Power Commission of Ontario, defended the commission's approach to hydro planning and to the distribution of electricity.

Supporting Perrin's contention that industry should get priority access to 'power at cost' instead of wasting it by supplying small Ontario towns, Carlisle demanded a larger allocation of power for Goodyear:[87]

> Our present plant is one twelfth of the plant we planned for in New Toronto, our Company is employing about three thousand people, and this new development will call for about six thousand five hundred people. I have been informed that we can get no increase of power. When I located in New Toronto, I took it up with some of your representatives and was assured of a continuous power and plenty of power. We have made an investment at the present time of $6,842,000. The additions mean two and a half million dollars more, so we have quite an investment. In planning this plant we made no provision for space for [sic] steam plant ... *We will*

need by January 1st, [1921], 6500 H.P. and we get a promise of 2300. That is one reason that I think we should first see that the manufacturing interests that employ the people of this Province and the concerns that are practically the backbone of commerce should be first considered, and their future extensions be provided for.[88] [my emphasis]

Perrin thought it a waste that small communities (he did not even mention small manufacturers) should have electric lighting and that, instead, more reliable power should be supplied to his factory and to Goodyear. Perrin observed that 'a short time ago I had occasion to pass through a number of hamlets and small villages in the Province. Now it is very nice to see them all lighted up and all that but it seems a waste and this power should be devoted to industrial [transmission] lines.'[89]

Responding to his fellow manufacturers, Beck explained the difference between the obligations of a cooperative public utility and a private utility with regard to supplying electricity: 'You [large-scale manufacturers] are getting power at cost, and I think we have [a]lways made an equitable adjustment as between the manufacturer, street lighting service, and the individual householder. You say the manufacturer should have some preference, but as I have already explained, this is not possible. It is a municipal affair, and we cannot say as a private company might say that we will not take on this town or that village, because the manufacturers pay us a better price and it is cheaper and more convenient to render one bill instead of 10,000 bills.'[90]

Since the power commission could not supply the requested power, and the Goodyear management did not want to wait for the industrial supply that would come on line from the Queenston plant, Goodyear signed a contract with the Toronto Power Company (also called the Electrical Development Company), which sold power for profit rather than at cost.[91] As indicated in Goodyear memoranda, the cost of power under the Hydro Commission contract was $22.75 per hp per year and that of the Toronto Power Company was more than triple the rate at $72.51.[92] Goodyear signed a contract in 1920, and, 'under its terms, the Toronto Power Company agreed to supply and hold in reserve for the Goodyear Company, 3000 HP ("Firm Power") during 24 hours of every day for a period of five years from 1st January, 1921, to 31st December, 1925.'[93] Just two months later, Goodyear no longer needed the power. In the company's defence, Carlisle explained to the commission, 'As you know, the American Goodyear became involved with losses of upwards of $70,000,000.00, causing us a loss through the contracts we had with them for foreign business of somewhat over $5,000,000.00. This loss made our company insolvent, and it was necessary to refinance and reorganize.'[94]

Carlisle argued that, for these reasons, 'the Goodyear Company was in no position to take the additional power specified in the contract with the Toronto Power Company.'[95] The value of the legal and collectible contract, a sum of $360,000, was scheduled as one of the assets of the Toronto Power Company, which the Hydro-Electric Power Commission was negotiating to purchase.[96] The compromise acceptable to the Goodyear Company was to suspend its contract with the Toronto Power Company and to obtain a power rate of $29.27 per hp per year from 1921 to 1926 (reduced from $72.51 per hp per year) from the commission.[97] As will be shown in later chapters, other Canadian utilities have similarly overbuilt their capacities at the urging of expansion planners in transnational corporations who thereafter reneged on their electricity orders.

Such turns of events demonstrate the serious problems that emerge when a public power company tries to meet the electricity needs of manufacturing in small towns, foreign-directed branch plants, and natural-resource processing. Attempts to supply public energy for such varied paths of industrial growth often result in power surplus capacities. Schemes to export such surplus capacity became a contentious issue during the second decade of the twentieth century.

Repatriation Crisis: Power Exported Is Power Lost

The federal government's reactions to the export of electricity varied. The government instituted controls in 1907 but allowed them to slacken prior to the First World War. This led to what is known as the repatriation crisis of 1917, when Canada was unable to reclaim electricity exports from the US in order to supply electricity for its own production.[98]

The federal government's earliest position appeared to have been that electricity should be treated as any other good; in other words, it could be exported at the discretion of the electric utility that owned the power. Then, unexpectedly, Canadian subsidiaries signed long-term export contracts – of up to a century – with their US parent utilities. Dal Grauer, a political economist and former professor of social science in Toronto, reviewed Ontario's export history in 'Export of Electricity from Canada,' an essay he published while he was chair of the BC Electric Company. In this essay, he indicates that 'the Ontario Power Co. in 1904 envisaged the export of 45,000 horsepower of electrical energy to the US for a period of 99 years.' The threat of Canadian water powers being absorbed by the US led to the passing of An Act to Regulate the Exportation of Electric Power and Certain Liquids and Gases by the federal government in 1907. Grauer reasons that this act essentially restricted exports out of concern that Niagara Falls power and other sites would not be available for future Canadian needs. Now, exports of electricity, as well as international power lines, came under federal jurisdiction. Every export licence needed a government permit; electricity diversion through export was not permanent; and licences were revocable when electricity was required by purchasers in Canada after approval by the governor-in-council (the Cabinet).[99]

US attempts to divert more electricity from the Canadian side of the international boundary had led the Laurier government to establish the National Commission of Conservation under the minister of the interior, Sir Clifford Sifton, in 1909. His commission reported that 'should water power be exported to the US, the vested interests which it would create there would prevent its subsequent withdrawal to meet future needs of Canadian industries.'[100] US companies in Canada now had to obtain an export licence to send electricity back to their home markets. The New York State Public Service Commission, however, showed little respect for Canadian restrictions on export. The commission wrote in 1914 that when 'affecting so important a subject as the means of continuing great industries ... the time has long since passed when governments proceed ruthlessly from pure national rashness or anger to destroy the settled accepted commercial relations.'[101] This US position raised a storm in Ottawa, and the Canadian Privy Council sent His Majesty's ambassador in Washington to the US government with a carefully worded minute outlining Canadian energy export laws and regulations, which included indication that export licences are revocable, valid for only one year at a time, and not permanently binding in the case of exports to US industry.[102] Nevertheless, as shown in Table 3.2, exports from Ontario had risen steadily until 1914, levelled off in 1915-6 (the two years after the protest), but returned to higher levels again in 1917-9. 'Having allowed export agreements to be made ... they were unable to repatriate firm power [once] exported.'[103] Canadian administration, at both the provincial and federal levels, was found to be wanting.[104]

The repatriation crisis intensified when Canada needed electricity for war production in 1917-8. During the war, the federal government's inability to enforce the repatriation of Canadian electricity through legislation became clear. The power comptroller, Sir Henry Dayton, found that export commitments and industrial demands had absorbed existing capacity. Grauer maintains that 'the real explosion [of anti-export sentiment], if it can be described as such, was heard in 1917.'[105] It occurred at a time when Canadians most needed electricity, when so much of the power in the Niagara area was used for war manufacturing that power needed to be rationed by the power comptroller. When the Imperial Munitions Board consulted the Niagara Falls hydroelectricity producers, it 'found that export commitments, together with the already inflated demands of industrial and other consumers in Canada, had absorbed practically all of their existing capacity.'[106] Adam Beck described the situation in the following way: 'Industries [in Ontario] either had their power cut off, or reduced to a point which entailed great financial losses; in many cases

Table 3.2

Annual quantity of electricity exported to the United States, 1908-20 (GWh)

Year	From Ontario	From Quebec	From other provinces	Total exports
1908	113			113
1909	358		1	359
1910	474		1	475
1911	536		2	538
1912	536		2	538
1913	656		6	662
1914	746		27	773
1915	605	29	22	656
1916	647	359	16	1,022
1917	779	429	17	1,225
1918	730	381	16	1,127
1919	731	396	16	1,143
1920	643	283	24	950

Sources: 1908-10, Canada, Department of Inland Revenue, publications in Dominion Bureau of Statistics Library; 1911-20, *Canada Year Book* (cited by Dal Grauer, 1961: 257).

almost complete paralysis of business was experienced. At this period the Hydro Electric Power Commission was supplying power for the operation of over 360 plants manufacturing munitions and war supplies, and these plants were using over 80 percent of the entire power supply in the Niagara district.'[107] In a letter to T.J. Hannigan and S.R. Clement he added, 'One can hardly find fault with our neighbours to the south for desiring to have such a valuable commodity to aid in building up their industries and communities, but it is scarcely to be expected that Ontario citizens can be induced to part with a commodity so essential to their own necessities and welfare.'[108]

The 1917-8 power shortage in Canada made clear to utility executives, industrialists, and residential customers the near impossibility of repatriating power. Leading politicians all 'spoke to much the same effect, namely, *that Canada should never again export firm power*'[109] [my emphasis]. In 1929, Prime Minister W.L. Mackenzie King summarized the rethinking of the use of hydroelectricity in Canada since the turn of the century: 'Public opinion in Canada ... is insistent that such power ... shall be utilized within the Dominion to stimulate Canadian industry and develop the natural resources.'[110] The experience of the Niagara Falls power export trap strengthened the perception of electricity as Canadian industrial energy but did not lead to a comprehensive national energy or industrial policy. Numerous federal government debates between 1925 and 1937 supported Dayton's view that 'power exported is power lost.'[111] Provincial premiers expressed similar views, as Armstrong found that George Ferguson, the premier of Ontario, and L. Alexandre Taschereau, the premier of Québec, held several meetings in early 1925 'at which they condemned electricity exports to the United States.'[112] Despite such pronouncements by federal and provincial governments, as Armstrong tells us, from the 1920s to the 1950s federal and provincial support for exports varied with provincial shortages or surpluses of electricity and with the size of proposed power projects on rivers suitable for navigation, canals, and water-power developments (e.g., the Ottawa and St. Lawrence Rivers).[113]

Conclusion

Generally, Canadians continue to assume that energy infrastructure expansions are followed by industrial development, as had been the case in Britain during its Industrial Revolution. At Niagara Falls, however, developers of a transnational hydroelectric infrastructure stifled industrial growth in Ontario and, instead, strengthened it in New York State. The growth that did occur was not of the type and quality anticipated by engineers, utility executives, and politicians. Ontario's dependence on technology transfers, importation of entrepreneurs, and US capital showed less industrial autonomy than had been the case in Britain. In other words, Canada has not followed the European industrial development model.

The Niagara experience also casts doubt on another currently strong assumption among Canadians and their provincial governments: that electric power development should be left to the private sector. Local governments at Niagara were soon forced to reverse their allocation of water power from the Falls to private, US-owned profit-making utilities in order to repatriate and nationalize the Niagara energy necessary for the progress of Ontario manufacturers.

As shown above, the private power companies' failure to transmit power to the small manufacturers in southwestern Ontario towns gave rise to the formation of the publicly owned Ontario Power Commission. It belatedly began bridging the transmission gap, so that, by 1910, small manufacturers could convert their factories from American coal-fuelled steam engines to industrial electric motors. This reversal of privatized development at Niagara Falls offered an object lesson to other provinces: do not select foreign ownership for the development of hydro utilities. Reliance on that form of ownership also contributes to imitative technology, difficulties in planning industrial electricity supply, exports to the US, and the repatriation crisis.

The federal government, which controls export policy, reversed its original view that power could be treated like other exports and began advocating that electricity exports be stopped. The threat of having Ontario water power absorbed by US industry resulted in the Exportation Act of

1907, which, however, was insufficient to repatriate power needed for Canadian war industries in 1917 and 1918. As a result of the repatriation crisis, a temporary 'consensus' underpinned power policy: electricity, Mackenzie King emphasized in 1929, 'shall be utilized within the Dominion to stimulate Canadian industry and develop natural resources.'

Would Canada's political elite and planners learn from the Niagara case that privatizing water-power sites may not be in the best interest of provinces, that adding new hydro plants may not lead to the kind of manufacturing expected, that the timing and size of new hydroelectric projects may not coincide with industrial need, and that exports to the US may not be in the interest of the country as a whole? Would Québec, like Ontario, pursue a provincial hydroelectric policy? As I will demonstrate, forty years later, members of Québec's political elite repeated history. At the beginning of the 1960s, Québec integrated its hydroelectric expansion program into its economic sovereignty strategy at the moment that Canada achieved the technical ability to build a national power network.

4

Power from the North and Neighbour: Distinct Interconnections (Québec)

Introduction

During the early 1960s, western Canadian planners – oblivious to the onset of the Quiet Revolution in Québec – argued that Canada's planned national power network would permit the flow of energy from the east and west towards the centre of the country.[1] They presumed that Québec would support the national power grid and would dam its mighty rivers, not only for its own electricity needs, but also for export to Ontario, New Brunswick, and, in a lesser amount, to Nova Scotia.[2] Planners assumed that the construction of relatively low-cost projects such as Churchill Falls would allow the deferral of the more expensive hydro and other energy projects, and the benefits from Canada's large northern sites would be integrated and administered by a central planning and operating authority.[3] The province of Québec has some of the best power sites in Canada, including Shawinigan Falls, river sites in the James Bay region, and on the Manicouagan and Outardes Rivers (Figure 4.1). While sites such as that at Shawinigan Falls and at Beauharnois on the St. Lawrence near Montréal were developed well before the mid-1940s, southern and northern river expansions started during the 1960s with the Quiet Revolution.

Subsequently, Premiers Jean Lesage, René Lévesque, and Robert Bourassa directed an enormous hydro expansion that did not begin to slow down until 1994 when Premier Jacques Parizeau cancelled the Great Whale hydro project in the James Bay area. Bourassa promoted the continental energy policy favoured by the US in the hope of developing Québec as the 'Kuwait of the North.' Within Québec, he proposed hydro development as the cornerstone of a 'new economy,' which was to bring increased autonomy from the political and economic control of English Canada; protests by Aboriginal peoples, environmentalists, and others were often dismissed as obstructing industrial progress. A detailed examination of this process and parallel developments in Québec – mixed and changing hydro ownership, hydro power as a determinant of Québec's industrial growth

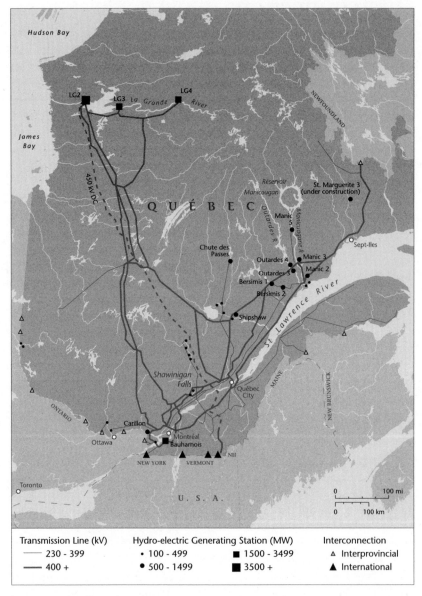

Figure 4.1 Québec's main system features

Sources: Government of Canada, Department of Energy, Mines and Resources, Energy Sector, *The National Atlas of Canada*, 5th edition, Electricity Generation and Transmission, 1983, MCR4069 and MCR4144; Hydro-Québec, TransEnergy 1998.

(assisted by a 'buy-Québec' or 'achat chez nous' government procurement policy), the surplus of power plants, and interconnection with Québec's southern neighbours – will demonstrate that Québec, like other provinces, acted in its own interest when it developed northern power in order to attract foreign industry and to support continental integration. The brief historical sketch that follows will set the background and will demonstrate that the provincial government of Québec first allocated water-power rights and their development to private interests and then reversed this process by nationalizing utilities in order to regain water-power rights and to develop hydroelectricity in the provincial interest – a pattern also evident in other provinces.

Reversal of Privatization

Québec governments from the 1890s to the 1960s had transferred the rights to develop the best river sites to private, mostly anglophone, hydro developers and relied on them to dam rivers, construct powerhouses, and transmit electricity. For most of the twentieth century, politicians and government planners expected their new hydro projects to bring new kinds of industries to Québec. However, Québec governments increasingly found, as, earlier, did those in Ontario, that often provincial development interests did not match those of private power companies. As it turned out, private hydro developers were more concerned with maximizing their profits from their utility investments than with fostering overall regional economic development: private utilities, for example, sold overpriced electricity to their urban customers, showed little interest in supplying smaller, more remote industries in rural areas, and employed too few francophones.

From the 1940s onwards, Québec governments reversed these earlier privatizations by reappropriating water power and by taking over generating facilities and transmission systems in order to gain greater control over the building of the 'new Québec.' The initial goal of reappropriating development rights over hydro power was to assure an adequate and better-priced supply of power in Montréal and its industrial hinterland. During the Quiet Revolution, with the help of René Lévesque, as will be discussed below, hydro development was intended to employ the francophone middle class and to foster indigenous industrial development; thereafter, it was to serve Premier Robert Bourassa's vision of building 'l'Alberta de l'Est.'[4] Bourassa's government acted to attain more energy autonomy from English Canada, globalized its invitation to industry, and planned to export energy and other natural resources (staples) to the declining industries in the New England states. Becoming 'a major energy hub' in the North American electricity market based on developing nine hydroelectric projects in the North is the revised concept central to Hydro-Québec's strategic plan approved in 1998 by the Lucien Bouchard government.[5] For instance, the 9 March 1998 announcement of Québec premier Bouchard and Newfoundland premier Brian Tobin to start hydroelectric power development talks about the Lower Churchill Falls development in neighbouring Labrador is part of this Québec strategy.[6] Looking back on the second half of the twentieth century, it becomes evident that the expansion of Québec's electrical system became part of ever more ambitious provincial development goals and occurred in several phases between the 1940s and 1990s.

Québec Government Intervention (1944)

On 15 April 1944, the provincial Commission Hydro-Electrique du Québec (Hydro-Québec) took over a major electrical utility, the Montreal Light, Heat and Power Company, on the grounds that it did not provide adequate public service. The Montreal Light, Heat and Power Company's record before the mid-1940s included a number of shortcomings: contentious pricing policies in Montréal, disregard for government regulations since the mid-1930s, restraint of industrial activities because of its high industrial rates, and insufficient investment in plants on the St. Lawrence and the lower Ottawa Rivers.[7] Other private utilities in Québec conducted their business in a similar way during the 1940s and 1950s. They aimed primarily at profit, did little to fill insufficiently profitable transmission gaps in the regional, particularly rural, hydro infrastructure despite government subsidization of such private electricity producers, and (as writers of hydro history in Québec, André Bolduc and Daniel Larouche, tell us) blocked francophones' access to many utility positions. Increasingly, Québécois came to think of utilities as 'les grandes entreprises d'électricité symbolisant la domination économique des anglophones au Québec' [the large power companies symbolizing the economic domination by Anglophones in Québec].[8] Such sentiments aided the nationalization of private electricity companies for the greater social and economic benefits of Québec during the 1960s and accompanied the mandate of Hydro-Québec to develop major new sources of hydroelectric power for the province.

Maîtres chez Nous (1963)

During the Quiet Revolution, politicians and Québec nationalists perceived hydro development to be essential for gaining more autonomy from English Canada. At the time not a broadly popular move, their strategy included purchasing additional power companies and damming more rivers for industry. Once most of the electrical system was unified under government ownership, industrialization could be encouraged, and economic benefits could be guaranteed through a 'buy-Québec' policy. Dissatisfied with the anglophone private-sector efforts, Jean Lesage, the premier of

Québec from 1960 to 1966, saw the Québec state as the most efficient instrument of economic liberation that Québecers possessed.[9] René Durocher, a history professor at the Université de Montréal, claimed that the Québec government's 'most spectacular accomplishment in economics was the nationalization of private electricity companies.'[10] The front-page headline in *La Presse* on 1 October 1962, during the provincial election campaign, affirmed that the provision of electricity was seen as the key to Québec's autonomy: 'L'électricité: clef qui nous rendra MAÎTRES chez nous.' [Electricity: key to making us masters in our own house.][11]

In a series of speeches leading up to the provincial election in February 1962, René Lévesque, then Québec's minister of natural resources, pointed to the failures of the remaining private power companies.[12] Private utilities had failed to provide opportunities for francophone upper-level management, despite their undisputed qualifications.[13] He argued that an enlarged Hydro-Québec could increase equity for French Canadians: 'Les Canadiens français étant économiquement faibles, il appartient au gouvernement du Québec d'adopter une politique qui favorise le progrès de la majorité francophone, sans pour autant défavoriser les anglophones.' [French Canadians being economically weak, it is up to the Québec government to adopt a policy that favours the progress of the Francophone majority, without putting the Anglophones at a disadvantage.][14] In economic performance, private utilities had not sufficiently equalized prices and had failed to provide industrial energy to strengthen economic growth in many regions of Québec. Lévesque emphasized that 'il faut, de toute urgence, fournir l'énergie à meilleur coût aux régions de l'Abitibi, de la Gaspésie et du Bas-Saint-Laurent' [one must with the utmost urgency provide energy at lower costs to the Abitibi, Gaspé, and the Lower Saint Lawrence River regions][15] as part of the Lesage government's resolve to stimulate industrial progress in these regions.[16] According to Lévesque, the reappropriation of water-power rights and the purchase of utilities would give francophones more employment opportunities in the energy sector and would fill the transmission gaps left by the private sector.

During the next fifteen years, from 1963 to 1978, Hydro-Québec quadrupled its size through buyouts of existing power facilities and by damming power sites on the Manicouagan and Outardes Rivers. On 30 April 1963, Hydro-Québec took over private electrical utilities, including the Shawinigan Water and Power Company, the Companie Québec Power, the Southern Canada Power Company, the Gatineau Power Company, the Companie de Pouvoir du Bas-Saint-Laurent, the Companie Électrique du Saguenay, and the Northern Québec Power Company as well as other smaller companies.[17] To finance this takeover, Hydro-Québec sold twenty-five-year bonds amounting to $300 million at 5 percent interest on the American market.[18] This capital allowed for a substantial increase in Hydro-Québec's generating capacity. To give some idea of the extent of Lévesque's 1963 takeover, the generating capacity involved is equivalent to the electrical load required to run 600 universities. Among major power users, universities need roughly 10-12 MW, pulp and paper mills 40 MW, and aluminum smelters 100 MW of generating capacity. Hydro-Québec's capacity before the takeover was 3,675 MW from its nine power plants; the 1963 nationalization of the remaining private utilities added 6,224 MW from forty-one power plants.[19] In the 1960s, Hydro-Québec also continued the Manic-Outardes complex, which, by 1978, augmented Hydro-Québec's capacity by a further 5,517 MW.[20]

The Manic 2 powerhouse, sixteen kilometres from Baie Comeau, began generating power in 1965, and by the end of November, the 735 kV transmission line (the first in the world to carry such voltage) transported power from Manicouagan via Lévis to Montréal.[21] While these southern system expansions were intended to develop francophone industries within Québec, the northern expansions – Churchill Falls and James Bay – were to serve the Québec government's strategy of inviting industrial investors from outside Québec, while strengthening interconnections with networks in the eastern United States and supplying natural resources from the James Bay region to the New England states.

Northern Expansions: Churchill Falls (1969) and James Bay (1971)

In 1969, Hydro-Québec bought nearly all of the Churchill Falls power from the Churchill Falls (Labrador) Power Company, a subsidiary of the British Newfoundland Company (Brinco). Brinco's corporate behaviour later influenced how the Bourassa government proceeded with its corporate approach to the development of James Bay. In 1953, Brinco had received extensive mineral, forestry, and water-power resource rights in Labrador and on the Island of Newfoundland from the Newfoundland government. This British-owned firm was based in Montréal, rather than in St. John's, and directed the extraction of Labrador's resources by incorporating each resource development venture separately. For instance, its rights to Labrador resource exploration ventures were incorporated as British Newfoundland Exploration Limited (Brinex), and its rights to develop Churchill Falls became the Churchill Falls (Labrador) Power Corporation. Brinco and its subsidiaries developed hydro projects in three stages: they first contracted to sell the electricity, then used their power contract to raise the capital for the hydro project and, thereafter, developed the site. From its Montréal offices, Brinco's Churchill Falls (Labrador) Corporation sold power to Hydro-Québec in 1969, raised the capital in New York, and supplied the first power from the Churchill project by 1971.[22] By 1972, the 5,225 MW powerhouse became to a large degree part of the Québec power system (Figure 5.1).[23]

Following Brinco's corporate approach to the development of Labrador, in 1971 the Bourassa government inaugurated the James Bay Development Company (JBDC or SDBJ) to promote resource extraction by foreign companies in the James Bay region, and the James Bay Energy Company to plan and build hydroelectric facilities.[24] With its James Bay scheme, the Bourassa government continued 'the conquest of northern Québec, its rushing, spectacular rivers, its lakes so immense they are veritable inland seas, its forests of coniferous trees concealing fabulous mineral resources of all kinds.'[25] According to Article 41 of Bill 50, approved by l'Assemblée Nationale du Québec on 14 July 1971, the JBDC was granted all remaining public rights to mining, hydraulic resources, forests, hunting, fishing, agriculture, colonization, and tourism within the James Bay region.[26] The JBDC, like Brinco, became the government's management agency for the development of the James Bay territory.[27] Funded with '$100 million capital underwritten by the province,' it was to cooperate with the James Bay Energy Corporation (JBEC or SEBJ).[28] On 21 December 1971, in Premier Bourassa's words, 'the development of the area's hydroelectric resources was entrusted to the James Bay Energy Company.'[29]

Initially, Hydro-Québec thought the northern project premature, but then, 'on January 21, 1972, Hydro-Québec agreed to underwrite 70 million shares of [the James Bay Energy Corporation], stock totaling $700 million, or 70 percent of [JBEC's] authorized capital.'[30] A newspaper report of Bourassa's discussion with financier David Rockefeller of New York stated that 'Premier Bourassa had already warned Ottawa not to impose any measures which would restrict in any way job-creating investments from south of the border.'[31] With the help of borrowed money from the United States, Hydro-Québec gained a major influence over the building of James Bay dams.

The JBEC dammed the La Grande River at three sites. By far the largest dam is LG2, which is located 117.5 kilometres upstream from the James Bay Delta of the river. La Grande Phase 1, which included the generating stations of 'LG2, LG3, and LG4, cost nearly $15 billion in total for an installed capacity of just over 10,000 megawatts' expansion which, together with Churchill Falls, doubled Québec's power capacity (Figure 4.1).[32] Overall, power from the North became part of the Bourassa government's strategy to advance US-Québec transborder electrical network integration.

Bourassa frankly hoped that James Bay's power and semi-processed resources (industrial staples) would benefit the US northeast – a region that in the 1970s had been under threat of deindustrialization due to the southward migration of its industries:[33]

Le Nord-Est américain a connu au cours de la dernière décennie une perte de poids relatif dans l'économie des Etats-Unis. Un grand nombre d'industries se sont relocalisées dans le Sud des Etats-Unis, qu'on appelle parfois le 'croissant de soleil' ...

Le déclin du Nord-Est américain nuit au commerce extérieur du Québec, puisqu'il s'agit de notre principal partenaire commercial parmi les régions américaines. Par ailleurs, il apparaît que les entreprises du Sud des Etats-Unis ont une faible propension à importer du Québec que [sic] les entreprises du Nord-Est américain, essentiellement en raison des coûts de transport. Par example, les entreprises d'Arizona achèteront leur bois, papier, aluminium et amiante en Colombie-Britannique plutôt qu'au Québec, alors que la situation inverse prévaut pour les entreprises fonctionnant près de l'Etat de New York.

Nous voyons donc qu'il est dans l'intérêt direct des Québécois d'aider les Américains de la Nouvelle-Angleterre à renforcer leur économie.

[The American North East has known for the last ten years a loss of relative power in the economy of the United States. Numerous industries have relocated in the South of the United States, that is sometimes called the 'sun belt' ...

The decline of the North East is damaging to Québec's foreign trade, as it is our principal commercial partner among the American regions. Moreover, it seems that the businesses of the southern part of the United States are less likely to import from Québec than the businesses from the American North East, essentially for reasons of transportation costs. For instance, Arizona businesses will buy their wood, paper, aluminium, and asbestos in British Columbia rather than in Québec, while the reverse prevails for the businesses located close to the State of New York.

Thus we can see that it is in the direct interest of the Québecers to help Americans from New England to strengthen their economies.][34]

The mandate to develop electricity as an industrial input within the boundaries of Québec, so typical of the Quiet Revolution of the 1960s and of Lévesque's policy, changed during the 1970s and 1980s as Bourassa directed the damming of northern rivers for foreign industry and export to the US. On the one hand, he planned to use James Bay hydroelectric energy to support manufacturing in the New England states; on the other hand, he planned to use it to attract foreign industrial firms to Québec.[35] Besides expecting growth in the resource sector, Québec governments assumed industrial diversification would follow based on the growth of supply industries for hydro projects and on the implantation of new industries attracted by cheap electricity. As events showed, the hydro scheme built under Bourassa's policies fostered neither the expected quantity nor the quality of such growth.

Linkages to Industry

Contrary to the anticipation during the 1960s through to the 1980s that new hydro projects in Québec would increase economic liberation, autonomy, and indigenous industrialization, new supplies of power have attracted or enlarged mostly old types of resource industries. Existing research, such as studies of Québec's industrial and technological development, federal government statistics, manufacturers' records, a Québec government report, and Hydro-Québec's industrial contracts, supports the claim that hydro developments failed to deliver on their promises. The prominent shortcomings of industrial growth from hydro development include its dependent and imitative nature, the continuation of weak inter-industry integration, the use of electricity for predominantly primary processing of natural resources (e.g., pulp and minerals), the refusal of foreign branch plants to allow their technology to be built for export from Canada, and the failure to capture sufficient electricity revenue from global players in industry. In addition, those new industries that were attracted to new supplies of power needed far less industrial capacity than was planned for them.

To discover whether hydro development is linked to an increase in new economic activities, one can trace whether the development of new supply industries resulted from utility purchasing preferences.[36] For instance, the choices made by Hydro-Québec and the James Bay Energy Corporation in the procurement of equipment for new hydro plants largely determined the local growth of industries producing technology for hydroelectric stations.[37] State corporations such as Hydro-Québec and the James Bay Energy Corporation played a 'nationalistic' role in the import substitution manufacture of generators. Both corporations had some success in increasing local hydro equipment manufacture, though they were dependent on foreign investment and technology.

Non-Québécois Industries

Although Ontario and Québec developed their own hydro systems over an eighty-year period, foreign dependency in this sector was actively encouraged by utilities. For example, Faucher and Fitzgibbins found that some firms were lured into Québec with the promise of guaranteed markets, 'and/or preferential pricing along with other advantages (in the form of government loans, subsidies, and fiscal incentives).'[38] By 1979, of forty major hydro-technology firms in Québec, half were foreign subsidiaries, many of which remained technologically dependent on parent firms.[39] Canadian Westinghouse (US-owned) and the Canadian General Electric subsidiary, Dominion Engineering Works (reliant on its US parent for its direction), were dominant supply companies for Hydro-Québec and the

Table 4.1

Turbines and generators in power plants on the La Grande River, James Bay

La Grande River power plants	Capacity of plant (MW)	Years of turbine installations and manufacturer	Years of generator installations and manufacturer
L G 2	5,328	1979-81 DEW, MIL	1979-81 CGE, ASM
L G 3	2,304	1982-4 DEW	1982-4 MIL
L G 4	2,651	1984-6 MIL	1984-6 MIL

Note: DEW = Dominion Engineering Works; MIL = Marine Industries Ltd.; CGE = Canadian General Electric; ASM = Alsthom Savoisine, Marine Industries.
Source: Statistics Canada, *Electric Power Statistics: Inventory of Prime Movers and Electric Generating Equipment*, cat. 57-206, 31 December 1986.

James Bay Energy Corporation (Table 4.1; Appendix 20). The exception was Marine Industries, then owned by the governments of Québec (65 percent) and France (35 percent), which used 'French technology (bought from Neyrpic, a subsidiary of the Alsthom group)'(Table 4.1).[40] Yet, even these locally owned firms that manufactured in Québec, rather than importing such equipment, relied on foreign initiative and technology.

Moreover, the 'achat chez nous' policy of procuring hydroelectric equipment purchased in Québec had lost its effect by 1970. A case study by the Centre de recherche en développement industriel et technologique (CREDIT) shows that 'even with the massive investment realized in the seventies, the industrial structure of the electrical equipment industry in Québec has not changed a great deal.'[41] The authors of this study, Jorge Niosi and Philippe Faucher, found in 1988 that 'the value of the goods and services purchased for hydro projects in Québec had not increased since 1971.'[42] In particular, Niosi tested the ability of 150 companies to export hydroelectric technology from Canada. He found that some Canadian-owned hydro equipment firms (not necessarily suppliers of Canadian hydro utilities) were interested in exporting. Yet, foreign-owned branch plants in the same industry had no export mandate from their head offices and were not even permitted to bid on Canadian-engineered hydro projects in the Third World and other countries.[43] The exception was Dominion Engineering Works: although owned by General Electric, it retained its world product mandate.[44]

New Energy: Same Old Industries
Contrary to popular assumptions that hydro power is a determinant of industrial diversification, manufacturers' records from the 1960s, 1970s, and 1980s reveal that the new supply of electricity from Manicouagan, Outardes, Churchill Falls, and James Bay did not significantly change the

structure of Québec's industry. During these three decades, approximately 80 percent of the manufacturers used the new supplies of electricity for the same 'old' semi-processed, extracted, and export-oriented commodities: pulp and paper, chemicals, and primary metals (Table 4.2 and Appendix 4).

The report, *L'Électricité: Facteur de developpement industriel au Québec* (by Québec's Direction générale de l'Énergie et al. in 1980), revealed an even higher concentration of electrical consumption in the older semi-processing industries in a few industrial sectors. The report concluded that about half of Québec's electricity was used by industry in 1977 and that 85 percent of

Table 4.2

Consumption of electricity in manufacturing industries in Quebec, 1962-84

Year	Total consumption in manufacturing (GWh)	Three major sectors			Sum of three sectors (%)
		Pulp and paper (%)	Chemical (%)	Primary metals (%)	
1962	17,656	52	10	21	83
1963	16,979	48	10	23	81
1964	18,264	45	10	25	80
1965	18,712	44	11	24	79
1966	19,679	43	11	25	78
1967	19,872	40	11	26	77
1968	21,110	40	10	27	77
1969	23,343	39	10	30	79
1970	23,322	40	10	27	77
1971	23,074	40	10	25	76
1972	24,504	42	9	25	76
1973	26,549	40	10	25	75
1974	28,475	41	9	27	77
–	–	–	–	–	–
1976	27,967	44	10	22	76
1977	28,693	42	9	25	76
1978	31,976	42	9	25	76
1979	33,261	40	9	26	75
1980	36,126	39	10	28	77
1981	37,532	41	–	27	–
1982	34,580	42	10	25	78
1983	36,510	44	10	24	78
1984*	42,277	50	9	21	80

* Manufacturing survey was discontinued in 1984. Since manufacturing has not increased in the recent decade in Québec and Canada, I would expect to find little change in the industrial diversification pattern during the 1990s.*Sources*: Statistics Canada, cat. 57-206, *Consumption of Purchased Fuel and Electricity by Manufacturing, Mining and Electrical Power Industries, 1962-1974.* Cat. 57-208, 1975-84.

the industrial total was used by just four sectors: 'L'essentiel de cette consommation industrielle est le fait de quatre groupes industriels (85% du total): le papier et les produits connexes, la première transformation des métaux, les produits minéraux non métalliques et les produits chimiques.' [For the most part this industrial consumption occurs in four industrial groups (85% of the total): paper and related products, primary transformation of metals, non-metallic mineral products, and chemical products.][45] The other types of industries that consumed the remaining 15 to 20 percent could be run by one of the ten generators at Churchill Falls (or on 554 MW). As Hydro-Québec's industrial contracts (over 5 MW) from 1984 show, manufacturers in the non-metallic mineral sector (glass, cement, lime, abrasives) purchased 3 percent of the manufacturing electricity, the textile and clothing industry purchased 2 percent, the food-and-beverage industry purchased 1 percent, and the transportation equipment industry purchased 2 percent (Appendix 6). In all, less than 8 percent of the total electricity contracted to manufacturers in Québec was directed towards secondary manufacturing. Another aspect of the truncated industrial development in Québec is the lack of integration of major industrial power users within the Québec economy when compared with the integration of corporations within the US economy. While energy-intensive industries in the US ship 90 percent of their end products to other productive sectors within the same country, only 38 percent do so in Québec.[46] This failure to use semi-processed goods within Québec as manufacturing input is indicative of the lack of integration between primary and secondary industries within the province. Nevertheless, growth in secondary industries was usually predicted before adding new power plants to Québec's system. It seems that government policies in the 1980s were not responsive to the failure of earlier projects to achieve their goals. In particular, in 1968, Hydro-Québec had argued that integrating the Churchill plant was 'necessary for the economic development of Québec and particularly for that of secondary industry.'[47] A review of Hydro-Québec's industrial contracts from the 1970s and 1980s indicates that, contrary to Hydro-Québec's expectations, neither Churchill Falls nor James Bay led to major new industrial growth.

Churchill Falls first delivered power to the Québec system in 1971, but thereafter, during the early 1970s, industrial consumption showed only minor increases. Hydro-Québec's industrial contract statistics show that the pulp and paper sector expanded consumption by only 18 MW from 1973 to 1976 (Appendix 5), a very modest growth rate considering that the average pulp mill requires 40 MW of power. At the end of the same three-year period, the chemical industry (with an average consumption of 15 MW per customer) required 25 MW more, whereas the electro-metallurgical industry (with an average consumption of 60 MW per customer) required

100 MW less of the Churchill Falls power. Power requirements in the textile industry dropped from 57 to 46 MW, and requirements in general manufacturing dropped from 177 to 111 MW (Appendix 5). By 1976, five years after the first generator at Churchill produced a fresh load of power, the consumption of electricity in industries had hardly changed. Both the number and kind of industries stayed virtually the same after power from Churchill Falls entered the system.[48]

Only slightly smaller than Churchill Falls (5,429 MW, the largest hydro station in Canada), the La Grande 2 powerhouse (5,328 MW) started delivering power from the James Bay region in 1979. The La Grande 3 powerhouse (2,304 MW) supplied power in 1982, and La Grande 4 (2,651 MW) started in 1984.[49] These enormous new capacities from the James Bay region were not put to new industrial uses in Québec. Hydro-Québec's statistics on its industrial customers (over 5 MW) show a continued concentration of consumption in a few industrial sectors. Of the total electricity consumed by its industrial customers, the same three sectors (Chimie, Pâtes et Papiers, and Métaux [including Sidérurgie], Appendix 6) used 84 percent in 1981 and 85 percent in 1984.[50] From 1981 to 1984, Hydro-Québec signed contracts with only three additional pulp and paper mills for a total of thirty-four contracts, and one extra chemical plant signed for a total of seventeen chemical industry contracts. The loss of contracts in the primary metal category left a total of twenty-one contracts in this industrial sector in 1984 (Appendix 6). Counting on abundant energy as a base and relying on investors from outside Québec to implant new industries proved a flawed strategy for the James Bay region.

Global Invitation

Strategies to solicit more foreign participation continued at a time when foreign ownership was already high in Québec. As previously mentioned, the James Bay Development Company was established on 14 July 1971 by an act of the provincial government and was empowered to 'create subsidiaries, share projects with government or paragovernmental associates, and enter into partnership with private companies' in order to develop other industrial ventures.[51] By the mid-1970s, the JBCD was associated with twenty-six private companies in eighteen mining exploration programs and conducted nine other projects by itself.[52] Contrary to what one would expect of a province that aspires to be master in its own house, the majority of the JBDC's partners were either English Canadian or foreign enterprises such as Eldorado, Brinco, and Cominco.[53] By 1974, for instance, the distribution of control in Québec manufacturing was estimated by André Raynauld, the chair of the Economic Council of Canada, as 'French Canadian 15.4%, English Canadian 42.8%, and foreign 41.8%.'[54] In line with an old bonus practice, already wealthy industries were still receiving special subsidies from provincial sources. Many of the mining exploration pro-grams, for instance, received funding from the James Bay Development Company.

In creating corporate liaisons and partnerships with private companies, the JBDC did not rely on Québecers or stay within the North American continent but, rather, visited mining and industrial firms in Europe and Japan. Fred Ernst, one of the five administrators of JBDC, travelled first-class to Europe in the fall of 1973 'aux fins de rencontrer les dirigeants des entreprises minières qui sont intéressés à participer au développement minier de la région de la Baie James' [in order to meet with the directors of the mining industry interested in participating in the mining develop-ment of the James Bay region].[55] Ernst also travelled with government industrial promoters to Japan in 1974 'aux fins de donner des conférences aux hommes d'affaires japonais qui sont intéressés par le projet hydroélec-trique et l'aménagement régional de la Baie James' [in order to confer with Japanese businessmen interested in the hydroelectric project and territor-ial and regional development of James Bay].[56] Charles Boulva, the president of JBDC, also invited industrialists from outside Québec and arranged a meeting with le Commissariat à l'Energie Atomique in Paris from 6 to 10 April 1974 to discuss the installation of a uranium enrichment plant in Québec.[57] In further pursuit of this uranium project, Boulva accompanied the premier of Québec to Paris on his official visit during the first week of December 1974.[58] In addition to overseas investors, mining firms from the US and Toronto were encouraged to start mining explorations in the James Bay area.

During the 1970s, Hydro-Québec still set the price for its industrial cus-tomers. Under the electricity surplus conditions of the late 1980s, how-ever, foreign industrial firms 'set' the price for (or negotiated extremely advantageous conditions with) Hydro-Québec in costly special-risk con-tracts, many of which were signed for twenty-five years. For example, Hydro-Québec set the industrial rate with the Canadian Reynolds Metal Company for contract #4,407-75 over the period of 1974-9 for 102 to 108 MW power at $7.36 (1978-9) kW per month; this amounted to $9.5 mil-lion per year at full capacity.[59] By 1989, 'Reynolds [for its $500 million expansion] a obtenu d'Hydro-Québec un contrat d'approvisionnement d'électricité dont le tarif est lié aux cours mondiaux de l'aluminium et la rentabilité de l'usine de Baie-Comeau' [Reynolds (for its $500 million expansion) received from Hydro-Québec a contract for supplying electric-ity the rate of which is linked to the aluminium world market, as well as profitability of the Baie-Comeau factory].[60] In other words, the price of electricity is limited by the profits of the branch plant and the global mar-ket conditions. A special-risk industrial contract, in which profitability in Québec and world prices determine the utility's revenue, was also signed with a Norwegian company for its 'l'Aluminerie de Bécancour (ABI) where Norsk Hydro invested $500 million.[61] During 1993, these special-risk industrial contracts, thirteen in all, brought enormous losses to the owners of Hydro-Québec – the people of Québec. *Le Devoir*, on 13 March 1994, reported the following:

La société d'Etat [Hydro-Québec], qui divulgera ses résultats financiers annuels dans quelques jours, a comptabilisé une perte de 324 million$ pour 1993 seulement, au chapitre de son programme de partage des risques et bénéfices. Les pertes prévues pour 1994 et 1995 sont respectivement de 271 et 237 millions$.

Selon Hydro-Québec et ses plus récentes prévisions, c'est-à-dire révisées en 1993, la valeur des 13 contrats signés en vertu de ce programme de partage des risques serait de 24 milliards$. La même quantité d'électricité vendue au tarif 'L', qui est le tarif industriel ordinaire, donnerait des revenus de 25,2 millard$. Cela voudrait donc dire un manque à gagner de 1,2 milliard$ dans le cas des 13 contrats à partage des risques.

[The state corporation (Hydro-Québec), which will divulge its annual financial statements in a few days, has counted a loss of $324 million for 1993 alone, under its program of sharing risks and profits. Losses forecast for 1994 and 1995 are respectively $271 and $237 million.

Hydro-Québec, in its most recent forecast, i.e. revised in 1993, stipulates that the value of the 13 contracts signed under this program of risk sharing would be $24 billion. The same amount of electricity sold under rate 'L,' the ordinary industrial rate, would yield a revenue of $25.2 billion. This would thus mean a shortfall of $1.2 billion in the case of the 13 risk sharing contracts.][62]

These special-risk contracts will remain in effect well into the twenty-first century unless they are challenged under provisions of the free trade agreement with the United States. They show how Hydro-Québec and the people of Québec shoulder an increasing risk in the provision of energy for transnational corporations in an increasingly global economy.

The Roots of Overdevelopment

The Québec government's 'industrialization by invitation' strategy not only brought expensive 'foreign masters' into the provincial house, but also led to the building of hydro projects before they were needed. According to public records, the Québec government authorized Hydro-Québec to build new power projects only for the needs of the Québec population.[63] Hydro-Québec's provincial mandate was to provide low-cost power for municipalities, for industrial and commercial enterprises, and for the residents of Québec.[64] While Québec's population grew by less than 15 percent between 1971 and 1986 (from 6 to 6.8 million inhabitants), the hydroelectric system capacity grew by more than 100 per cent between 1971 and 1984 (when the capacity of the northern expansion is compared with the total capacity in Québec) (Table 4.3).

By the mid-1980s, the additional generating capacity of the Churchill Falls powerhouse and James Bay plants resulted in all of the James Bay plants being surplus to provincial needs during the summer months. To appear to stay within the provincial mandate (in force until the mid-1980s), provincial governments usually claimed that dams in excess of provincial power requirements were being constructed to supply the anticipated industrial market, and any surplus electricity could be exported 'temporarily' to the US until it was needed in Québec. To legitimate such actions, hydro officials would forecast power shortages before launching hydro projects and would admit to 'unplanned' surpluses after the projects were installed. These surpluses went far beyond the normal surpluses of the phase-in periods for large power projects.

Churchill Falls Surplus (1970s)

In 1974, Mark Zanis, a sociologist from McGill University, questioned the planning practices of Hydro-Québec – especially the contradiction between shortage predictions and surplus realities. At the NEB's export licence hearing, he commented: 'As far back as the Churchill Falls deal in 1966, there was talk about shortages in Québec. In fact, at that time, there was mention of a shortage in 1971, and here we are again in 1974, and we have heard about James Bay, there is going to be a shortage, and, suddenly, we just wind up with surpluses. It's either poor planning on the part of Hydro

Table 4.3

Northern hydro expansion, Churchill Falls and James Bay, 1971-96

Year of initial operation	Name of plant	Rated capacity (MW)
1971	Churchill Falls (NF)	5,429
1979	James Bay La Grande 2	5,328
1982	James Bay La Grande 3	2,302
1984	James Bay La Grande 4	2,651
1985	Total northern expansion	15,710
1985	Total hydro capacity in Québec	24,928
1996	Total James Bay	15,237
1996	Total installed capacity in Québec	34,731

Sources: Canada, Energy, Mines and Resources, *Electric Power in Canada 1988*, p. 32; *Electric Power in Canada 1986*, p. 61; *Electric Power in Canada 1996*, pp. 65-72.

Québec or that these things are deliberately created.'[65] The Hydro-Québec official denied both poor supply forecasting and deliberate planning for export, claiming that 'nous ne créons pas délibérément des excédents d'énergie pour des fins d'exportation, absolument pas' [we are not creating surplus energy for exports, absolutely not].[66]

When the Churchill energy was supplied to the Québec system by 1974, Hydro-Québec did not need the additional power for its old or new industrial customers. Its industrial customers consumed virtually the same amount in 1973 (20,894 GWh), one year before all the generators were connected, as they did in 1976 (21,078 GWh), two years after the fresh supply of electricity became available (Appendix 5). In September 1976, the 'unplanned' surplus generating capacity for every October from 1977 to 1984 was revealed by Hydro-Québec before the NEB as equivalent to more than half of all Churchill Falls generators (5,429 MW) (Table 4.4). Although a refined hydraulic optimization analysis of seasonal reservoir operations and storage levels would show a more nuanced variation in available capacities, the surplus data presented here are given to expose the dramatic contrast between public pronouncements of capacity shortages and surplus realities.

James Bay Surplus (1980s)

The James Bay projects were legitimated in the same fashion as was the Churchill Falls annexation: predictions of shortage prior to the launching of the scheme and actual surplus upon its completion. Québec premier Robert Bourassa warned that, if James Bay were not built by 1985, then a 15,000 MW deficit would result (equivalent to the capacity needs of more

Table 4.4

Hydro-Québec estimated power surplus (MW)

October* surplus capacities, 1977-84 (Churchill Falls plant on line)

1977	1978	1979	1980	1981	1982	1983	1984
3,342	2,205	2,311	2,773	2,982	3,356	3,195	3,506

March surplus capacities, 1984-90 (some James Bay plants on line)

1984	1985	1986	1987	1988	1989	1990
4,683	4,849	5,114	6,136	5,556	4,781	4,049

July surplus capacities, 1984-90 (some James Bay plants on line)

1984	1985	1986	1987	1988	1989	1990
10,328	10,936	11,692	12,580	12,345	11,957	11,634

* October was indicated as the most critical month in most years when capacities are needed both in the United States and Canada (NEB Report, September 1976).
Sources: *National Energy Board Report to the Governor in Council: In the Matter of an Application under the National Energy Board Act of Quebec Hydro-Electric Commission*, September 1976, Appendix 5. National Energy Board, *Reasons for Decision: In the Matter of an Application Under the National Energy Board Act of Hydro-Québec*, January 1984, Appendices: VI and VII.

than one thousand universities), and he questioned whether Québecers 'would want the wood-burning stove and the oil lamp restored in order to cope with these [power deficit] problems.'[67] He was 'convinced' that 'besides ensuring firm sources of supply to existing industries,' the James Bay expansion was needed for 'the establishment of new industries.'[68] In 1981, Bourassa reserved 2,500 MW from James Bay for the 'implantation' of new industries: 'Un cinquième de la puissance et près du tiers de l'énergie de la Baie James se trouvent implicitement réservés à la croissance de la demande dans le secteur industriel. Il y a donc un bloc de 2 500 MW dédié aux implantations industrielles.' [A fifth of the power, as well as close to one-third of the energy, of James Bay are implicitly reserved for the growth of the demand in the industrial sector. Thus there is a bloc of 2,500 MW devoted to establishing industries.][69] Yet, after the three James Bay power-houses, La Grande 2, 3, and 4, were connected to the provincial system, the new industries that resulted subsequent to the James Bay development needed a capacity of 176 MW – nowhere near the 2,500 MW forecast by the premier.[70] Instead of being attracted to Québec's fresh supply of electricity, industrial investors stayed away.

Québec's James Bay surplus capacity far exceeded the allotment for industry and revealed an enormous capacity in excess of Québec's needs. Simply stated, during the summer, the equivalent of all the James Bay plants (10,287 MW) was surplus to Québec's needs, and during the winter, half these plants were surplus to Québec's needs, as Hydro-Québec's March and July surplus estimates for the years 1984-90 clearly indicate (Table 4.4). Poor industrial forecasts and unpredictability as to where and when global industries would implant their manufacturing facilities were not the only roots of Québec's capacity surplus. The damming of Québec's mighty rivers beyond provincial mandates was also intended to produce electricity for US export.

Directing Québec's Surpluses

Although in the 1970s, as discussed in Chapter 2, Québec participated in provincially initiated plans to develop a national power network, the province thought it most profitable to negotiate bilateral electricity trade arrangements (e.g., Churchill Falls) with neighbouring provinces and US states rather than national or regional system integrations within Canada. Hydro-Québec was supported by Canada's National Energy Board in its increase of electricity exports to utilities in US states. From 1974 to 1988, starting two years after Québec received the first Churchill Falls power, fourteen Hydro-Québec export licence applications were processed and approved by the NEB.[71] Two case studies concerning the licensing processes, including hearings for the 'diversity power' contract during the post-Churchill period and for the export to the New England Utility during the post-James Bay surplus period, will reveal how the NEB (with the approval of Cabinet) legitimated surpluses to be used as *continental energy*.

Labrador Power to Montréal, Montréal Power to New York

Québec began paving the way for the export of surplus electricity as early as the 1960s. In 1962, Jean Lesage, the premier of Québec, refused to take part in the federal government's national power network meetings, insisted on buying all the power from Churchill Falls, and refused wheeling electricity from Labrador across Québec.[72] Although Hydro-Québec, as I shall explain further in the next chapter, also participated financially in the development of Churchill Falls, it nevertheless proceeded to buy all the 'trapped' electricity at Churchill Falls from Brinco's Churchill Falls (Labrador) Corporation at very favourable rates for sixty-five years.

Québec then arranged to resell the Churchill Falls surplus to US utilities. Before any export, NEB regulations in the 1970s demanded proof that the energy would be surplus to Canadian needs, and it required utilities to offer surplus energy to neighbouring utilities. In 1974, following its export agenda, Hydro-Québec reluctantly went through the NEB formalities, choosing to offer the surplus electricity to its Canadian neighbours in the form of the US 'diversity power' contract, which it signed with the Power Authority of the State of New York (PASNY) on 29 March 1974 in Montréal.[73]

This export contract involved the interconnection of the Beauharnois plant (nationalized in 1944) on the St. Lawrence with a US utility. In part, because of the incompatibility problems between Québec's transmission system and that of its US neighbours, about half the generating capacity (800 MW) of the plant (1,574 MW) was disconnected by the publicly owned Hydro-Québec from the Québec system and made 'effectively part of the US system.'[74] The electricity sold was labelled as 'diversity power,' which was 'defined as power exchanged or made available in order to benefit from the diversity between two peak loads [peak periods of power demand]. In this case, the diversity is seasonal and is between the annual peak loads of the two systems, the Québec peak occurring in the winter [heating] and that in southern New York in the summer [air-conditioning].'[75] As the NEB report notes, 'in a true seasonal power diversity exchange between systems, both parties should be able to defer the installation of new generation ... In the present case, the contract will enable PASNY to defer 800 MW of new generation, whereas Hydro-Québec will defer none.'[76] Since virtually no power was imported from the US in the 1980s and a relatively small amount in the 1990s (Table 4.5), the real reason for the export to the US was the scrapping of a nuclear project in New York 'because of financing, siting and environmental problems' and the advantageous contract terms.[77] At the time, Hydro-Québec announced that formerly firm electricity

Table 4.5

Québec's international electricity trade, 1983-96 (GWh)

Year	Exports to US	Imports from US	Net exports
1983	10,128	8	10,120
1984	11,250	8	11,242
1985	9,581	3	9,578
1986	12,674	35	12,639
1987	16,401	0	16,401
1988	11,863	86	11,777
1989	5,458	1,001	4,437
1990	3,403	1,188	2,215
1991	5,959	730	5,229
1992	8,877	1,388	7,489
1993	13,009	684	12,325
1994	17,337	1,304	16,033
1995	16,854	838	16,106
1996	15,327	540	14,787

Sources: Canada, Energy, Mines and Resources, *Electric Power in Canada 1986*, Table A5, p. 57; *Electric Power in Canada 1988*, Table A5, p. 77; Natural Resources Canada, *Electric Power in Canada 1994*, Table A5, p. 155; *Electric Power in Canada 1996*, Table A5, p. 149.

contracts with its Canadian neighbours were not to be renewed (or tailored to their needs) and that surpluses from Québec or Labrador power plants would be offered as tailored-for-the-US package contracts.[78]

At the hearing to grant the diversity power export licence in 1974, government representatives from neighbouring provinces opposed the export on the grounds of interprovincial interest. New Brunswick declined the offer of Québec's surplus, but neither Ontario nor Newfoundland was offered this US package contract.[79] Ontario had only been notified of this export but had not been offered surplus electricity by Hydro-Québec, and the Ontario ministry of energy contended during the 1974 hearings 'that Ontario Hydro had not been offered the proposed export quantities and that therefore those quantities could not be found surplus to Canadian requirements.'[80] Furthermore, just when the provinces had renewed the initiative to plan a national power grid (see Chapter 2),[81] the Ontario ministry of energy argued at the hearing that 'the power transfer possible over the line [to Utica, NY] could reduce incentive to expand regional and national interconnections in Canada, particularly in relation to the Province of Ontario.'[82] Aware of the global energy crisis in oil at the time, the ministry added that Ontario may, 'by the early 1980s, be unable to meet the needs of its customers in the Province of Ontario.'[83] The NEB, in its decision to grant Québec the right to export, chided Ontario for failing to initiate a contract: 'Ontario could itself have made a bid for the power but it did not.'[84] The attorney general of Newfoundland, who had also been neglected by Québec, asked 'why the diversity power had not been offered to Newfoundland.'[85] Hydro-Québec replied that 'there is no inter-connection with the Island of Newfoundland and that, because of this, Newfoundland would presumably have no use for the power.'[86]

Together with provincial government representatives and utility officials, industrial power interests from Ontario also opposed the exports. The Association of Major Power Consumers from Ontario argued that 'surplus hydraulic energy [should] be kept in Canada and utilized to displace energy generated by fossil fuels,' such as oil, coal, gas, and nuclear power, and it 'urged the establishment of an Eastern Canada Power Pool.'[87] The power consumers made this proposition at the time of the energy crisis in the mid-1970s and while initiatives to form national and regional inter-provincial grid systems were under way. Ontario's major power users opposed integrating with the eastern US power pool and so cooperating with utilities within US states and, instead, favoured the formation of utility pools within the eastern provinces of Canada.

The price differential between the Churchill Falls power and the Beauharnois power indicates terms that were profitable for Québec and advantageous to PASNY. Churchill Falls power, severely underpriced at 2.8 mills per kWh (equal to $.0028 kWh), was brought to Montréal, and Montréal (Beauharnois) power was exported to PASNY at 8.2 to 9.2 mills per kWh.[88] Thus, Hydro-Québec charged triple the rate of Churchill Falls electricity for its export to PASNY. However, PASNY's price was not high; in fact, it was less than half of 22.3 mills per kWh, the price Hydro-Québec charged for its export to Consolidated Edison of New York (Con Ed).[89] A summary of the key features of the 'diversity power' contract (with some NEB amendments) shows some other unusual arrangements favouring US network integration:

Seventeen-year contract:
- 800 MW of power (and more in the future) from the Beauharnois plant near Montréal to Utica, New York (via Massena)
- 'Up to 3,000 GWh per year of energy for five export years [of export] from 1 June 1977 to 31 December 1981' and renewable for a total term of 13 years, no later than 31 October 1991.[90]

Sixty per cent of the cost of this US power line is paid for by this Québec export:
- During the term of the contract, the difference between the low price of 8.2 to 9.2 mill per kWh and the 20 mill per kWh, common in this U.S. market, would pay for most of the 134 miles of transmission line (sixty

per cent of $226 million) from Massena near the Canadian border to Marcy substation near Utica, NY.[91]

These beneficial terms were initially even more advantageous to PASNY, until the NEB changed the licence required for the contract: it was reduced from 25 to 13 years and it stipulated that, rather than the export paying for 100 percent of the PASNY transmission line, it would pay for 60 percent.[92]

But why would the NEB allow such a disadvantageous sale to proceed? One reason was the expectation that once transmission lines were interconnected with the US they would, according to the board, 'undoubtedly have potential for future additional benefits' to Canada's economy.[93] One benefit that such a Canada-US interconnection of power systems was expected to provide was the early development of northern Canadian power resources for export, a major feature of the federal government's national-continental power policy, as stated on 8 October 1963 in the House of Commons.[94] The NEB approved interconnection with the US for the sake not only of 'future additional benefits' from exports, but also for the expected benefits of US portfolio capital investment in Québec dams. Possibly, by paying for two-thirds of the US power line, Québec utility officials hoped that future exports and hydro development capital would be easier to obtain.

As a matter of routine, the board ahistorically evaluated each application on its own merit, primarily within a narrow definition of public and national interest; that is, in the 1970s and 1980s, the board verified both that the electricity to be exported was surplus to Canadian needs and that the price of exported electricity in the local US market was above the price in the Canadian market. Such a price differential, this NEB practice assumed, would give a competitive advantage to Canadian industries and would attract industrial investments from the US to low-cost power in Canada; by providing cheap Canadian electricity in the US, the advantage to Canadian industries and the incentive for energy-intensive industries to move north would otherwise be undermined. In the case of the 'diversity power' contract, to establish that the electricity destined for export to the US was surplus to Canadian requirements, the board instructed Hydro-Québec to offer the contract to Newfoundland and Ontario, both of whom rejected the offer. Since 'no Canadian requirement [had] been established for this type of [US-tailored] sale,' the board reasoned that the power was surplus to Canadian needs. Hydro-Québec then convinced the board to waive the price criterion, which states that an export price shall not be less than the price to Canadians for equivalent electrical service. The board also noted that, since 'Hydro-Québec does not propose selling firm power to neighbouring provinces in the future, it did not or could not provide up-to-date [price] comparisons.' The board noted, however, that the export price of 8.2 to 9.2 mills per kWh for PASNY was less than Ontario's 'cost of producing electricity from coal, given in previous hearings as 12 to 12.5 mills per kWh.'[95] The price of the local US market fitted NEB regulations because Québec's export price carried most of the amortization (debt repayment) costs of the US power line.

One further consequence of the diversity power contract is that the supply of electricity to Montréal tends to become unstable in the winter months. With most of the generators at the Beauharnois hydroelectric plant disconnected from the Québec system, Montréal relies on power from Labrador, despite winter icing and storms causing technical problems on the transmission line. Meanwhile, power generated near Montréal is exported, with less risk of interruptions, to the US.

In sum, as part of the annexation of Churchill Falls, Québec first denied Newfoundland permission to transport Labrador electricity across its territory; then Hydro-Québec signed a 'diversity power' contract with the Power Authority of New York, offered the US contract as an entire take-it-or-leave-it package to some neighbouring provinces, interpreted the unsuitability of the US contract to the provinces as proof that the power was surplus to Canadian needs, and, finally, applied for and received a licence to export power. With the James Bay surplus, Hydro-Québec went even further in making interconnection benefits available to US buyers (as did other provincial utilities with regard to their northern electricity). Québec's plants, as those of other provinces, could have been designed to meet the needs of integrated national or regional power networks within Canada (as discussed between Québec and other provinces in the 1970s) rather than merely to offer US-tailored contracts to neighbouring provinces as part of the NEB export licence application requirement.

James Bay Power chez E.U.

During the 1970s, Hydro-Québec continued to offer package contracts to neighbouring provinces, often after being reminded to do so by the NEB. Then, in 1985, Hydro-Québec directly challenged the authority of the NEB by contracting to sell electricity to New England without first offering it to the Canadian provinces.

As indicated above, since 1984, capacities equivalent to about one-quarter to half of the three James Bay hydro stations (total: 10,287 MW, power-houses La Grande 2, 3, and 4) have been indicated as surplus during the winter, and all of them have been indicated as surplus during the summer.[96] Therefore, Hydro-Québec negotiated the '1985 New England Utilities Contract' (NEU Contract) based on this mounting James Bay surplus.[97] This firm export contract (not interruptible for Canadian needs) was signed (with later NEB approval) between the NEU and Hydro-Québec in October 1985 and involved, in part because of transmission system incompatibility, disconnecting some of the generators in the largest powerhouse in Québec and linking them, by separate transmission facilities, directly into the United States.[98] This contract concerned the export of 'up to 9,000 GWh annually [enough to operate 30 pulp mills for one year], 70,000 GWh in total, of firm energy to the New England Utilities for a period of from 10 to 14 years beginning 1 September 1990 and ending 31 August 2004.'[99] To reduce system instabilities for the US utilities, the NEB report states that 'generators at LG 2 [James Bay, La Grande River powerhouse 2] would be isolated from the Hydro-Québec system and would feed directly to Sandy Point in New England ... This scheme would effectively isolate 2,000 MW of exports from the system.'[100]

In May 1987, Hydro-Québec failed to offer the surplus energy to 'neighbouring utilities in Ontario, New Brunswick, and even Newfoundland.'[101] The NEB found that 'Hydro-Québec had not provided sufficient evidence to demonstrate that the electricity it proposed to export was surplus to reasonably foreseeable Canadian requirements and that the export price was not less than the price to Canadians for equivalent service.'[102] The NEB rejected Hydro-Québec's position that 'the Board's surplus test should only be applied in the narrow sense of the requirement of Hydro-Québec's own domestic customers plus its firm commitments to supply other utilities or systems.'[103] The NEB also rejected Hydro-Québec's understanding of the 'national' intent of the National Energy Board Act.[104] The NEB viewed the act as indicating that 'its statutory [national] mandate is to ensure that the power exported is surplus to *requirements in Canada*'; Hydro-Québec, by contrast, argued that the NEB's role is only to ensure that the power exported is surplus to the *'requirements in any particular province [Québec].'*[105] With this interpretation, Hydro-Québec saw no need to offer its surplus to neighbouring provinces. But with both Brian Mulroney's Conservative government and Bourassa's Québec Liberal government supporting the continentalist US-Canada Free Trade Agreement (FTA) during the late 1980s, a solution by the about-to-be-deregulated NEB was soon found.

The NEB considered the suggestion that the application could be 'cured' if Hydro-Québec offered the surplus electricity to neighbouring provinces, even if the price and surplus conditions were not fully met.[106] As the NEB reported in February 1988, Hydro-Québec offered the terms of the '1985 NEU export deal' to the neighbouring provinces, but, predictably, neighbouring Canadian utilities could not match all the conditions or elements of the export deal with the US utility.[107] The NEB found, for instance, that the Newfoundland and Labrador Hydro requirements 'would be substantially different than the terms and conditions provided for under the export contract.'[108] Despite disagreeing with Hydro-Québec's position that 'the offer must be on exactly the same terms and conditions as provided under the export contract,' the NEB, on 22 January 1988, approved the licence, with minor variations from the NEU export contract,[109] thereby contributing to the legitimization of Hydro-Québec's continentalization program.

In the 1980s, such exports constituted initial steps towards Bourassa's even more ambitious plans for two further large-scale water projects, one expected to cost $25 billion, the other $100 billion.[110] The former constituted the 'Power from the North' scheme of doubling Québec's electricity capacity once again; that is, of developing additional power plants amounting to 30,000 MW capacity in northern Québec and Labrador primarily for export to the US. According to Bourassa, this development would partially realize Hydro-Québec's available energy resources, almost all hydroelectric, which at the equivalent of 57,337 million barrels of oil (valued at $1,095 billion) are 'substantially larger' than 'Canada's total proven oil and gas reserves [equivalent to 23,704 million barrels] as well as the combined proven oil and gas reserves of the ten largest United States energy companies [equivalent to 48,369 million barrels, or $463.6 billion].'[111] Once this first project was developed and these energy values realized, the $100 billion project, the 'Grand Canal Concept,' or 'Recycled Water from the North' plan, could be launched.[112] In this scheme, water that has passed through James Bay turbines would be diked in James Bay and flushed as 'fresh' water through the Great Lakes and (along one of many canals) via the 'revitalized Mississippi' into the Gulf of Mexico (solving pollution problems along the way).[113]

In 1994, however, reality finally hit home. His Parti Québécois government pressured by the cancellation of power contracts by eastern US utilities, by the mounting debt of Hydro-Québec, and by the environmental resistance of the Cree in northern Québec, Premier Parizeau quashed the more 'modest' $13 billion Great Whale hydro project and shelved plans for large hydro developments. Since 1997, however, with long-term contracts more difficult to negotiate, Hydro-Québec has expected to sell surplus electricity to the US on a shorter-term basis. In part, such expectations of exporting to the deregulated northeastern US electricity market, as well as the election of a new premier and the retention of the Parti Québécois state interventionist policy, led to the reinvention of the idea of Québec as the 'l'Alberta de l'Est,' or the 'Kuwait of the North.' Thus by the end of the 1990s, Québec's premier Lucien Bouchard's energy policy envisaged making Québec 'a major energy hub' (with one of the new supply spokes expected to extend to Labrador) in a restructured North American electricity market.

Québec as a Major Energy Hub by 2002

Although in the 1990s more than 90 percent of the electricity from Québec's generating stations was sold to customers in *Québec*, the 10 percent sold to markets outside Québec – particularly electricity sold to the US

– became subject to US energy policy and disproportionately influenced the restructuring and provincial regulation of Hydro-Québec.

Québec sold decreasing amounts of electricity to New Brunswick and Ontario: from 1985 to 1994 sales to New Brunswick declined from 5,985 GWh (GWh: 1 million kilowatt hours) to 2,174 GWh, and sales to Ontario declined from 8,685 to 1,418 GWh – while sales to the US increased from 9,585 to 16,835 GWh.[114] However, in the 1990s, Hydro-Québec has found new, long-term, high-capacity contracts with US utilities increasingly difficult to negotiate. The US department of energy reported in March 1994 that in 1992 the New York Power Authority cancelled or opposed two contracts: a 1,000 MW purchase from Hydro-Québec worth about $13 billion and a twenty-year $5 billion contract.[115] The department added that US importers in 1995 drew off only 6 percent of the installed hydroelectric capacity of Québec.[116] Thus, this relatively small proportion of capacity allocation to international electricity export has had a disproportionate influence on the restructuring of Hydro-Québec, as similarly small proportions have influenced hydroelectric utilities in other provinces. Nevertheless, Hydro-Québec wishes to retain its supplementary revenue from exports. One can assume that, because, of its $7.7 billion in revenues (more than 90 percent from its Québec customers), $400 to $600 million is usually derived from US exports and can be paid towards the interest on Hydro-Québec's $38 billion long-term debt, the utility desires to maintain and increase its electricity exports.[117] However, the export market remains uncertain.

By the year 2002, according to the Bouchard government-approved utility's *Strategic Plan* for the period 1998-2002, Hydro-Québec plans to increase its sales by 6,000 GWh but will also have to find new buyers for 8,000 GWh from terminated electricity contracts outside Québec. In total, the utility will need to find new buyers for 14,000 GWh.[118] To accomplish this, the utility plans to make direct sales to industrial customers, wholesale buyers, and resellers of electricity in markets outside Québec; that is, in the US and possibly in neighbouring provinces.[119]

In order to make itself acceptable to the US energy authorities as a supplier of electricity, Hydro-Québec restructured, or deintegrated the administration of, its transmission system. While in the Lesage era its transmission lines were not accessible for the trans-provincial delivery of electricity, Hydro-Québec's *Strategic Plan* for the years 1998 to 2002 states that in Québec 'the wholesale market and the electricity transmission system have been open to third parties since May 1, 1997' as part of a new energy policy.[120] Two elements of the new energy policy, an open transmission system and Québec's vision of becoming 'a major energy hub,' need to be addressed.

In 1996, the Québec government unveiled a new energy policy entitled

Energy at the Service of Québec, and many of its objectives are important for future exports. While other provinces also changed their transmission policy and proposed new forms of regulation, Québec, unlike them, has not proposed privatization and cutbacks in public electricity generation. The Québec government's new energy policy proposed the following:

Hydro-Québec shall remain the exclusive property of the Québec government;

...

Hydro-Québec shall become the cornerstone of an industrial strategy aimed at making Québec a major energy hub;

...

and Québec's regulatory framework shall be modernized through the Act respecting the Régie de l'énergie which stipulates, among other things, that the generation, transmission, and distribution of electricity are regulated activities.[121]

The government of Québec created the *Régie de l'énergie* in December 1996, and Hydro-Québec's division TransÉnergie intended to 'broaden access by its customers to transmission systems peripheral to Québec.' Since, under free trade agreements, US federal and regional authorities set the reliability criteria for integrated transmission systems, TransÉnergie plans to lobby the Northeast Power Coordinating Council and the North American Reliability Council to be allowed to increase the capacity of existing interconnections by 250 MW to 400 MW.[122] But to what degree – given its history of blocking transmission from Labrador – does Hydro-Québec intend to open its transmission system by following US requirements?

In the US, energy authorities advocate and regulate the opening of power lines to *retail* power marketers who sell electricity from various sources, whereby, with new metering technology being able to distinguish between different time-of-day consumption (which, in turn, allows the electricity provider to identify the power suppliers), in the future 'transmission and distribution systems will act like public road networks, over which suppliers of electricity will be able to deliver electricity they produce or purchase to customers who will have chosen them.'[123] It will be the suppliers who will pay the tariffs for access to power lines.[124] In Québec, the transmission system was opened to the provincial *wholesale* market, which Hydro-Québec identifies as eleven electric power distributors (municipal systems and a regional cooperative system).[125] In addition, according to the *Strategic Plan*, since 1 May 1997 other distributors and market intermediaries have been allowed to purchase electricity outside Québec, and independent power producers have also been allowed to sell their electricity outside Québec.[126] However, by October 1997, no outside producer supplied electricity directly into the Québec wholesale market; this, the *Strategic Plan* explains, is due to the low electricity rates in Québec.[127] Also because of the low electricity rates in Québec, Hydro-Québec claims that, in Québec, market pressures to open the transmission system to the retail market (which includes industrial, commercial, institutional, residential, and agricultural consumers) are not as significant as they are elsewhere in North America. Therefore, 'Hydro-Québec does not intend to promote the opening of the province's retail electricity market.'[128] It does, however, plan to export its surplus electricity to the US. Hydro-Québec's subsidiary, Hydro-Québec Energy Services (US), applied to the Federal Energy Regulatory Commission (FERC) in the US, and since November 1997 'has been licensed to sell electricity at market based prices in the United States.'[129] Since FERC has become very influential, what is the role of Canada's federal energy authority, the NEB, with regard to such exports to the US?

Hydro-Québec had already applied on 24 May 1994 for flexible export permits to take advantage of selling its surplus electricity to US customers. On 7 December 1994 the NEB approved Hydro-Québec's two sixteen-year electricity permits, which are valid from 1 December 1994 to 31 December 2010. They allow exports of up to 4,300 MW of power: one permit authorizes up to 30,000 GWh of interruptible exports (minus firm exports); the other allows firm exports of up to 20,000 GWh (minus the interruptible exports). The board explains that 'these permits allow the company to sign short-term export contracts for a maximum duration of five years to any customer, without having to obtain prior approval of the Board.'[130] The application was dealt with, not by oral public hearings, but by 'a written proceeding.' Since long-term contracts are becoming rare, this licence allows Hydro-Québec to supply electricity to the US spot market. However, resistance to building projects for such exports continues.

The Cree, who are the people most affected by the building of hydro projects for export (Hydro-Québec officials and politicians in the 1970s publicly claimed such projects were needed for consumption within Québec) intervened, as they did during the NEB hearings over Hydro-Québec's licence application at the US Federal Energy Regulatory Commission in November 1997. In a press release on 13 November 1997, Matthew Coon Come, the grand chief of the Cree, objected to the licence, stating after the FERC licence had been granted that 'we did not sign the James Bay and Northern Quebec Agreement so they could export Great Whale River electricity to the United States and play on the energy spot market.'[131] Two other criticisms were levelled against this FERC application. One was that FERC based its decisions on very narrow parameters – just two measures of market power ('total installed capacity and uncommitted [surplus] capacity'). The Cree representative, Brian Craik, pointed out that FERC traditionally limits market shares in these measures to 20 percent but that

Hydro-Québec studies had shown them as up to 38 percent, failing even to mention the substantial error in surplus capacity estimated by Hydro-Québec: 'Hydro-Québec's surplus capacity analysis [was] based on their surplus (peak) capacity in the winter (3,480 MW), but the American peak is in the summer ... [when] Hydro-Québec has some 10,000 MW of surplus capacity. Our expert testimony demonstrated that this would result in market shares of over 60 per cent.'[132]

Trying to oppose the application using FERC's own traditional criteria of preference for allowing competitive new and small generating capacities into regional transmission groups, the Cree had little effect. While not refuting the Crees' arguments, FERC did not even mention 'the substantial errors' the Cree pointed out concerning the market-share analysis submitted to FERC by Hydro-Québec.

Hydro-Québec's subsidiary, Hydro-Québec Energy Services (US), argued that it is the responsibility of the Régie alone to protect the interests of ratepayers in Québec from losses incurred through exports. However, the Mouvement Au Courant, in levelling the second major critique against the FERC licence, argued that, in its decision to grant the export permit, FERC relied on Hydro-Québec's assurances that the 'Régie de l'énergie plays a regulatory role equivalent to that of FERC,' whereas in reality, as Bill Namagoose, executive director of the Grand Council of the Cree, pointed out, 'Régie remains, for the moment, an empty shell ... until the government sees fit to give it regulatory powers over Hydro-Québec.'[133] Furthermore, with public hearings for export licences in Canada becoming rare and US regulatory agencies not being responsible for the interests of Canadian electricity consumers, in frustration Craik asked, 'What do the Americans care if Québecers end up shouldering the losses incurred on the U.S. spot market?'

As mentioned above, the Québec government's new policy of *Energy at the Service of Québec* also envisaged Hydro-Québec's becoming 'the cornerstone of an industrial strategy aimed at making Québec a major energy hub.' The company has already started acquiring substantial shares in North American energy companies: 'In 1997 ... Hydro-Québec acquired a 41% stake in Noverco, the holding company which owns Gaz Métropolitain Inc. The company also acquired an indirect interest in IPL Energy which operates the world's longest petroleum pipeline system, and is Canada's largest natural gas distribution company.'[134] In addition, Hydro-Québec's *Strategic Plan* predicts that from 1998 to 2002 'Hydro-Québec's investments in generation, transmission, and distribution will exceed $13 billion,'[135] that the utility will attract industries to Québec that need up to 500 MW in capacity, that export sales will increase by 6 TWh (billion kilowatt hours), and that overall 40 TWh in new sales will develop within the next ten years. Given these optimistic expansion predictions, which are reminiscent of those of the 1960s, Québec is also negotiating with New-foundland about Phase II of the Churchill River – the $12-billion Lower Churchill project, which includes the construction of dams, powerhouses, and $2 billion for a federally subsidized 800 MW transmission line from Churchill to St. John's, Newfoundland.[136] After having reduced its total workforce from 27,234 in 1992 to 19,500 in the year 2000, Hydro-Québec also plans to hire people and 'complete the development of Québec's hydro-electric potential, in line with the development of the past 30 years.'[137] On one hand, Hydro-Québec appears to follow an internal corporate restruc-turing process (including layoffs) possibly to make itself appear attractive to moneylenders; on the other hand, the utility has proposed a renewed hydro expansion strategy in line with the Parti Québecois globalization adjustment policy of 1993, as identified by Daniel Salée and William Cole-man.[138] In the context of that policy, Hydro-Québec's *Strategic Plan* coin-cides with the modified Parti Québécois neo-liberal adjustment policy, which continues to support a strong interventionist state, building on the benefits from nationalizing electricity, employment retention and creation in the construction sector, and an orientation to the export market through making 'trade agreements to protect access to these markets.'[139]

Conclusion

Like other provinces, Québec first allocated its water-power resources for development to the private sector and then nationalized them. After the 1944 takeover of the Montreal Light, Heat and Power Company and the 1962 formation of Hydro-Québec, René Lévesque, as minister of natural resources during the Quiet Revolution, nationalized private utilities, in part because they failed to provide sufficient electricity for industrial development in rural areas, such as the Bas-Saint-Laurent, Abitibi, and the Gaspésie.[140] Other nationalization objectives included the integration of the system, the encouragement of industrialization, and a guarantee of increased economic benefits for the people and the economy of Québec. However, in the 1970s, Premier Bourassa's 'conquest of the North' (after the 'annexation' contract at Churchill Falls in 1969 and the development of James Bay in 1971) went beyond the industrial needs of Québec, and in the 1980s Québec increasingly accommodated the US national energy pol-icy of 'sharing' the benefits from Québec's renewable hydroelectric energy resources.[141]

Through a 'buy-Québec' policy, Hydro-Québec and the James Bay Energy Company did strengthen provincially owned and private turbine and gen-erator manufacturing and industry branch plants supplying hydroelectric equipment. Yet, despite industrial 'masters' from other parts of Canada, the US, Europe, and Japan being invited into Québec, few of them estab-lished new industries to take advantage of the new supplies of electricity reserved for them, and secondary industry did not develop as predicted, with about 80 percent of the province's electricity between 1962 and 1984 continuing to be used to process pulp and paper, primary metals, and chemicals. The acceleration in installing generating capacity and the global industry's lack of interest in James Bay electricity contributed to making electricity surpluses that Québec could not absorb. Therefore, in export contracts with US utilities Hydro-Québec agreed, in part because of incompatibility in transmission technology, to transmit Labrador power to Montréal and to export Montréal power – from the Beauharnois plant – to Utica in New York State. Similarly, Hydro-Québec dedicated some generators in its La Grande 2 plant to the direct transmission of power to Sandy Point in New England.

In the mid-1990s, Québec politicians and planners – having used Hydro-Québec *not* merely as a utility that generates and supplies electricity, but as a highly politicized provincial development and export agency – cancelled further expansions. For example, Premier Parizeau called off the $13 billion James Bay expansion, to some extent because of the failure to achieve long-term contracts with utilities in the northeastern US. Nevertheless, in 1996 Premier Bouchard's government passed a new energy policy, and in 1998 Hydro-Québec has proposed a new expansion plan – a strategy that it hopes will make Québec into a 'major energy hub' in North America. Another 1998 initiative involves the lucrative power sites on the Lower Churchill River in Labrador. However, to understand the background of these plans and the negotiations with Newfoundland in the late 1990s, it is important to look back and to examine the bitter historical lessons of the Churchill Falls project.

5

The Churchill Power Trap (Newfoundland)

Fortunately Premier Smallwood had the imagination and audacity to enlist Winston Churchill's support as the first step towards persuading the Rothschilds to form a syndicate to launch a costly and risk-ridden venture into the unknown. That was in 1953 ... For out of that great enterprize [sic] will flow an annual output equal to 40,000 million kilowatt hours of energy. That is very nearly the equivalent of the total production of electricity in Ontario in 1965 from all sources. It is a prodigious achievement ... What the Churchill energy will add to Canadian industrial output is immense to a degree beyond calculation.[1]

– *St. John's Daily News*, 3 July 1968

Newfoundland Premier Brian Tobin is threatening to pull the plug on the giant Churchill Falls hydroelectric project in Labrador unless Québec agrees to renegotiate the controversial 65-year deal under which it resells the power at a tenfold profit.[2]

– *Globe and Mail*, 21 September 1996

Newfoundland Premier Brian Tobin and Québec Premier Lucien Bouchard lay the groundwork for the new hydroelectric development in Labrador ... The [3200 MW Lower Churchill Falls] projects announced total $9.7 billion, with an understanding to study a further project for Muskrat River evaluated at $2.1 billion.[3]

– *Vancouver Sun*, 10 March 1998

Introduction

In 1894, A.P. Low, on an expedition for the Geological Survey of Canada, identified the potential of Churchill Falls (formerly Hamilton Falls) as a source of useable energy when he reported the availability of 'several millions of horsepower.'[4] In the late 1950s and early 1960s, as shown in Chapter 2, utility officials, federal politicians, advisors on energy policy, and engineers envisaged Labrador's hydroelectricity-generating facilities as part of a national electricity grid. Premier Smallwood of Newfoundland, like other premiers, did not or could not rely on the emergence of industrial entrepreneurs from among the 500,000 people of Newfoundland and Labrador and, instead, combined the allocation of rights to raw material extraction and the rights to develop hydroelectricity and assigned them to private, often overseas, developers or development companies.

Although some of the specific events differ (e.g., Québec's refusal to permit transmission of Labrador electricity across its territory), hydro-related development patterns similar to those in other provinces (e.g., privatization reversals, expected but unrealized hydro-related industrial development, electricity surpluses, and the restructuring implications of the export of these surpluses to the US) are also evident in the Churchill Falls case. The water resources of Labrador, including Churchill Falls, were initially entrusted to the British Newfoundland Company (Brinco) for private development and were then nationalized by the Newfoundland government. The industrial benefits from the project accrued disproportionately to Québec, and all power surplus to Labrador was sold to Hydro-Québec by way of the Churchill Falls power contract (so Québec, in turn, could export its surplus electricity to the US). Furthermore, Québec's 1965 unwillingness to allow transport of Labrador electricity across its territory was challenged neither by Canada's National Energy Board (NEB) nor by the federal Cabinet, but in the late 1990s the US Federal Energy Regulatory Commission (FERC) required Québec to open its transmission system to public and private utilities (including those in Labrador) in order to retain Hydro-Québec's electricity export status in the US. To understand the present initiatives, such as the proposed 1998 development of the $13 billion Lower Churchill Falls project together with Québec, we first need to examine how, since the 1950s, Newfoundland has tried to combine water-power development with the development of other natural resources.

In 1952, three years after joining the Canadian federation, Newfoundland trusted, and therefore invited, Brinco, a consortium of British banking and industrial firms (including N.M. Rothschild), to combine timber, mineral, and water resources for industrial development in Labrador. Newfoundland's 'first act relating to the Upper Churchill [the river above Churchill Falls] is an act cited as the Government-British Newfoundland Corporation Limited-N.M. Rothschild & Sons (Confirmation of Agreement) Act 1953 (the 'Brinco Act'). This act, as its name implies, approved the execution and delivery of an agreement (the 'Brinco Agreement') between the lieutenant-governor in Council (the 'Government') and N.M. Rothschild & Sons, who had brought about the incorporation of Brinco.'[5] In 1958, Brinco created and incorporated a subsidiary, the Churchill Falls (Labrador) Corporation (CFLCo), which showed little interest in new local industry but signed a power contract with the Québec Hydro-Electric Power Commission (Hydro-Québec) even before raising US capital and draining the Falls to generate electricity. Original CFLCo and NEB records

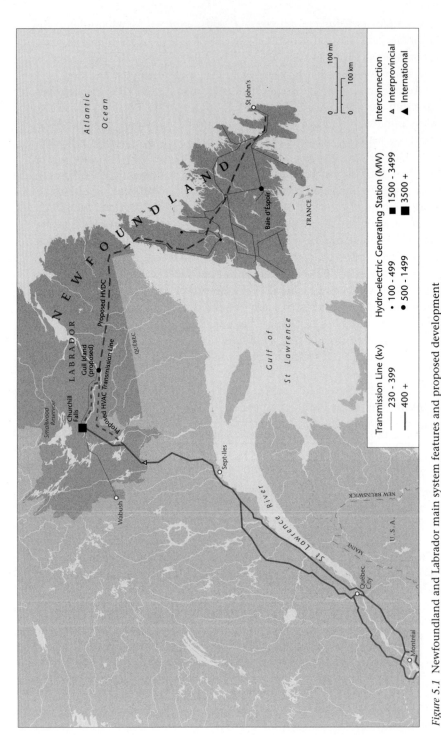

Figure 5.1 Newfoundland and Labrador main system features and proposed development

Source: Government of Canada, Department of Energy, Mines and Resources, Energy Sector, *The National Atlas of Canada*, 5th edition, Electricity Generation and Transmission, 1983, MCR4069 and MCR4144; Government of Newfoundland, 'New Churchill River Developments Plus HVDC Infeed to Island,' 1998.

reveal that Newfoundland lost industrial benefits to Québec, lost 'the right to transmit or export power,' and found itself unable to resolve sovereignty over Churchill Falls. In 1974, the Newfoundland government, incensed by Brinco's intent to alienate more of Labrador's water power from local industry and sell it to Québec, reversed its hydro privatization and bought back the people's water resources.

Inviting Brinco

In 1952, determined not to become overly dependent on US business interests for the development of Newfoundland's extractive industries, Premier Smallwood told the British prime minister, Sir Winston Churchill, that Newfoundlanders would prefer a return to a stronger link with Britain: 'He told Churchill it was American capital that was beginning to exploit Labrador's iron ore; but the people of Newfoundland had a tremendous pride in their British roots and did not want to see their province virtually annexed by United States' investment. He mentioned the East India Company and the Hudson's Bay Company as historic models for what he had in mind.'[6] After informing Churchill of his ideas, Smallwood gave the financiers, Anthony Rothschild and his nephew Edward Rothschild, the sales pitch. He offered them rights to explore minerals, to cut timber, and to develop water resources in Labrador and Newfoundland; in return for these rights, Brinco guaranteed it would invest $1.25 million every five years for the exploration and evaluation of the potential of such resources (Table 5.1).[7]

At first, seven United Kingdom banking and industrial firms joined Brinco: N.M. Rothschild and Sons, the Anglo-American Corporation of South Africa, the Anglo-Newfoundland Development Corporation, Bowater Corporation of North America, Frobisher Limited, the English Electric Company, and the Rio Tinto Company.[8] Essentially a consortium of private companies, Brinco was expected to explore and develop Newfoundland with government backing.[9] In 1953, the Newfoundland government allocated to Brinco timber, mineral, oil, and gas concessions in over 77,000 square miles, which included the Churchill Falls and all the hydro sites on the rivers and watersheds, both in Labrador and Newfoundland: 'The Brinco Agreement gave Brinco an option exercisable at any time within 20 years from May 21, 1953, to take an exclusive right and concession to harness and make use of all of the rivers and watersheds in Newfoundland and Labrador including the Churchill River and to be vested with all hydro-electric and hydraulic rights to the same.'[10] One of the hydro sites was the Baie d'Espoir site on the Island of Newfoundland. Among several others in Labrador was the Lower Churchill Falls (Gull Island) site, about 130 miles downstream from the Churchill Falls (Figure 5.1). All were designated for private development.

Soon, the Newfoundland government lost decision-making power over the export of Churchill Falls electricity when Brinco appropriated the right to export power without the government initially realizing what it had agreed to. The original 1953 Brinco Agreement stipulated that 'the Corporation shall not export any electric power from Newfoundland or Labrador without the previous consent of the Government';[11] but export was allowed according to the lease of the full rights to the Upper Churchill waters granted to CFLCo by the Newfoundland government's Lease Act. Passed on 13 May 1961 by the Government of Newfoundland, the Lease Act contained, in a deal involving power sites on the Island of Newfoundland, a ninety-nine-year lease granting 'to CFLCo the full right and liberty to use exclusively all usable waters of the Upper Churchill and its catchment area [above Churchill Falls] with the right to flood.'[12] This Lease Act allowed Brinco's CFLCo subsidiary the right to generate and transmit electric power in the province from the harnessed Churchill Falls and without qualifying consent to export from the province such power.[13] This clause effectively constituted the Government of Newfoundland's losing its export rights to Brinco.

Selling the Power, Raising the Capital, Developing the Site

Brinco used the most profitable means of extracting electricity from its water-power resources. For instance, the conglomerate took no risks of maintaining expensive unused capacity: in the Churchill Falls venture, it sold the power before raising the capital and developing the site. As indicated in Table 5.1, the letter of intent signed in 1966 to secure the Churchill Falls power contract preceded the financing and project launch in 1969. Power from the smaller sites was sold to existing resource-processing industries in Labrador, whereas Churchill Falls was primarily intended for customers outside the province. For each individually incorporated site, Brinco undertook initial surveys of geographic suitability, profitability, and market potential; if sites were found to be unfavorable, then Brinco sold them back to the Newfoundland government. In particular, Brinco pulled out of its plans to develop the Baie d'Espoir plant because it was unwilling to finance the $24 million transmission line to its nearest major industrial customers, the Bowater and Anglo-Newfoundland companies, which controlled more than half of the Island's productive forests. In exchange for selling back to the government all its formerly publicly owned hydro sites on the Island of Newfoundland and charging for survey and engineering fees, in 1961 Brinco was granted a ninety-nine-year lease on the water powers in the drainage basin of the Churchill River,[14] including the Churchill Falls and the Lower Churchill River (Gull Island) site.[15] As noted previously, the largest and most profitable site (then known as Hamilton Falls) had been incorporated in 1958 by Brinco to become the

Table 5.1

Newfoundland's reversal of hydro resource privatization

1952	Smallwood, Newfoundland's premier, invites the British Newfoundland Company (Brinco) to develop natural resources extending over 70,000 square miles in Newfoundland and in Labrador and to invest $1.25 million every five years.
1953	Privatization for twenty years of the all hydroelectric and hydraulic rights to all the rivers and watersheds in Newfoundland and Labrador including the Churchill River, which becomes part of the Government-British Newfoundland Ltd.-N.M. Rothschild & Sons (Confirmation of Agreement) Act 1953 (the Brinco Act).
1958	Brinco incorporates the Hamilton Falls Power Corporation, the predecessor company of the Churchill Falls (Labrador) Company (CFLCo) to plan the Churchill Falls project.
1961	CFLCo and the government of Newfoundland sign the Lease Act of 13 May 1961 that permits CFLCo the right to export power from Churchill Falls without the consent of the Newfoundland government and grants a ninety-nine-year lease on the water power in the drainage basin of the Upper Churchill River.
1965	Lesage, Québec's premier, states that the transmission of Labrador electricity across Quebec to other provinces will not be allowed.
1966	CFLCo and Hydro-Québec sign a letter of intent on 6 October 1966 to secure the Churchill Falls power contract; preliminary work starts in 1967.
1968-9	Brinco's CFLCo obtains North American financing (not British investment) by selling equity shares totalling $125 million to Hydro-Québec and by borrowing $840 million from US and Canadian sources in 1969.
1969	CFLCo and Hydro-Québec sign the power contract of 12 May 1969 for nearly all the power from Churchill Falls for 65 years at about one-quarter cent per kilowatt hour; launch of the major phase of construction.
1972	The Churchill Falls plant starts delivering electricity into the Hydro-Québec system, electricity which is resold at high mark-ups to industry and US utilities.
1974	Brinco plans the next project, the Lower Churchill (also called the Gull Island) and woos Hydro-Québec with another power contract.
1974	Moores, Newfoundland's premier, buys back the 'property of the people' and informs the legislature that the Churchill Falls company and all of Brinco's remaining water rights in Labrador have been bought back for $160 million.

CFLCo. Its mandate included the development of the drainage basin of the Upper Churchill River into a reservoir and of the powerhouse at Churchill Falls.[16]

Despite the fact that Canadian engineering and construction firms had more than seventy years of experience in building hydro projects in Canada, Brinco contracted with foreign companies (such as Bechtel Construction of San Francisco) and retained Canadian experts and engineers only as subservient partners.[17] In their offices located outside Newfoundland and Labrador, most of which were located at 1980 Sherbrooke Street West in Montréal, Brinco and its group of companies negotiated supply contracts, power contracts, and financial arrangements with Hydro-Québec (which was also in Montréal). Thus, Premier Smallwood's 'chosen instrument,' Brinco, directed Labrador's hydro development from Québec rather than from Newfoundland.

As in other ventures, Brinco drew its key customer into supplementing the financing of Churchill Falls before starting work on the site. After prolonged negotiations, Brinco's CFLCo subsidiary signed a letter of intent in 1966 with Hydro-Québec, which then purchased $125 million worth of equity and mortgage bonds.[18] Memoranda in Brinco's files indicate that in 1969 an '$840 million' debt 'was arranged with the United States and Canadian financial institutions and banks'; the total project cost was $946 million (Appendix 8).[19] With Québec's credit and the US loan, the financing amounted to $965 million, and the total project cost near the time of completion was estimated at $946 million – $19 million below the financing raised in North America. Preliminary underground work at the site began in 1967. Following the installation of the new 735 kV transmission line, the first power was delivered in 1972 into the Québec power system (Table 5.1). Keeping the construction time short meant that the plant would earn money at the earliest possible date; that is, within five years of the start of construction.

Disputes over Permitting Transmission

The Churchill Falls power station is located on the Labrador side of the boundary with Québec and about 300 kilometres inland from the coast of Labrador (Figure 5.1). Maurice Duplessis, Québec's premier in the 1950s, wondered, because of the boundary dispute from earlier in the twentieth century, if 'the Hamilton River [renamed in Winston Churchill's honour as Churchill River] was in Newfoundland at all.'[20]

Just as the connection of Niagara Falls plants to US utilities gave rise to international boundary conflicts, joining the Churchill Falls plant to the Hydro-Québec system rekindled the unresolved Labrador Boundary Dispute:

> The territorial limit between Québec and Newfoundland in the Labrador peninsula, at over 3,500 km long, is the longest interprovincial boundary. It has not yet been surveyed and marked on the ground. A dispute concerning the ownership of Labrador arose in 1902 when the Québec government protested Newfoundland's issuing a timber licence on the Churchill River.

Two years later Québec asked Ottawa to submit the controversy to the Judicial Committee of the Privy Council in London.

...

Newfoundland traced its claim of ownership to the commission issued to Gov. Thomas Graves in 1763, which extended his jurisdiction to the 'Coasts of Labrador.'

...

The judicial committee [of the Privy Council in London] refused to accept Canada's contention that 'coast' meant a strip of land one mile (1.6 km) wide along the seashore. It found that the evidence supported [pre-confederate] Newfoundland's inland claim as far as the watershed line or height of land. The court's decision in March 1927 settled the boundary in its present location. When Newfoundland joined Confederation in 1949, its boundary in Labrador was confirmed in the Terms of the Union (now the Newfoundland Act), enshrined in the Constitution Act, 1982. A 1971 Québec royal commission decided that Québec's case against the 1927 boundary was not worth pursuing; by 1987, although the province did not consider the issue to be settled, the dispute appeared dormant.[21]

Given this background, the development and generation of power in Labrador and its transmission across the provincial boundary with Québec posed special problems for Newfoundland. From the 1960s onwards, the Government of Newfoundland at times supported the integration not only of Churchill Falls, but also of any future Labrador hydro plants into a national power grid.[22] At the same time, the Québec government did not allow Newfoundland to transport electricity across Hydro-Québec's power lines to other provinces; thus the Québec utility became the only extra-provincial buyer of Labrador electricity.

On 12 May 1969, Hydro-Québec signed a power contract with CFLCo allowing about 90 percent of the Churchill Falls capacity of 5,429 MW to be used by Hydro-Québec. The contract specified an extremely cheap, declining rate of less than 3 mills per kilowatt hour (2.7 to 2.4 plus adjustments) for forty years and for automatic renewal involving another twenty-five years at about 2.4 mills (less than a quarter of a cent).[23]

Controversy over this contract began to broil in 1976 when the Newfoundland government asked to buy power from Churchill Falls in order to meet its estimated industrial growth needs. By an Order-in-Council dated 6 August 1976, the government requested, via the Supreme Court in Newfoundland, that CFLCo supply its new parent company, Newfoundland and Labrador Hydro, with electricity.[24] The key issue became the existing long-term contract between Hydro-Québec and CFLCo. The Newfoundland Supreme Court judgment concluded that 'there is a wide gap between what the Government really seeks and what it can get. For example, no matter how much energy it requests, it can receive no more than is surplus to the present commitment for sale [by] CFLCo plus whatever remains of the 300 MW recapture allowance.'[25] It is uneconomic and technically problematic, however, to transmit only 300 MW across the iceberg-prone ocean straight to the Island of Newfoundland.[26]

National issues were aired at export hearings in the 1980s, during which Newfoundland emphasized that the NEB should have a 'national' mandate and argued that, under the British North America Act, Newfoundland should have the right to transport energy across provinces. In the early 1980s, the federal government and Newfoundland held discussions on the possibility of 'incorporating a wheeling amendment into the National Energy Board Act.'[27] Part of the discussions included 'wheeling opportunities through Hydro-Québec's system for future developments' in Labrador.[28] The outcome of these initiatives resulted in extending regulatory powers to the NEB over select interprovincial transmission lines:

> Bill C-108, *An Act to amend the National Energy Board Act* was proclaimed on 1 February 1983. This Act included a provision to designate particular interprovincial transmission lines for treatment as if they were international power lines – giving the Board jurisdiction over the regulation of such lines and over the expropriation of land for their construction.[29]

In 1992, the NEB reported that no such designation had taken place.[30]

In September 1986, Marcel Masse, the minister of energy, mines and resources, asked the NEB to review and report on the simplification and reduction of the federal regulation of electricity.[31] In the NEB's public hearings in November 1986 to discuss this matter, Québec claimed that the board's role was to encourage interconnections with the US and to open access to the American market, whereas Newfoundland claimed that the NEB's mandate should be to encourage interconnection between systems in Canada.[32]

During the 1986 NEB Regulation of Electricity Export hearings in Ottawa, Newfoundland and Labrador Hydro argued: 'We have repeated throughout our evidence ... that we must look at Canada first, we must look at Canada as if it is a single market, as if it is not divided by provincial boundaries, when we are attempting to establish where there is Canadian need [for provincial surplus power].'[33] Also at these hearings, David Mercer, president of Newfoundland and Labrador Hydro, referred to the Borden Royal Commission Report, which preceded the formation of the NEB, to illustrate that the National Energy Board Act arose, in part, out of concern over energy for industrial use within the boundaries of Canada. He further contended that 'it is advisable to encourage, where possible, the development in Canada of processing industries relating to energy and sources of energy as opposed to making such benefits available outside of Canada.'[34] He added that permission to export should be withheld if the granting of a licence would 'interfere' with the 'energy requirements of those parts of Canada within economic reach of the producing province.' Further, if 'provincial government objectives' are 'at variance with national objectives,' then the provincial 'producer's revenue objective would first have to be reconciled with the national objective.'[35] In contrast to Québec's view of the NEB as an export regulating agency, the Newfoundland government's position was that Parliament passed the National Energy Board Act 'to restrict access to foreign markets in the energy sector.'[36]

Newfoundland maintained that, since gas and oil pipelines are allowed to cross provinces within Canada, the same practice should apply to electricity. Its representative argued that 'energy, in forms other than electricity, moves between producing and consuming provinces under federal jurisdiction derived from Section 92 (10) of the British North America Act. Moreover, the establishment of a National Power Grid would extend this principle to electricity.'[37] During the hearings, Jacques Guevrement, the executive vice-president of external markets for Hydro-Québec, advocated a regulatory environment for oil and gas that differed from that for electricity crossing Québec in transmission lines (i.e., wheeling) because, in his view, 'electricity is less a natural resource and is more a manufactured product' and, thus, NEB regulations should not apply.[38] Québec's refusal to allow the transport of electricity across the province had also come up twenty years earlier.

From the 1960s onward, Québec officials have resisted wheeling and its regulation by the federal government, invoked the Labrador Boundary Dispute in the Churchill Falls power contract, and convinced Québecers of industrial energy shortages while simultaneously applying to export surpluses to the US. For instance, in 1965, when Jean Lesage, Québec's premier, was 'asked if he would give Brinco a right-of-way for the power he replied, 'very emphatically,' that 'he would never agree to that.'[39] He told reporters that it was 'an absolute condition' for continued negotiations [over buying Churchill Falls electricity] that Hydro-Québec buy all the power. 'Never under any conditions' he repeated, 'will we let others build a transmission line across Québec ... and never will we agree to simply rent out the transmission facilities.'[40] Lesage acted accordingly, thereby virtually landlocking Labrador's surplus electric power. The NEB commented that 'without a wheeling or transmission access option being available to it [Churchill Falls], Newfoundland entered into financing and sales agreements with Hydro-Québec.'[41] In May 1965, as stated in the second chapter of this book, the point was made in the Pearson Cabinet that, given these events, 'it was important that negotiations be carried on directly between the two power corporations concerned [Brinco and Hydro-Québec], without interference from the provincial governments.'[42] Two years later, in March 1967, Brinco, as the owner of Churchill Falls (Labrador) Hydro, agreed with Hydro-Québec that the power from Churchill Falls would be delivered into the Québec grid for sixty-five years.[43] With the NEB's permission and federal Cabinet approval, Hydro-Québec then exported its electricity surplus during the 1970s, 1980s, and 1990s.

In its submissions to the NEB, Québec continued to insist that it would negotiate only with individual utilities, treating energy transactions as commercial transactions between neighbouring jurisdictions. Whether they are states of the eastern US or provinces of Canada, Québec proposed to treat its business partners in these jurisdictions simply as external markets.[44] In 1986, without help from Ottawa, and without federal enforcement of the national interest, no wheeling agreement had been reached between Newfoundland and Québec.[45]

One Hydro-Québec official argued before the NEB that Québec had a distinct definition of surplus electricity, one that differed from that of the NEB. During the review hearings about NEB regulations in November 1986, Hydro-Québec proposed to redefine the concept of surplus from 'existing surplus plant capacity' to 'undeveloped hydro resources,' possibly to propose a change in the NEB regulation that required a test to determine whether electricity is surplus to Canada's needs before allowing its export to the US (see Table 5.2). Guevrement stated: 'Le surplus pour nous à Hydro-Québec, ça a différentes notions. Nous avons 40 000 mégawatts de ressources non développées ... Cette notion de surplus m'apparaît comme très differente dans l'esprit de differentes personnes.' [For us at Hydro-Québec surplus has different connotations. We have over 40,000 megawatts of undeveloped resources ... The definition of surplus seems to me to be very different for different people.][46] Guevrement defined surplus as sequentially exploitable river energy of 40,000 MW to be extracted from northern Québec and Labrador for the US market – a return to the 1890s Niagara Falls idea of extracting energy resources for a manufacturing base elsewhere (i.e., treating them as staples).

In the 1970s, Québec increasingly took steps towards exploiting electricity as an export commodity. In the years before applying to the NEB for export permits, planners of Québec's hydro system had argued that electricity from Churchill Falls was needed for industry in Québec. For instance, the letter of intent, dated 6 October 1966, between Hydro-Québec and CFLCo securing the Churchill Falls power contract stressed that 'power is necessary for the economic development of Québec and particularly for that of secondary industry' and that 'it is necessary to foster by all possible means the industrial potential of Québec.'[47] However, a few years after Churchill Falls first delivered electricity in 1972, Hydro-Québec did not use much of the electricity to power Québec's industry but, instead, exported power to the US, for the utility had more power than was needed for the development of industries in Québec. On 28 November 1973, while additional generators were being installed at Churchill Falls from 1972 to 1974, Hydro-Québec signed the 'diversity power contract' with the Power Authority of the State of New York, a contract which relied on the construction of a 765 kV international power line with a carrying capacity of 2,300 MW – nearly half of the Churchill Falls capacity.[48] In part because of the expected increase in Hydro-Québec's surplus in the 1980s and 1990s, the NEB approved the interconnection between Québec and the US and the export of power.[49]

In the 1983 licence hearings, when Newfoundland expected the NEB's help with regard to having Hydro-Québec offer Newfoundland at least some of its surplus electricity, the NEB instead allowed Hydro-Québec to exclude Newfoundland as part of the Canadian public interest (Table 5.2). The NEB reported in January 1984 that Newfoundland and Labrador Hydro had applied to it in 1983 to repatriate (in lieu of Churchill Falls exports to Québec) 'up to 1200 MW of power' of Québec's planned export and to block the export hearings.[50] The NEB allowed the hearings on 19 and 20 September 1983 but dismissed Newfoundland's request, foregoing

Table 5.2

Hydro-Québec electricity export hearings and interventions by Newfoundland

1973	Hydro-Québec signs power contract with the Power Authority of the State of New York, while new supplies of electricity from Churchill Falls enter the Québec system.
1976	During the Hydro-Québec export licence hearings, Newfoundland attempts, through legal action, to recapture some of the power output of Churchill Falls, but is unsuccessful.
1982	Newfoundland and NEB discussions about wheeling through the Hydro-Québec system result in the NEB Act (1 February 1983) amendment allowing designation of interprovincial transmission lines for treatment as if they were international power lines. By 1992 such a designation had not been assigned.
1983	During export licence hearings to examine surplus exports of 2,699 MW for the period 1984-99, Newfoundland tries to obtain 1,200 MW of Québec's surplus power but is unsuccessful.
1984	NEB grants an export licence to Québec in January 1984, after deciding that offering surplus power to Newfoundland was not a requirement for permitting exports.
1986	During the NEB Hearing about *The Regulation of Electricity Exports* in November 1986, Newfoundland argues for including a national role and Québec for merely a continental role of the NEB.

its usual practice of asking that electricity surplus be offered to neighbouring provinces as evidence that 'energy proposed to be exported is surplus
to reasonable foreseeable requirements in Canada.'[51] The NEB had a broad
conception of the term 'surplus'; and its decision shows that it accepted
other evidence for the surplus test, as it concluded that there '[was] no
requirement in the [NEB] Act that an offer [of the export] be made' and
that Hydro-Québec had complied with the regulations.[52] Therefore, it
granted Québec an export licence in 1984, contending that the export was
in the public interest, even without the offer to Newfoundland (Table 5.2).[53]

Newfoundland and Québec have had uneven bargaining positions
because until the late 1990s Hydro-Québec treated its transmission facilities
as sovereign and the Churchill Falls power contract as non-negotiable.[54] In
1996, for instance, Hydro-Québec 'resold the [Churchill Falls] power at a
tenfold profit.'[55] Until Newfoundland's premier Brian Tobin threatened in
October 1996 to force a reopening of the Churchill Falls contract, Québec
appeared unwilling to negotiate questions of interprovincial transmission
from future hydro plants in Labrador. Its plan was to use its renewable
northern energy resources for the 'joint economic benefit' of the US and
Québec.[56] The continentalization of surplus electricity brought with it the
necessity for Canadian provinces to comply with American regulations
when planning to export to the US.

While the NEB has had or has exercised essentially no power over
wheeling electricity in Canada, FERC, in contrast, mandates the wheeling
of transmission groups in the US and of electrical utilities in Canada that
join such transborder regional groups, so that transmission policies of
provincial regulatory commissions must increasingly comply with FERC
regulations. Although the NEB could call on the federal Cabinet to desig-
nate this an issue of significant national interest, by 1998 it had still not
requested that new public hearings be held on this issue. As it stands, 'new
FERC rules require American utilities to open their transmission grids to
US and Canadian competitors at rates set by regulators.'[57] Transmitting
through its system electricity from Labrador destined for export to utilities
in the US, Hydro-Québec could be required to allow 'reciprocal' access to its
system (which it did not allow under NEB regulations). Konrad Yakabuski,
who has analyzed this regulation issue, argues that 'with an open transmission system' Newfoundland could theoretically compete in the Québec
market against Hydro-Québec; this potential competition, he suggests, is
in part behind Québec's new interest in teaming up with Newfoundland.[58]
Therefore, with reference to the Washington-based FERC, Premier Tobin,
while at a meeting of Canada's eastern premiers with New England state
governors in Rhode Island in June 1997, mused 'God bless FERC' and
added that 'it is ironic that federal regulatory renewal in the United States
has brought new life to an old project in Canada.'[59] Nine months later, on

9 March 1998, Québec premier Lucien Bouchard and Newfoundland premier Brian Tobin announced a plan to negotiate a $12 billion project on the Lower Churchill River in Labrador, which is largely reliant on exports to the US (discussed in more detail below). The Tobin government in Newfoundland is proceeding on this set of negotiations carrying a legacy of bitter historical lessons from Smallwood's Upper Churchill (Churchill Falls) project. Possibly now resolved through having to conform with US regulations, the intensity of the transmission dispute between Québec and Newfoundland appears to have been a historically unique event. Another occurrence, however – the reversal of privatization – seems to repeat a pattern evident in other provinces. However, an examination of the background of this incident shows that specific events did differ.

Repatriation and Ownership

As stated above, Newfoundland's control over electricity exports from Churchill Falls was lost to Brinco, but why was the province not able to repatriate or recapture electricity supply from the Churchill Falls plant, as Ontario had at Niagara Falls in the early decades of the twentieth century?

The amount of Churchill Falls electricity that the Newfoundland government can 'recapture' for its consumers, according to the power contract between CFLCo and Hydro-Québec (12 May 1969), 'shall not exceed ... 300 MW,' and Newfoundland's priority access is restricted by 'not less than three years written notice to Hydro-Québec.'[60] The other capacity amount available to Newfoundland from Churchill Falls is 230 MW, an amount which was part of the contract to replace the capacity of the Twin Falls near Churchill Falls and to deliver it to industry in Labrador. According to the Supreme Court of Newfoundland, the power exported to Québec from Churchill Falls cannot be recaptured until the year 2041.[61] This contract and the Lease Act oversight, as already described, led Newfoundland, one of the poorest provinces in Canada, to allow Brinco to usurp its export rights and to recapture claims for sixty-five years (Table 5.1).

Furthermore, as part of this same private contract between CFLCo and Hydro-Québec, Newfoundland risked its ownership of the Churchill Falls facilities, especially given the possibility that, in the future, the Labrador Boundary Dispute might be settled in Québec's favour, for the British Privy Council decision of 1927, which supported Newfoundland's claim to own the inland watershed of Labrador as well as the coast, has never been fully recognized by Québec.[62] The power contract between CFLCo and Hydro-Québec circumvented the issue by allowing for the possibility that Québec actually owns part or all of the Churchill Falls facilities on either side of the 'Delivery Point,' where electricity enters Québec. The power contract stipulates that 'any facilities of the Plant which may at any time be found to be located on the Québec side of the Delivery Point shall be and are hereby acknowledged always to have been the property of, and to have been purchased for the account of, Hydro-Québec.'[63] Should the court recognize the validity of this clause, the Newfoundland government could lose its stake in Churchill Falls, which it currently owns and operates through CFLCo.

As shown so far, Brinco, the parent company of CFLCo, weakened Newfoundland's interest by taking actions that compromised sovereignty and export control. Under similar circumstances, when US utilities signed export contracts for up to ninety-nine years, the Ontario government had supported the reversal of the Niagara Falls privatization. Brinco's corporate conduct, subsequent to the signing of the sixty-five-year power contract, led also to the reversal of Labrador's hydro resource privatization (Table 5.1).

When, in 1974, Brinco was ready to start its next hydro scheme, the Lower Churchill River site (also called Gull Island), which has less than half of Churchill Falls' potential power (2,300 MW), Brinco once again set about selling the power, raising the capital, and then developing the site, and once more intended to woo Hydro-Québec as a customer. It argued that 'a large, financially-secure consumer of power is required as party to any Gull [Lower Churchill] power contract in order to achieve the same types of support which occurred with Churchill Falls. At present, the only foreseeable customer that might satisfy these requirements is Hydro-Québec.'[64] This time the Newfoundland government intervened.

Reversal of the Churchill Falls Privatization

As we have seen, Brinco first gained rights to Newfoundland's natural resources and hydro power development at the invitation of Premier Smallwood in the early 1950s. Brinco's approach to development not only led to Newfoundland's loss of provincial decision-making power over hydro projects, but it also failed the province in other ways. Brinco usually picked the best hydro sites and developed them as capitalist ventures separate from other raw materials available in the region (rather than combining energy and natural resources for local industrial development) and made large power contracts with wealthy customers while leaving Newfoundland's unprofitable electrical network expansions to the government. The Newfoundland government was forced to rethink its earlier policy of relying on transnational corporations (corporations in which ownership, management, sales, and production take place within several national jurisdictions), such as Brinco, to develop the hydroelectric resources in Labrador and Newfoundland (Table 5.1).

Brinco Takeover

By the 1970s, the Newfoundland government had decided to interrupt Brinco's no-risk, maximally profitable development of the province's water resources. On 28 March 1974, Frank Moores, Newfoundland's premier, 'told the House that the province and the company had reached "an amicable negotiated position" under which the province would buy the Churchill Falls company and all Brinco's remaining water rights in Labrador for $160 million.'[65] The Conservative premier asserted that 'the resources of Newfoundland are the property of our people,' and in the context of these events he argued that 'it follows that control over these resources must rest with the people's government.'[66] Unlike Brinco, which separated hydro energy development (for export) from that of other natural resources, his government 'intended to establish a comprehensively planned approach to the development of the province and its resources.'[67] He added, 'indeed we are of the firm opinion that government ownership of Brinco is a necessary and fundamental step toward ensuring that this province will continue to be able to attract private enterprise in the future.'[68] However, taking back control over the development of hydroelectric infrastructure so that it can serve to attract private enterprise leaves much of the development of (forward) industrial linkages to chance. In contrast, Québec had taken more direct control over negotiating (backward) industrial linkages from the development of Churchill Falls. During the late 1960s, and possibly in return for Québec's 13 percent contribution to the financing of Churchill Falls[69] and for purchasing most of its power, Québec's hydro-technology industries had received a boost.

Linkages to Industry

Brinco favoured Québec's interests in the development of the Churchill Falls site by allowing Hydro-Québec to impose nationalistic 'buy-Québec' manufacturing conditions on the purchase of supplies. In this way, Newfoundland's participation in the growth of supply industries was subordinated to the interests of Québec. The Churchill Falls power contract of 12 May 1969 went far beyond the normal price and length of service clauses by giving preference to Québec in several areas. The contract stipulated first preference to manufacturers and contractors from Québec, and only second preference to those outside Québec. Preference for turbines and generators manufactured in Québec was also made clear:

> 20.1 Preference concerning Québec Equipment, Services and Personnel
> CFLCo [Churchill Falls (Labrador) Corporation] will, in the procurement of materials, services and equipment and in the employment of personnel, extend or cause to be extended by its contractor, sub-contractors and agents preference to Québec labour, personnel and services and to materials and equipment manufactured in Québec.
>
> 20.2 Preference to other Canadian manufacturers
> In the purchase of items of equipment, such as turbines, generators, transformers and the like, CFLCo will, subject to the prior preference stipulated in Section 20.1, give the next preference to other Canadian manufacturers.[70]

Québec's procurement position was, therefore, strengthened at the expense of Newfoundland's – supplies from Newfoundland amounting to only $19 million.[71] While explaining the breakdown of these purchases in 1971, four years after the start of the project, Brinco's general manager, Roy Legge, argued that 'Newfoundland [subsumed under 'other Canadian manufacturers'] has benefited also from a preference clause, written into all Churchill Falls contracts, favouring materials manufactured or distributed and serviced in the Province.' As a result, so he claimed, 'purchases from Newfoundland manufacturers and suppliers for the Churchill Falls project have been considerable. Alcan's Stephenville plant supplied 4,714 miles of aluminum conductor worth 11 million dollars ... North Star Cement of Corner Brook have [sic] supplied 97,686 tons of cement worth 2.5 million dollars. Newfoundland Steel Company has supplied tons of rebar [reinforcing bars] worth 2.3 million dollars.'[72] In isolation, Mr. Legge's figures of millions for aluminum conductors and millions for reinforcing bars and cement from Newfoundland may be impressive, but even when inflated with purchases of such items as beer, groceries, and paint, they amount to a minuscule fraction of the project cost. All in all, the $19 million worth of purchases (Appendix 7) from Newfoundland amounted to 2 percent of the total of $946 million in project costs (Appendix 8).

In contrast to the 2 percent from Newfoundland, 'approximately 75 percent of all equipment, materials and supplies' originated from Montréal or further west in Canada.[73] As an illustration, more than two-thirds of the manufacture of ten turbines and generators, valued at $50 million, was carried out in Québec.[74] An examination of the negotiations concerning manufacturing locations and other issues, which surfaced while the $50 million turbine and generator contract for Churchill Falls was finalized in 1967, will show just how choices regarding strengthening supply industries were made by Hydro-Québec and the Brinco group.

Generators and Turbines for Churchill Falls

Criteria such as cost, quality, and quantity continued to be important in the selection of turbines and generators, but during the mid-1960s in Québec, in the midst of the Quiet Revolution, the manufacturing location became a crucial issue. Under Québec's nationalistic manufacturing conditions, English Electric, a founding member of Brinco and a supplier of turbines for Brinco's smaller Twin Falls project, was passed over for a contract because it manufactured its turbines in Netherton, England. The Churchill Falls (Machinery) Consortium, consisting of Marine Industries, in partnership with Canadian General Electric and its Québec subsidiary, Dominion Engineering Works, became a contender. Marine Industries had previously manufactured turbines for the Hydro-Québec Manic project on the Outardes River north of Baie Comeau in the mid-sixties. Unlike English

Electric, Marine Industries had established a production facility to build equipment for the Manic project, was owned 65 percent by the Québec government and 35 percent by the French government, and manufactured under licence to Neyrpic, a French company.[75]

Manufacturing benefits from the Churchill Falls generating station became an issue two years before the 1969 power contract was signed. On 9 June 1967, the $50 million turbine and generator contract was reviewed in a meeting between CFLCo, Acres Canada, Bechtel, and the Churchill Falls (Machinery) Consortium. Their main concern was with the Québec manufacturing content of the generating equipment. They asked 'how much of this equipment will be manufactured in the Province of Québec – what percentage?' and demanded explanations if the answer to the question was under 90 percent.[76] J.E. Pontbriand, representing the Québec government's Marine Industries, reported that turbines and generators would largely be manufactured in Sorel, Québec. J.D. Lewis, representing Canadian General Electric and Dominion Engineering Works, also confirmed that while most of the manufacture of turbines and 50 percent of its generators would be done in Québec, the rest would be built in Peterborough, Ontario. Don McParland of CFLCo made it known – particularly to Canadian General Electric – that if they could not manufacture some components in Québec, then they would have to give their reasons. On 28 June 1967, a new turbine and generator letter of intent was signed.[77]

Two days after the letter of intent had been signed, the turbine and generator contract was awarded to the Churchill Falls (Machinery) Consortium of Montreal.[78] In its press release, CFLCo summarized its adherence to Hydro-Québec's stipulation of prior preference to Québec and Canada in manufacturing the ten turbines and generators for Churchill Falls, stating that, overall, 'more than 75 per cent ... will be manufactured directly in the Province of Québec' (Appendix 9).[79] To speed up the project – in part to keep the $189 million in interest costs low and to achieve early profits from generation – tenders for other turbine and generator manufacturers were circumvented and a consortium was formed. It consisted of Marine Industries Limited (MIL), Canadian General Electric Company Limited (CGE), and CGE's subsidiary, Dominion Engineering Works Limited (DEW). Between 1967 and 1974, these companies, which did their manufacturing in Québec, installed their equipment in Newfoundland and Labrador powerhouses (Table 5.3). Nevertheless, with regard to technical considerations, CFLCo remained dependent on the Acres Canadian Bechtel Consortium for design, engineering, construction management, and equipment selection: the ultimate approval of Canadian equipment rested with Bechtel in San Francisco.[80] This generating equipment, used in the cavernous powerhouse at Churchill Falls, consisted of some of the most powerful equipment then in existence. With most of the Churchill

Table 5.3

___

Turbines and generators at Churchill Falls and Baie d'Espoir, 1967-85

	Capacity of plant (MW)	Years of turbine installations and manufacturer	Years of generator installations and manufacturer
Baie d'Espoir,	613	1967-70 CAC	1977 DEW, 1967-77 CGE
Churchill Falls	5,429	1971-4 DEW, MIL	1971-4 CGE, MIL

Note: DEW = Dominion Engineering Works; CGE = Canadian General Electric; MIL = Marine Industries Ltd; CAC = Canadian Allis-Chalmers.
Source: Statistics Canada, *Electric Power Statistics: Inventory of Prime Movers and Electric Generating Equipment,* cat. 57-206, 31 December 1986.

Falls power available only to Québec, including its industries (and allowing Québec to export the surplus from other plants), Newfoundland planned to use its revenue from Churchill Falls, however modest, to subsidize its electricity rates for manufacturers on the Island of Newfoundland. And it tried to create what Hirschman identifies as a fiscal, rather than a physical (e.g., electricity supply), linkage to new manufacturing activities.[81]

Industrial Customers
On the Island of Newfoundland, the government's attempt to use part of the anticipated revenue from Churchill Falls as a fiscal linkage merely provided cheap electricity for invited forest products industries. Despite the addition of Churchill Falls, industries in the province continued to be based on the primary processing of resources, particularly minerals and forest products. In Labrador, Brinco built hydro projects only for large 'financially-secure' consumers, thereby helping the expansion of already existing mining operations without bringing any new manufacturers to the region. Neither the hydro-related industrial development strategy for Labrador nor the strategy for the Island of Newfoundland created new and innovative secondary industries.

Labrador
As we have seen, Brinco's development of hydro projects failed to start new industries in Labrador. Brinco sold off hydroelectricity and natural resources separately, without combining them in order to develop secondary industries. To explain its lack of industrial development, Brinco claimed that 'the markets at the time were not in Labrador and many attempts to develop a local load [industrial electricity consumption] failed.'[82] John Dales, extrapolating from his 1957 historical study of the

relationship between hydroelectric and industrial development, had argued that a contrary pattern would emerge, that diversification had been most successful where power companies had to develop their own local industrial markets (Chapter 1). Newfoundland had expected Brinco to create its own market for industrial electricity by risking capital investment in the value-added manufacture of resources. To develop its own resource prospects over 77,000 square miles, Brinco established a wholly owned subsidiary, British Newfoundland Exploration (Brinex), to serve as its exploration corporation.[83] However, Brinex developed no new industries; 'with the exception of copper deposits on the Island, at Whales Back, discovered in 1961, of which mining commenced in 1964 and [was planned for closure] in June 1972, the efforts of Brinex have located occurrences of minerals or deposits ... which ... have not been feasible to develop.'[84] Brinco simply followed a strategy of sequential extraction of resources by selecting only the most profitable ventures first.

Brinco's corporate approach to developing hydroelectric power reveals that, unlike new steam power – one of the determinants of the British Industrial Revolution – Churchill Falls electrical power was not a determinant in the growth of new industries in Labrador. Rather than integrating the new supplies of hydroelectricity with the transformation of northern resources in an entrepreneurial fashion (as is often expected from the private sector and is still the basis of a continuing belief in innovative production located in private enterprise), Brinco sold electricity to the 'large, financially-secure' iron ore mining companies and Hydro-Québec.

The low-cost industrial electricity consumed in Labrador was largely applied to the minimal processing of manufacturing inputs exported to the US. For instance, the minerals from the Iron Ore Company's open-pit mines at Carol Lake, near Labrador City, are first processed in a concentrator (which separates rock particles from iron grains until reaching about 66 percent iron content) and then shipped to a pelletizing plant (which converts the concentrate to half-inch-round balls for furnace reduction) in Labrador City. Yet, not all the Labrador iron ore was even processed in the region, because, for example, the Wabush Iron Company, using electricity available in Québec, set up its $50 million pelletizing plant at Sept-Iles.[85] In this way, some of Labrador's iron ore was pelletized in a Québec resource shipping port.[86] Such instances illustrate that Newfoundland was not effective in enforcing manufacturing commitments from the mining industry.

Representatives of US firms, for their part, made exaggerated claims about 'their rights' to continentalize energy and other natural resources. For instance, in a letter dated 12 December 1959, Mr. Humphrey (of the US-owned Hollinger-Hanna, corporate owners of the Iron Ore Company in Labrador) tried to claim water-power rights to Churchill Falls and planned to export power generated at the Falls to New York. In his response to Hollinger-Hanna, Premier Smallwood clarified the nature of Brinco's control over hydro resources:

> To me the situation is very clear. You have the right to take iron ore out of the Province. Others have rights to the same. The British Newfoundland Corporation has, among other things, the right to develop the hydro resources of Labrador. The Government of Newfoundland regard [sic] you and other mining concession holders basically as miners of iron ore. The British Newfoundland Corporation are [sic] the chosen instrument of Newfoundland for the development of the hydro resources of Labrador. The position is that the Government see [sic] your relationship to the British Newfoundland Corporation as that of a possible customer.[87]

Smallwood's letter also reveals his willingness to separate the development of hydro resources from the development of Labrador's natural resources for the sake of US export.

The Island of Newfoundland

In anticipation of the revenue expected to flow from the 1969 power contract between CFLCo and Hydro-Québec, Premier Smallwood established a discount price for industrial electricity on the Island of Newfoundland, even though statistics in 1954 and 1966 had already demonstrated that low-cost electricity plays only a minor role in secondary manufacturing.[88] For example, the cost of electricity amounts to less than one-half of 1 percent of the total cost of production for most industries producing finished goods.[89] Low-cost electricity is a more significant factor in the natural-resource-based sector because for semi-finished goods the electricity component constitutes 4 to 8 percent of the total production cost (Appendix 11).[90] Nevertheless, government advertising in 1968 invited potential customers[91] to take advantage of power from Baie d'Espoir in Newfoundland and Churchill Falls in Labrador, and offered substantial blocks of power at the discounted rate of 2.5 mills per kWh at either coast.[92]

From 1975 to 1978, subsequent to the building of these hydro projects, the semi-finished goods sector, wood, pulp and paper, and chemical manufacturers consumed an increasing percentage of all electricity purchased by manufacturing industries in Newfoundland (Table 5.4). After low-cost industrial power became available, consumption in extractive industries intensified, with consumption in the pulp and paper sector increasing from 816 GWh to 1,050 GWh during the same period (Appendix 10). A significant proportional increase also occurred in the processing of chemical products, which are mostly used in the processing of pulp (Table 5.4).

The additional hydro projects did not change the concentration of use, for nearly 90 percent of manufacturing electricity is used in the forest-based and chemical supply sectors (Table 5.4). Some of the manufacturers that consume a smaller percentage of the total are those involved in fish processing (about 4 percent) and food-and-beverage producers (about 6 percent) (Table 5.5).

Dependent Development Continues

Unlike industrialized countries in Europe and East Asia (e.g., Switzerland, Germany, or Japan), where self-reliant entrepreneurs in the manufacturing sectors and the hydroelectric sector emerge from within the society, in Canada – albeit a country industrialized in the 1850s and included among the seven most industrialized countries in the world – a peculiar dependence on entrepreneurs from outside the nation has continued late into

Table 5.4

Consumption of electricity in manufacturing industries, Newfoundland, 1975-8

Year	Total consumption in manufacturing (GWh)	Three major sectors			Sum of three sectors (%)
		Wood (%)	Pulp and paper (%)	Chemical (%)	
1975	1,130	0.5	72*	13	86
1976	1,468	0.4	66	25	91
1977	1,639	0.3	–	32	–
1978	1,883	0.3	56	34	90

*Statistics from category 2710.
Source: Statistics Canada, *Consumption of Purchased Fuel and Electricity by the Manufacturing, Mining, and Electrical Power Industries*, cat. 57-208, 1975-84. Only three complete years were reported during this period.

Table 5.5

Consumption of electricity by fish products and food-and beverage production, Newfoundland

Year	Consumption in manufacturing (GWh)	Manufacturing categories			
		Fish products (GWh)	Fish products (%)	Food and beverage (GWh)	Food and beverage (%)
1975	1,130	48.8	4	71.8	6
1976	1,467	56.1	4	80.1	5
1978	1,883	83.7	4	113.5	6

Source: Statistics Canada, *Consumption of Purchased Fuel and Electricity by the Manufacturing, Mining, and Electrical Power Industries*, Catalogue 57-208, 1975-84. Only three complete years were reported during this period.

the twentieth century. For example, by 1968, the control of the $946 million Churchill Falls project (i.e., the control of Brinco itself) was held by British and US firms. Rio Tinto-Zinc Corporation of London and Bethlehem Steel Corporation, Pennsylvania, formed a Canadian holding company that owned about 50 percent of the stock in British Newfoundland Corporation of Montréal. Brinco held about 70 percent of the shares of Churchill Falls, with Hydro-Québec and the Newfoundland government owning the balance.[93] Predominant ownership of the Churchill Falls hydroelectric franchise returned to the government of Newfoundland in 1974 when it bought back the remaining shares of CFLCo and all water rights from Brinco. However, foreign ownership and foreign-directed development continued in mining and forestry. The Iron Ore Company and the Wabush Iron Company became US-owned and were directed from holding companies in Cleveland. The Melville Lake and Stephenville pulp and packaging material (linerboard) complex, using some Labrador forest fibre and subsidized electricity, provides a similar case. The development of this complex involved chiefly British and European firms, which co-signed this forest-product venture with the Newfoundland government in 1968.[94] The *St. John's Evening Telegram*, on 11 June 1968, listed the signatories and the roles of the firms they represented in financing, marketing, and manufacturing:

> Officials of Lazards, Greyhound and the Newfoundland government [arrangers of the financing of the linerboard complex] were represented at the signing ceremonies Friday.
>
> Also on hand were officials of Walmsley (Bury) Ltd., of the United Kingdom, who will build the 350 inch wide linerboard machine, one of the largest in the world. The machine will be 900 ft. long and capable of producing 1,000 tons of linerboard daily.
>
> Two 80,000 ton multi-purpose bulk cargo ships are to be built by Harland and Wolff of Belfast, Ireland. To be delivered in December 1970, they will transport wood chips from Labrador to the mill.
>
> Construction of the structures and installation of machinery will be done by McAlpine of Newfoundland Ltd., and Sir Robert McAlpine and Sons Ltd. of England.
>
> Other firms taking part in the signing of documents Friday were W. Heinzel and Co. of Austria, paper manufacturers and linerboard marketing agents in Europe; G.T.R. Campbell London Co. of Montreal, naval architects; Javelin Bulkcarriers of London, ship owners, a subsidiary of Canadian Javelin Ltd.; and E. and B. Cowan of Montreal, pulp and paper consultants, designers of the Melville project.
>
> In charge of the complex will be H.Y. Charbonnier, executive vice-president and general manager of the company.[95]

This account demonstrates that, although the province has a population of more than 500,000 people (among whom, one would think, there are individuals with industrial and entrepreneurial ability), Newfoundland continued to be technologically and economically dependent on an English company for the production of linerboard machinery, on an Irish company for the building of bulk cargo ships, on an English company for the construction of buildings and the installation of machinery, and on the professional skills of marketing agents from Austria and naval architects from London.

To summarize: contrary to public claims, Brinco's building of the Churchill Falls generating plant had little transformative effect on industrial development in Newfoundland. Its group of companies compromised Newfoundland interests in developing supply industries while strengthening Québec's hydro equipment industries. In Labrador, hydroelectricity as an input served to strengthen existing mining operations, and, on the Island of Newfoundland, low-cost electricity promoted the forest-products sector and chemical production – all of these operations were concerned with primary processing. Both the Newfoundland government and Brinco continued to opt for foreign ownership and technology in related manufacturing activities. Few new industries emerged from within Newfoundland and Labrador society itself.

Because the capacity of Churchill Falls exceeded the needs of Newfoundland, CFLCo executives advocated 'the sale to Québec of essentially all of the output from the Churchill Falls Power Development which is surplus to the needs of Newfoundland.'[96] The new hydroelectric power from Churchill Falls failed to be a determinant of industrial transformation in Labrador and Newfoundland; rather, the small amount of Churchill Falls electricity available in Labrador, the large amount contracted for delivery to Québec, and the industrial power subsidized (from anticipated revenues from Churchill Falls) and available on the Island of Newfoundland were all associated with a subordinated, truncated, and dependent industrial development process.

The Lower Churchill by 2007

Brinco's corporate legacy of incorporating each hydro site as a separate company has remained, albeit under new ownership. In 1998 Newfoundland and Labrador Hydro continued to be the parent company of incorporated hydro properties that were individually developed, such as the Churchill Falls (Labrador) Corporation, or that have remained undeveloped, such as the Lower Churchill Development Corporation and the Gull Island Power Company. The Churchill Falls (Labrador) Corporation became a subsidiary of Newfoundland and Labrador Hydro in 1974 when the government of Newfoundland and Labrador purchased 65.8 percent of the issued share capital from Brinco; Hydro-Québec had already acquired the remaining shares.[97] The rights to power sites below Churchill Falls were incorporated as the Lower Churchill Development Corporation on 15 December 1978; in 1996 it was owned 51 percent by CFLCo and 49 percent by the Government of Canada. It had a total share value of $30 million, was established to develop all or part of the hydroelectric potential of the Lower Churchill Basin, had an option with respect to the Gull Island Power Company assets, and held the hydroelectric development rights to the Lower Churchill, which were scheduled to expire on 24 November 1998.[98]

Discussions about the joint development of the Lower Churchill power sites continued, Newfoundland nearly reaching agreement with Québec on several occasions. In 1974, one agreement was nearly closed. However, after Brinco planned to sell all the power from the proposed generating facility to Hydro-Québec, the Moores government of Newfoundland reappropriated all hydro sites in Labrador and nationalized CFLCo. On a later occasion in 1991, the premier of Newfoundland, Clyde Wells, was close to reaching an agreement between Newfoundland and Hydro-Québec jointly to develop the Lower Churchill River at Gull Island and Muskrat Falls. The agreement foundered, however, according to Chuck Furey, energy minister in Brian Tobin's government in 1998, in part because of Québec's insistence that Newfoundland not exercise its right (provided for in the 1969 Churchill Falls power contract) to tax the project in 2016.[99] Then, on 9 March 1998, Newfoundland premier Brian Tobin and Québec premier Lucien Bouchard announced in Churchill Falls, Labrador, their plans to negotiate the building of a $12 billion 3,200 MW complex on the Lower Churchill River, to be completed by the year 2007 and to include dams at Gull Island Rapids and Muskrat Falls and transmission lines to Québec, as well as a federally funded $2 billion HVDC transmission line (partly by underwater cable) that would carry 800 MW from the new generating facilities to St. John's, Newfoundland (Figure 5.1).[100]

Why this renewed initiative for joint development? Lynn Verge (a minister in the Peckford government, the leader of the opposition when Clyde Wells was premier of Newfoundland, and author of position papers on the 1991 Lower Churchill negotiations with Québec) said, with regard to the 1998 negotiations between Brian Tobin and Lucien Bouchard, that the opportunity is better now since the insistence of the US on deregulation (a misnomer, because FERC actually increased regulation over transmission-line access) 'forces Québec to open a corridor for the sale of power to northern New England. We now have the option of dealing directly with the US; so our bargaining position has improved since 1991.'[101] As stated above, the FERC's rules require that regional transmission group members (e.g., Hydro-Québec and utilities in the northeastern US and Canada) must be given 'reciprocity' of access to Québec's transmission grid.[102] As indicated in Chapter 4, Hydro-Québec's *Strategic Plan* for the years 1998 to 2002 indicates that the electricity transmission system has been 'open to third parties since May 1, 1997.'[103] The Island of Newfoundland, however, is not interconnected with the Québec transmission system at Churchill Falls; hence Newfoundland's initiative for building the $2 billion transmission line – a substantial cost for the province.

Are there valid reasons to expect federal funding for the $2 billion transmission line from the Lower Churchill to St. John's? Chuck Furey, energy minister in Brian Tobin's government, argued that 'giving Newfoundland a break on the Lower Churchill would "correct a 30-year-old injustice" on the Upper Churchill [Churchill Falls] ... created at least in part because Ottawa would not grant the right to sell electricity through another province – rights it eventually granted to western provinces for natural gas.'[104] By January 1998, according to Furey, Premier Tobin and Prime Minister Jean Chrétien have had informal discussions about the loan, but a formal request had not been made.[105]

One of the many unresolved issues surrounding the Lower Churchill hydroelectric development is that it is located in territory with outstanding claims by people of the Innu Nation of Labrador. The resistance of the approximately 1,500 members of the Innu Nation to the 570,000 non-Innu of Newfoundland is based on past injustices and the fear that provincial governments would proceed with the Lower Churchill development without their agreement. Back in the 1960s, Aboriginal land claims were rare, environmental legislation weak, and governments could flood vast areas of land without harvesting trees, without compensating people who had traditionally used the land, and without mitigating the effects of what they were doing.[106] This situation, however, changed during the 1970s when the Cree in neighbouring Québec halted the construction of James Bay in the courts of Québec and came to a compensation agreement with Hydro-Québec. On 17 January 1998 in Labrador, Katie Rich, representing the Innu, referred to previous injustices and potential legal actions:

> We were never consulted or even informed about what would happen to our land when Churchill Falls was built.
>
> We have never been compensated for the damage that was done by the flooding. Our people lost not only territory and possessions when Mishikamau and Mishikamas were flooded to create the Smallwood Reservoir; they also lost a part of their history and identity as Innu people ... The Innu will take legal action, if necessary, to prevent Newfoundland from entering into any contract should the province attempt to proceed without the approval of the Innu.[107]

Six weeks later, on 3 March 1998, a few days before the announcement of negotiations with Hydro-Québec, Brian Tobin said to his evening television audience, 'I'll be seeking to brief as quickly as possible at the time of the announcement the representatives of the Innu Nation and to engage the Innu Nation, as appropriate, in dialogue about participation by the Innu Nation in the negotiations in the next year.'[108] On 9 March 1998, the announcement that Bouchard and Tobin had broken a twenty-five-year impasse with an agreement that laid the groundwork for new hydroelectric development was staged as one of the most important economic investment announcements in years.[109] In the town of Churchill Falls, about 1,200 kilometres northeast of Montréal, fourteen airplanes full of politicians, officials, businesspeople, and journalists flew in for the announcement. About 100 Innu blocked the premiers' motorcade and disrupted the carefully planned media event so that the premiers had to make their announcement in a nearby hotel and a high-tech news conference had to be cut short. One placard said: 'HYDRO IS A NO GO WITHOUT INNU CONSENT.' Innu protester Paul Rich stated that 'any big project like this that wants to go ahead, it first has to deal with the compensation for the [original] Upper Churchill project.'[110] It is unlikely that the project can proceed as long as this social injustice issue remains unresolved.

Several other issues cast uncertainty upon the viability of the Lower Churchill project. The additional 2,400 MW capacity from the Lower Churchill complex may not be needed in Québec upon the project's completion in 2007. In 1996 Québec already had a surplus capacity and has had the lowest load factor since the 1960s (i.e., average consumption is only 60 percent of the peak consumption – the second lowest in the country). In addition, Québec's growth in electricity consumption is slowing down (between 1960 and 1996 the average annual growth rate was 2.9 percent; in 1995 and 1996, it was 1.2 percent); therefore, much of the electricity from the Lower Churchill Falls will likely be designated for export to the US.[111] Unlike the 1970s, when Churchill Falls electricity became available to Québec and the province was able to negotiate long-term contracts with utilities in a growing US market, in 2007, when the Lower Churchill Falls electricity is planned to come on line, the deregulated US market may, as now, be characterized by short-term contracts. Furthermore, with open transmission lines and trade 'reciprocity,' should utilities in the New England states and New York develop a surplus, US utilities may be able to sell their electricity to industrial consumers in Québec.

Luc Boulanger, executive director of the group representing thirty-five large industrial hydro consumers in Québec, said that 'Québec industrial rates may already be too high' when compared to places such as Washington State and the southern US, where large industrial customers have been able to negotiate rates better than the three cents per kilowatt hour rate in

Québec.[112] Boulanger also questions the prediction in Hydro-Québec's *Strategic Plan 1998-2002* that, on top of having to find buyers for the electricity from the Lower Churchill complex in the US, electricity sales will grow 25 percent over the next ten years. For instance, Jack Valentine, spokesman for the New York Power Pool, which represents the seven largest investor-owned utilities in New York State, indicated that demand in the last five years was pretty flat and predicted that demand will increase by only 1 percent per year.[113] In addition, with no fixed rates and with the prospect of only short-term and spot-market electricity exports, building the new Lower Churchill complex appears more risky now than it did in previous decades.

Conclusion

In the 1960s, when energy prices were moderate and expansionary growth in the northeastern US led to brown-outs (electricity supply interruptions), the long-term, fixed-price Churchill Falls contract may have seemed a sure way to amortize the debts of the projects and to provide sufficient return. However, this contract became unreasonably underpriced during the oil crisis of the 1970s, when energy prices rose and the US decided to secure more of its energy supply from Canada. Added to the Québec system, the contractually secured capacity from Churchill Falls allowed for long-term export arrangements. When the inequality in accrual of benefits from this contract became evident to politicians in Newfoundland, they failed in their attempts to have the contract reopened because federal jurisdiction over interprovincial integration of networks remained weak. To this day, wheeling of transmission remains unregulated in Canada, while in the US it is regulated by FERC and applies to regional transmission groups as well as to Canadian utilities (should they wish to sell electricity in the US).

New to the Canadian Confederation, Newfoundland became caught in the Churchill Falls power trap by inviting the Brinco conglomerate to be its development instrument, by Québec's blocking electricity transport from Labrador to other provinces, and by the unequal sixty-five-year Churchill Falls power contract. Even the reversal of hydro-resource privatization was insufficient to release Newfoundland from this entrapment.

Similar to Ontario's initial policy regarding Niagara Falls, Newfoundland's policy regarding Labrador allowed foreign firms to produce electricity for export. This policy resulted in Newfoundland's loss of political autonomy over its electricity use within the province and within Canada. By leaving hydro development up to Brinco, the Newfoundland government lost its ability to plan hydro policy comprehensively for at least sixty-five years. It gave up 'the right to transmit or export power' from Churchill Falls, its recapture allowance stayed at 300 MW, and it was not able to secure certainty over ownership of Churchill Falls facilities. Thus,

Hydro-Québec is the major beneficiary of the Churchill Falls contract: it receives electricity at one-quarter of a cent (2.5 mills) per kWh, which in 1994 was resold for up to three cents (30.6 mills) – more than tenfold the supply price.[114] Jean-Thomas Bernard, an energy economist and professor at Laval University, said of Hydro-Québec's income from Churchill Falls, that 'the money they make exceeds their total profit,' and the *Globe and Mail* estimated in 1996 that 'Hydro-Québec now generates a profit of anywhere from $536 million to nearly $1 billion a year on the deal.'[115] In addition, the Labrador boundary dispute remains unresolved.

The Government of Newfoundland had also granted Brinco extensive rights to timber, minerals, and water resources in the hope that it would develop secondary industries. Brinco, however, sold the power to 'large, financially-secure consumers,' raised the capital, and then developed primarily the hydro sites. In the case of Churchill Falls, for sixty-five years it sold the power to Hydro-Québec, raised the capital ($965 million), and built the project (5,429 MW). In 1974, the Newfoundland government bought back its water resources, including the Lower Churchill site, in order to stop Brinco from developing additional hydro sites for the joint benefit of its shareholders and of Hydro-Québec.

Given that Churchill Falls was developed as an energy project, what of the promise that it would provide momentum for new industrial development? Or, more specifically, what kind of industrial linkages were strengthened or captured as part of the Churchill development? The findings demonstrate that the Brinco group of companies strengthened Québec's supply industries while Newfoundland's supply industry linkages remained very weak. Furthermore, Hydro-Québec added 'buy-Québec' clauses to the Churchill Falls power contract. With regard to hydro equipment, these clauses allowed French technology manufactured in Québec to be substituted for English technology. In Labrador, Brinco used the new power from Churchill Falls to serve the continentally integrated iron ore industry. It only pelletized the ore; it did not use the new supply of electricity for secondary manufacturing. On the Island of Newfoundland, the reduced electricity rates did not lead to the development of secondary industries; rather, 90 percent of the manufacturing electricity was consumed in traditional forest, chemical, and fish-processing industries. In addition, while using hydro as an agent in the formation of backward, forward, and fiscal linkages, the government and Brinco continued to rely on foreign ownership and foreign technology.

Interprovincial trade of electricity is, in part, a federal responsibility. However, the national-continental power policy, interpreted and applied by the NEB as export policy,[116] led Newfoundland to claim in the mid-1980s that the NEB's role should be national, whereas Québec insisted that it should be continental. Newfoundland supported the view that its provincial surplus should be available to Canada as a single market (across provincial boundaries) and should restrict access to foreign markets, whereas Québec supported energy sovereignty, with its surplus being available for the continental market.[117]

In 1998, in the absence (or limitation) of an NEB mandate regarding wheeling jurisdiction and in the context of the Canada-US FTA, new federal US electricity regulations apply to electricity exports from Québec. Since 1 May 1997, to suit US regulations, Hydro-Québec has opened its 'wholesale market and the electricity transmission system to third parties.'[118] Unless purchased by Québec, Labrador electricity has so far had no access to the Québec, Ontario, Maritime, and US markets. In 1998, in light of meeting the US open transmission requirements, Québec and Newfoundland have relaunched discussions concerning developing the $12 billion Lower Churchill complex in Labrador. When exported, such electricity will become part of a continentally shared resource and such exports will be subject to providing 'reciprocity' of access to US utilities to customers in the Québec and Labrador markets. These interprovincial initiatives on the Lower Churchill project have proceeded without the agreement of the people of the Innu Nation, who claim title to this resource and want compensation for the damage to their livelihood brought about by the Churchill Falls development. Given the present power surplus in Québec, the Lower Churchill Falls project does not need to be built at this time. Should this project proceed, however, substantial benefits ought to accrue to both the Aboriginal and non-Aboriginal peoples of Labrador and Newfoundland.

6
Nelson River Power (Manitoba)

The objective of the February 15, 1966 [federal-provincial] Agreement was to facilitate development of the potential natural resource of the Nelson and Churchill Rivers [in Manitoba] into a major supply of low-cost hydro-electric energy. The achievement of this objective would, in turn, meet provincial and national objectives in the field of economic and industrial development. The supply of low-cost electrical energy would help the development of southern Manitoba, and would be a major potential export to other provinces and to the United States.[1]

– G.M. MacNabb, chairman of Federal-Provincial
Nelson River Review Committee, 31 March 1970

Introduction

Both the Winnipeg River, the southern inflow to Lake Winnipeg, and the Nelson River, its northern outflow, contain numerous power sites. Manitoba's two major rivers drop sufficiently at a number of rapids and waterfalls to drive electricity-generating equipment (Figure 6.1). Such water-power sites, administered before 1930 together with other natural resources under federal authority and after 1930 under provincial authority, were licensed to be developed by private and public utilities on the southern Winnipeg River and the northern Nelson River. After the 1950s, when the Winnipeg River was completely harnessed, Manitoba Hydro built the first Nelson River plant (the Kelsey generating station) for the energy requirements of the International Nickel Company in 1961.[2]

To increase industrial development within Manitoba and to make its east-west transmission the first major link in the national power grid, the federal government agreed to build a power line between the Nelson River and southern Manitoba. Manitoba Hydro became responsible for diverting part of the water from Manitoba's Churchill River into the Nelson River and for developing the Kettle, Long Spruce, and Limestone sites on the Nelson (Figure 6.1). Manitoba and federal officials planned to use hydro energy from the Nelson as a catalyst for the national power network and for secondary industry, but they fell short of both objectives. Manitoba Hydro claims to have achieved up to 90 percent of local manufacturing benefits from constructing hydroelectric projects – a claim that will be further examined below. In addition, industrial energy demand became difficult to predict, for manufacturers in Manitoba used low-cost electricity chiefly for export-oriented primary commodity production, and the

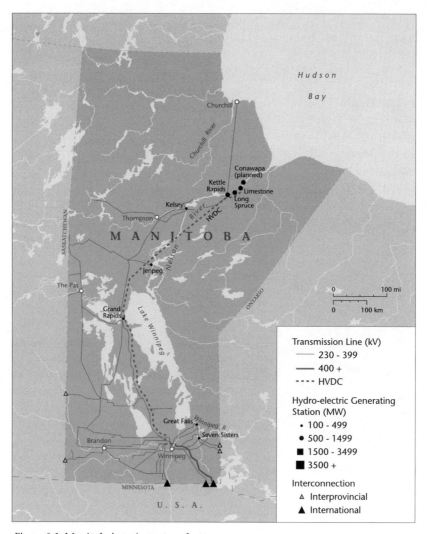

Figure 6.1 Manitoba's main system features

Source: Government of Canada, Department of Energy, Mines and Resources, Energy Sector, *The National Atlas of Canada*, 5th edition, Electricity Generation and Transmission, 1983, MCR4069 and MCR4144; Manitoba Hydro, 'Major Transmission Lines,' 1998.

'unplanned' electricity surplus was exported to US mid-continental utilities.[3] Whereas electricity trade and transmission network integration with neighbouring provinces has remained weak, exports to the US since 1976 have increased more than tenfold. In similar fashion to what happened in other provinces, in Manitoba something went wrong with the initial privatization of hydro resources, with the kind of industrial development that the addition of new hydroelectric projects were expected to bring, with the timing and size of the provinces' electrical supply expansion, and with the national and regional initiatives to link these major projects to a trans-Canada power grid. This chapter analyzes how these sometimes unexpected and problematic outcomes occurred.

Public Hydro

Manitoba is endowed with many hydro resources, such as waterfalls, rapids, and canyons in southern and northern rivers. Either private or public utilities could develop such river sites. Under which form of ownership – private or public – did Manitoba develop its southern water powers before expanding its electrical system northwards to integrate power supply from the Nelson River sites? In Manitoba, as in other provinces, rights to use and develop water-power sites were initially allocated to both private and public utilities. In Manitoba, however, the federal government planned and administered the water-power resources into the early decades of the twentieth century. Nelles points out, for instance, that 'the rivers, along with lands, forests, and mines in Manitoba had been administered since Confederation by the federal government' and its Department of the Interior until 30 May 1930, when, through the Manitoba Natural Resources Act, federal control over natural resources was transferred to Manitoba.[4] From 1911 to 1916, officials from the federal Water Powers Branch of the Department of the Interior, under the direction of Superintendent J.B. Challies, came to Manitoba and prepared detailed field surveys to assess the hydro-electric potential of southern and northern rivers in Manitoba. At the time of these surveys, the Manitoba legislature requested that H.A. Robson, the utilities commissioner, examine the possibility of establishing a province-wide hydroelectric system under provincial, rather than private, ownership. The issue to be examined was 'the question of the development of publicly-owned hydro-electric power within the Province, with a view to securing for all sections of the Province, rural as well as urban, the benefits and conveniences enjoyed by the citizens of Winnipeg as the result of the expenditure of public money for power development and distribution.'[5] Robson addressed his reply of 3 February 1914 to J.H. Howden, the attorney general of Manitoba, and made his case by drawing on the findings of two reports – one a survey by J.B. Challies of the Water Power Branch and the other a southwestern transmission and distribution survey by W.E. Skinner.[6]

Robson first discussed the report prepared by J.B. Challies, which identified the extensive power potential of the waterfalls, rapids, and canyons on the Winnipeg, Nelson, and Saskatchewan Rivers.[7] From the Challies report he listed the sites on the Winnipeg River already developed, such as the Pinawa Falls plant (1906) owned by the Winnipeg Electric Railway Company and the Pointe du Bois plant (1911) owned by the City of Winnipeg. He also included the undeveloped sites, such as the Slave Falls and Seven

Sisters site.[8] Using Chicago electrical consumption and population growth projections, Robson estimated that the growing population of Winnipeg alone would need the entire energy capacity that the Winnipeg River would be able to generate.[9] Therefore, setting his sights on northern Manitoba, Robson noted that the federal survey described the hydraulic potential of two northern rivers, the Saskatchewan River with its Grand Rapids site and the Nelson River with its Kettle, Long Spruce, and Limestone Rapids sites.[10] Furthermore, in his memorandum to the attorney general, Robson assessed the possibility of installing a transmission and distribution system, as proposed by the W.E. Skinner transmission report.[11] His evaluation did not consider 'Winnipeg and its environs' but suggested that northern development and long transmission lines would be premature. It also advocated that all sites be conserved for now and that power be purchased from existing hydro plants as an interim measure:

> While a general undertaking [of power generation and transmission] could not at present be justified, it is by no means certain that time will not come soon for its economic adaptability. For that reason, all water powers should, as herein stated, be conserved. It should also be said that provisional measures are quite within the range of possibility, and indeed of early adaptation. The purchase of power for public schemes is found in a notable instance elsewhere [e.g., in Ontario], and might be feasible here, thereby postponing for a long period highly expensive river work and transmission system.[12]

Referring to the report by Superintendent Challies, he pointed out 'that the Winnipeg River is unique in that the variation between maximum and minimum flow is much less than that experienced generally in water power rivers,' in part because it is regulated by the outflow of the Lake of the Woods.[13] Thus, he advocated the 'urgency not only of the preservation of these public assets for public uses, but the urgent necessity for continued – in fact, daily – vigilance with regard to the preservation of the flow of waters' from the east.[14] In southwestern Manitoba, the rather sluggish Assiniboine and Souris Rivers did not offer adequate hydro-power sites; therefore, studies were conducted to assess the westward transportation of electricity generated at Winnipeg River plants.

Robson reviewed one such study by Skinner, who proposed a southwestern transmission line from Pine Falls to Winnipeg, making a loop from Portage la Prairie to Brandon, then to a point near Minto, and finally through Roland and Morris back to Winnipeg. He compared the costs of locally generated electricity with those of electricity transmitted from the Winnipeg River and found that the difference was not really worth the effort. From this study, Robson estimated that the cost for operating small steam and gasoline generators would be cheaper, since one kilowatt per hour could be produced for 4 to 13 cents (compared to the transmission of electricity to towns and villages for 15 cents and to rural agricultural areas for 19 cents).[15] Building long transmission lines solely for rural agricultural use was not financially feasible.[16] However, members of the Legislature assumed that regional electrification would also foster industrial development, as it had in Winnipeg. If this were the case, Commissioner Robson argued, then the intermittent costs of transmission would be less formidable: 'It is not too sanguine an expectation that the electrical impulse throughout the Province would increase local manufacturing and thereby augment population, and, not only so, but the demand for power so arising would provide a market implementing the rural demand to such an extent that such revenues would be produced that the financial side of the undertaking would, as growth proceeded, become less formidable.'[17] Instead, being cautious, he proposed that, initially, the province could purchase surplus power from the City of Winnipeg and transmit it to other areas of the province. He added that, unless the city were approached, it would be impossible to predict on what terms such an arrangement might be made.[18]

Two years after Robson's 1916 memorandum, the Middle West Utilities Company of Chicago tried to determine whether rural electrification could be carried out as a profit-making venture. The plan was to purchase power from the City of Winnipeg Hydro Electric System, take over smaller rural power plants and utilities, and transmit power via Portage la Prairie to Brandon and to smaller communities such as Neepawa.[19] The Middle West Utilities Company had subcontracted this survey to the L.E. Myers Company, whose staff carried out the field work. Its survey report covered the lighting and industrial power needs of more than fifty small towns and, in assessing the feasibility of such an undertaking, concluded that the possible profits – a net balance of only $37,738 – were hardly worth the effort of transmitting power from Winnipeg to Portage la Prairie, Brandon, Minnedosa, Souris, and many other smaller towns in Manitoba.[20]

In Brandon, for instance, the privately owned Brandon Electric Light Company dominated electricity production.[21] In 1916, Brandon Electric charged 10 cents per kWh, an overinflated price at the time. Brandon Electric rates not only brought complaints from power users but also hindered small firms that were trying to convert their production facilities from steam to electric machinery. To reduce costs, some Brandon industries used steam to supplement their use of electric motor energy;[22] others were not running their operations at full capacity. The Myers survey report of 1916 concluded that 'the Electric Company is standing in its own light by not making reasonable power rates.'[23] Its overcharging reduced its own market.

While in 1916 the development of generating facilities and transmission networks outside Winnipeg was not sufficiently profitable to support the small regional markets, about twenty years earlier a major private company, the Winnipeg Electric Railway Company, and, about a decade earlier, a public utility owned by the City of Winnipeg, were incorporated and planned to supply the lucrative urban market. Having started its operations in 1892, by 1906 the Winnipeg Electric Railway Company 'had a monopoly on the transit service, electric light, power, and gas distribution in the Winnipeg area.'[24] Thus, at the turn of the twentieth century, the private Winnipeg Electric Railway Company acquired the best hydro sites and supplied the lucrative Winnipeg market.[25] This private utility obtained federal water-power sites and developed three hydroelectric generating facilities on the Winnipeg River: the Pinawa (1906), Great Falls (1913), and the Seven Sisters (1931), which made Pinawa obsolete. In his history of political tensions over hydroelectric development in the first decade of the twentieth century, H.V. Nelles found that in Winnipeg it was mainly businesspeople who were intent on breaking the private utility monopoly and who started and sustained a power movement until a public utility was established.[26] Businesspeople, so Nelles argues, wanted competition because 'if Winnipeg could obtain energy at cost, they convinced themselves, it would quickly rival cities in the east as a manufacturing centre.'[27] The establishment of the municipally-owned City of Winnipeg Hydro Electric System (City Hydro) in 1906 did curb the private monopoly and brought down electricity rates (see Table 6.1).[28] Alderman John Wesley Cockburn held, but had not developed, the private water rights to the Pointe du Bois site on the Winnipeg River.[29] From him, City Hydro obtained the right to construct the Pointe du Bois Hydro-Electric Generation Station and, in 1911, started supplying public power to the Winnipeg market.[30] Nelles observes that 'once the turbines began spinning at Pointe du Bois and the average domestic and commercial power bill had been cut in half, the businessmen were content to let the movement collapse, their work done.'[31] While most of the support for the initiation of a public utility owned by the City of Winnipeg came from businesspeople, most of the support for a public utility owned by the province originated from provincial politicians and civil servants.

Entrepreneurs did not come forward to develop rural and regionally interconnected electricity in Manitoba. Perhaps entrepreneurs could have shown an interest when, in June and October 1913, Manitoba's Public Utility Commissioner, H.A. Robson, offered opportunities for public participation through press invitations published in newspapers throughout the province. Though the June advertisements elicited very few responses, those who did respond expressed the desire for hydroelectric network expansion within the province, but their opinions were 'almost always

Table 6.1

Privatization and its reversal: hydroelectric resources in Manitoba, 1867-1966

1867-1930	The federal government's Water Power Branch of the Department of the Interior administers federal water-power rights on Manitoba rivers.
1882-1906	Winnipeg Electric Railway Company incorporates privatized water power rights to Winnipeg River sites and serves the City of Winnipeg with electricity and other utilities.
1906-11	City of Winnipeg Hydro Electric System (City Hydro) incorporates in 1906; acquires rights to Pointe du Bois site on the Winnipeg River, builds plant, and supplies power in 1911.
1914	Manitoba's Public Utilities Commissioner, H.A. Robson, recommends (3 February 1914) that private water rights to undeveloped river sites be appropriated and conserved for future 'publicly owned hydro-electric power within the province.' He highlights the possibilities of further hydro development on the Winnipeg River, Saskatchewan River, and Nelson River from the federal government's Water Power Branch report.
1919	The Manitoba Power Commission is established to generate, purchase, transmit, and distribute electricity to rural municipalities and later individual customers.
1931	Winnipeg Electric Company (name changed from Winnipeg Electric Railway Company in 1924) retains rights to and develops Seven Sisters Hydro-Electric Generating Project on the Winnipeg River and signs an agreement to supply Manitoba Power Commission with electricity.
1949	Formation of the Manitoba Hydro-Electric Board as a provincial government agency to formulate a coordinated hydro development policy (Manitoba Hydro Electric Development Act passed).
1953	Manitoba Hydro-Electric Board acquires and controls (4 April 1953) the Winnipeg Electric Company and thereby its generating and distribution system.
1961	Provincial government forms Manitoba Hydro by merging two provincial utilities: the Manitoba Power Commission and Manitoba Hydro. In this way formerly privatized hydro resources are appropriated by the province for distribution to the inner area of Winnipeg held by the municipally owned Winnipeg Hydro.
1966	Development of the Nelson River begins.

accompanied by a doubt as to the financial possibilities' for installing such a system.[32]

In 1919, the provincial government of T.C. Norris, 'under pressure from insurgent farmers, established a small provincially owned system in the southwest.'[33] When the provincial government passed the Electric Power Transmission Act, the Manitoba Power Commission, as it was called, came into existence with a mandate to 'generate, purchase, transmit, and distribute electrical energy' throughout Manitoba. Under the act, all municipalities in the region, many of which operated small and costly independent power systems, could apply for service through the minister of public works. The Manitoba Power Commission initially contracted to purchase the power from the City of Winnipeg and, later, from the Winnipeg Electric Railway Company under the Seven Sisters Agreement of 1931.[34]

The controversy leading up to the development of the Seven Sisters water-power site offers some insights into the federal government's role in allocating licences for hydro development in Manitoba. To start with, the federal government negotiated leases with private and public utilities to the Pinawa Channel (Winnipeg Electric Street Railway, 1904) and to the Pointe du Bois site (Winnipeg Hydro, 1906) years before the Inland Water Branch in Ottawa prepared hydraulic inventories or detailed federal plans for Manitoba's water-power resources.[35] Robson indicates that, by 1913, some private investors had obtained rights to water power that they failed to develop. Therefore, as Manitoba's provincial utility commissioner, Robson called for taking some of these rights back from private interests. As he stated in his memorandum to the attorney general: 'It is respectfully recommended that such action be taken as will prevent the grant of any further rights, and that in cases where rights so granted have not been exercised, and are merely dormant, lawful action be taken to resume control and terminate any supposed rights, so that the water powers involved may hereafter be held for public uses.'[36] However, a dozen years later, not without initial protest, the federal government did not heed such calls for provincial preference when granting further water-power rights to Winnipeg's private utility.

In 1925, the Winnipeg Electric Company (the new name of the Winnipeg Electric Railway Company) applied for the Seven Sisters site, and, as Nelles describes it, 'Premier [John] Bracken lodged an immediate protest, asking that the location be held in reserve for the provincial government's system.' Bracken became convinced that the new Seven Sisters site (which replaced the very inefficient Pinawa plant) would have been too costly to develop solely for the small market outside Winnipeg and, therefore, only authorized the private development after obtaining a 'rock-bottom $13.80 per horsepower for all of the power requested by the provincial rural electric system.'[37] In August 1928, the Water Power Branch routinely approved the Seven Sisters lease on the Winnipeg River, while City Hydro proceeded with its development of Slave Falls.[38] Negotiations undertaken to transfer federal control of the remaining lands, resources, and water rights to Manitoba were completed in 1930.

In subsequent decades, since it considered the investment in hydroelectric power insufficient, the involvement of the provincial government in public hydro development continued to grow. In 1949 the provincial government established the Manitoba Hydro-Electric Board in order to coordinate the future hydro expansion policy between the private Winnipeg Electric Company, the municipal City Hydro, and the provincial Manitoba Power Commission. Then, in 1952, with the approval of the government, the Manitoba Hydro-Electric Board acquired the Winnipeg Electric Company.[39]

The Manitoba Hydro-Electric Board and the Manitoba Power Commission became Manitoba Hydro on 1 April 1961. While Manitoba Hydro was responsible for power supply and transmission in the suburbs and the rest of the province, City Hydro remained responsible for distribution in Winnipeg. The mandate of Manitoba Hydro, as set out in the Manitoba Hydro Act of 31 March 1961, was confined to the supply of adequate power for use within the boundaries of the province: 'The intent, purpose, and object of this Act is to provide for the continuance of a supply of power adequate for the needs of the province, and to promote economy and efficiency in the generation, distribution, supply and use of power.'[40] Anticipating that they would soon develop generating capacities for needs outside the province and be able to transport large supplies of electricity from the North over newly developed long-distance transmission lines, the principal interest of the Manitoba power planners lay in the immense hydroelectric potential of the Nelson River system.[41]

The Nelson River, which flows northeastward from Lake Winnipeg to Hudson Bay, drops significantly at eleven sites (Figure 6.1). Only five of the eleven potential power sites of the Nelson River had already been developed or were under construction by the late 1980s. The Kelsey Generating Station, the first to be built on the Nelson, has a capacity of 224 MW, with potential for a 240 MW extension.[42] As will be discussed in more detail below, it was operating in 1960 to supply International Nickel's mining and smelting operations at Thompson. The other four Nelson plants were developed subsequently, in accordance with a scheme originating from the federal-provincial agreement to make Manitoba's east-west transmission lines the 'one major link' in Prime Minister Diefenbaker's proposed national power grid (as discussed in Chapter 2) and for export to the US.

So far, in developing the background of public hydro, I have shown that in the early decades of the twentieth century, as the Meyer survey indicates, the development of the generating facilities and transmission networks outside Winnipeg were estimated not to be financially profitable. In 1913, the utility commissioner believed that if electrical supply would result in more industrial development in small towns, then underwriting the costs would be justifiable. In the urban market of Winnipeg, it was primarily businesspeople who supported the initiation of public power by the municipality. This was so that competition would level prices charged by the private monopoly and lead to an advance in manufacturing (as had been the case in eastern Canadian cities). Starting with electrical supply outside Winnipeg, the influence of the provincial government began to increase, whereas the private and public urban utilities shared the lucrative urban market. With the takeover of Winnipeg Electric by the Manitoba Hydro-Electric Board in 1953, the earlier privatization of water power began to be reversed. The formation of Manitoba Hydro in 1961, and the launching of the Nelson River hydroelectric projects in 1966, continued the trend of hydroelectric developments coming increasingly under provincial control.

Nelson River: A Test Case for the 'National' Power Policy

At the Premiers Conference of 6 August 1962 in Victoria, Manitoba's premier, Duff Roblin, supported the development of the Nelson River in order to supply power for the proposed national power grid. He stated: 'We are definitely interested in the concept of a national grid. We do not visualize it springing into existence from Atlantic to Pacific overnight. We think the first step is to *develop and maintain one major link,* and, if it works to the satisfaction of all concerned, we could develop it across the country' (my emphasis).[43] Roblin pointed out to his fellow premiers that, although Manitoba already had interprovincial electricity linkages with 'Saskatchewan, based on coal' and with 'Ontario, based on hydro,' the development of the Nelson River needed more extensive cooperation. Arguing that the integration of Nelson River electricity into the national power grid was 'beyond our resources to handle by ourselves,' he called for continued federal 'financial and technical contributions' to Manitoba.[44] After several years of study, on 15 February 1966 the federal and provincial governments, together with Manitoba Hydro, came up with an agreement to develop the hydroelectric potential of the Nelson River system.

> The objective of the February 15, 1966 Agreement was to facilitate development of the potential natural resource of the Nelson and Churchill Rivers into a major supply of low-cost hydro-electric energy. The achievement of this objective would in turn meet provincial and national objectives in the field of economic and industrial development. The supply of low-cost electrical energy would help the development of southern Manitoba, and would be a major potential export to other provinces and to the United States.[45]

However, this scheme went beyond the 'needs of the province' in three ways: by helping to supply, develop, and maintain one major link of the national grid; by exporting to the US mid-continental grid; and by building for industries not yet in the province.

To meet these objectives, all three parties (the province, the federal government, and the utility) agreed that Canada would 'design, construct and place in service, at Canada's expense, the required transmission facilities' by 30 November 1971. Atomic Energy of Canada was assigned to act as the federal government's agent in the construction of the transmission line. The Kettle Generating Station would be connected to Winnipeg by power lines able to carry up to 3,200 MW, which constituted most of the future supply of Nelson River electricity (4,000 MW).[46] The main transmission line, 555 miles long and costing $175 million, would be rented from Canada starting on 1 April 1972: 'the rental charge is designed to repay Canada in full over a 50-year period.' Furthermore, it was agreed that Manitoba was 'to design, construct and place in service, at Manitoba's expense, generating facilities on the Nelson River.'[47] Thus, while the federal government installed the power line, Manitoba was responsible for the installation of the power plants. In keeping with this plan during the 1970s, Manitoba Hydro added the Kettle Generating Station (1,272 MW), the Long Spruce Generating Station (980 MW), the Jenpeg control structure and power plant (126 MW added as a design afterthought), and began construction of the Limestone project (1,330 MW).

Linkages to Industry

In Manitoba, as in other provinces, the 'if you build it they will come' mentality was problematic. Despite evidence from other provinces that new hydro-generating stations did *not* lead to a variety of new industries, Manitoba utility executives and politicians repeated the by now familiar refrain that hydro development brings industrial development. In 1987, Marc Eliesen, chair of Manitoba Hydro, observed: 'Manitoba's abundant low priced hydro-electric resources are offering a sound basis on which to attract new industries to the province, particularly those which rely significantly on electricity in their manufacturing and processing activities.'[48] Except for a period during the 1970s when Manitoba Hydro's insistence on Manitoba content led to a partial capture of local manufacturing linkages from Ontario and Québec, an examination of the new industrial development that followed the Nelson River projects shows something else. Two developments in particular stand out in Manitoba. One is that Russian generators and turbines were installed on a Nelson River plant and that Manitoba electricity was used for the processing of US strategic Cold War materiel.[49] Generating equipment had been previously purchased from central Canada, and the Cold War demand for Manitoba nickel led to overestimates of industrial electricity needs. The other is that Manitoba claimed to supply its Limestone project with 90 percent Manitoba-produced building and equipment content (in contrast to Newfoundland, which captured only 2 percent of supplies and services at the Churchill Falls project in

Labrador). But how was such a high degree of Manitoba content possible when, historically, core equipment (turbines and generators) for powerhouses came from central Canada? This claim demands closer examination.

Manitoba's Manufacturing Conditions

Branch firms of General Electric in Ontario and Québec manufactured 70 percent of all turbines and generators installed in Manitoba's powerhouses. Between 1923 and 1986, US-owned Canadian General Electric (CGE, Ontario) supplied thirty-three generators for hydro plants in Manitoba, and its subsidiary, Dominion Engineering Works (DEW, Québec) supplied thirty-eight turbines (Appendix 12). During the 1970s, Mitsui (Japan) supplied twelve generators, and the Leningrad Metal Works (from Russia) supplied six turbines and six generators (Appendix 12). To capture the contract, Energomachexport of Russia offered the six 28 MW turbine-generator sets at $15.6 million, whereas the nearest Canadian bid was $27.6 million.[50]

While the Kelsey plant (built for the International Nickel Company) generated power using Canadian General Electric and Dominion Engineering Works equipment that had been installed between 1960 and 1972, the Mitsui (MITS) generators were installed at the Kettle Rapids plant between 1970 and 1974, and the Leningrad Metal Works (LMW) equipment was installed in the Jenpeg control structure and generating station between 1977 and 1979 (Table 6.2). As a result of Manitoba Hydro's insistence on Manitoba content, Manitoba captured some component manufacture

Table 6.2

Turbines and generators by power stations (over 100 MW) in Manitoba, 1923-92

Site	Capacity of plant (MW)	Years of turbine installations and manufacturers	Years of generator installations and manufacturers
Great Falls	132	1923-8 DEW, SMS	1923-8 CGE
Seven Sisters	150	1931, 1949-52 AC, DEW, SMS	1931-52 CGE
Kelsey	236	1960-72 DEW	1960-72 CGE
Grand Rapids	437	1965-68 JI, CAC	1965-8 CGE
Kettle Rapids	1,224	1970-4 DEW	1970-4 MITS
Long Spruce	980	1977-9 DEW	1977-9 CGE
Jenpeg	186	1977-9 LMW	1977-9 LMW
Limestone	1,280	1985-92 CGE	1985-92 CGE

Note: AC = Allis Chalmers, CAC = Canadian Allis-Chalmers, CGE = Canadian General Electric, DEW = Dominion Engineering Works, JI = John Inglis, LMW = Leningrad Metal Works, MITS = Mitsui, SMS = S. Morgan Smith.
Source: Statistics Canada, *Electric Power Statistics: Inventory of Prime Movers and Electric Generating Equipment*, cat. 57-206, 31 December 1986.

from central Canada. Despite a bid to manufacture turbine and generator components in northern Manitoba, Winnipeg was eventually selected as the manufacturing site.[51]

The venture to establish local manufacturing in northern Manitoba was known as The Pas Machinery Manufacturing Project. The object was to use the Bertram machinery plant in the northern town of The Pas 'to manufacture an appropriate mix of parts of hydraulic turbines, hydro generators and hydraulic gates for the Long Spruce and Upper Limestone hydro projects.'[52] However, the Bertram Study Group calculated that, when compared to manufacturing in Winnipeg, manufacturing at The Pas would result in a $12.5 million deficit over an almost eight-year project period.[53] The group reported a long list of reasons for the higher costs of manufacturing in the North: 'lower production efficiency, higher northern wage rates including overtime, training programs, personnel turnover, relocation costs, start-up costs and the higher servicing costs due to plant location.' The workforce required to manufacture turbine, generator, and gate components would consist of 140 people, of whom only '40 could be recruited and trained locally,' with the balance being brought in from Winnipeg, industrial Canada, and possibly Europe. The study group also took into account jurisdictional problems between various unions, a shortage of steel, and the necessity of building 75 to 100 houses for the new workforce at a cost of $3 to $3.5 million. Finally, on 13 December 1973, Manitoba Hydro chair L.A. Bateman informed Edward Schreyer (premier of Manitoba), Max Drouin (president and general manager, Dominion Engineering Works Limited in Montreal), and Paul McConnel (manager, Large Motor and Generator Products Section, Canadian General Electric Limited in Peterborough, Ontario) of the following resolution:

[That] Manitoba Hydro does not wish to acquire any interest in the Bertram plant or to participate, directly or indirectly, in any scheme for the development of the facilities of the plant for the manufacture of hydraulic turbines, generators, and related equipment.
...
That Manitoba Hydro continues to encourage local companies to develop their existing and proven facilities in order to be able to manufacture a growing amount of Manitoba Hydro's heavy equipment requirements in Winnipeg, where there is an established pool of skilled workers.[54]

Unlike the situation of Québec's Marine Industries Limited, in Manitoba the manufacture of such equipment was left to the private sector. While Manitoba Hydro had claimed, possibly exaggeratedly, that manufacturing content increased at Long Spruce to between '50 and 55 percent,' the claim that 90 percent of the $1.73 billion Limestone project was Manitoba content needs to be more thoroughly researched.[55] Preliminary findings can only partly explain how Manitoba Hydro arrived at such high manufacturing content estimates.

Limestone

The US-owned Canadian General Electric Company Limited of Toronto was awarded a $100 million contract to supply Limestone's ten turbines and generators.[56] The mechanical gates and cranes were manufactured by Dominion Bridge in Winnipeg and by Ecolaire's branch plant in The Pas.[57] Ecolaire, of Kitchener, Ontario, ensured that, for a premium of $600,000, part of the equipment called for in the contract would be manufactured in Manitoba. The firm set up a manufacturing plant in The Pas, but after the gates were completed, this type of local manufacture was discontinued. Such extra premiums paid for manufacturing in northern Manitoba were not repeated.[58] Besides the gates by Ecolaire, General Electric subcontracted other mechanical components to Manitoba manufacturers. For instance, Canadian Anglo Machine and Iron Works manufactured the ten generator frames (17 m in diameter, 1.5 m high, and 90,720 kg in weight) and other 'turbine requirements on a subcontract basis for General Electric of Canada.'[59] But this, together with the local Manitoba manufacturing commitments by Canadian General Electric, leaves much Manitoba content unspecified.[60] What proportion of the Limestone contract was supplied by Manitoba companies and what remains unaccounted for? The report of the Manitoba Energy Authority offers a partial answer, as does the fact that Manitoba manufacturing content was circumvented by industrial offset agreements.

The Manitoba Energy Authority's *Annual Report* of 1984 predicted that an industrial benefits agreement with Canadian General Electric, in lieu of the Limestone turbines and generators being manufactured in Manitoba, would 'generate $150 million in long term direct and indirect economic benefits to Manitoba.'[61] At the time, the details were not specified. In its 1985 annual report, the CGE industrial offset agreement was hailed as 'one of the year's major achievements.'[62] A $100,000 research grant to the Manitoba Microelectronics Centre and the Manitoba High Voltage Direct Current Research Centre and a $10 million investment in viable operations in Manitoba were announced. Other offsets included sourcing a minimum of 15 percent of turbine and generator labour components in Manitoba, creating 100 new jobs in Manitoba's high-technology industry, and providing $2 million to stimulate northern Native business. Nevertheless, taken together, the benefits indicated by the Manitoba Energy Authority's *Annual Report* neither amount to $150 million nor convincingly make up for the increase in the estimate from 50 percent to 90 percent Manitoba content in the two most recent Nelson River projects.[63] In effect, the creation of a large variety of new manufacturing activities directly related to the Limestone project was circumvented. Overall, however, Manitoba was more successful in capturing manufacturing benefits through procurement strategies than was Newfoundland and, as will be shown, British Columbia.[64]

Hydro Power as a Determinant of Industrial Development

In the late 1980s, in a variation of the familiar theme of merging industrial optimism with hydro development, Marc Eliesen, Manitoba Hydro's chair, asserted that 'low cost electricity supply maintains and attracts new investment to the province, helping to encourage a stable, diversified economy.'[65] Yet even as early as the 1970s, there was evidence that such claims could not be sustained. The addition to the Manitoba power supply from Nelson River plants at Kettle Rapids (1,224 MW) by 1974 and Long Spruce (980 MW) and Jenpeg (186 MW) by 1979 failed to diversify the economy. Apart from the food-and-beverage sector, Manitoba's industrial firms continued to use electricity primarily in traditional semi-manufactured goods production rather than for diverse manufacturing.

Information on long-term industrial contracts is often confidential and difficult to obtain. However, industrial-contract information provided by

Table 6.3

Peak power and energy demand by major industrial customers in Manitoba, 1973-4

	Number of major customers	Estimated peak power (MW)	Energy (GWh)	Energy as a percentage of total industrial
Mining	3	209	1,488	34
M.H. Pipelines	1	78	442	10
M.H. Paper	2	42	330	7
M.H. Chemical	2	43	254	6
M.H. Steel	3	51	179	4
M.H. Oil Refin.	2	19	110	3
M.H. Railways	2	10	34	
Remaining M.H. industrial	4,557		1,201	27
Winnipeg Hydro industrial	4,011		341	8
Total industrial	8,583		4,379	100

Source: Manitoba Hydro, 'Application to the National Energy Board for a Certificate of Public Convenience for an International Power Line and a Licence to Export Power,' vol. 1, by the Manitoba Hydro Electric Board, Winnipeg, Manitoba, 1975, Exhibit 3 was entered on the NEB hearing day 13 January 1976. See p. 11 for the table indicating the industrial load served. Railways were included in the category of main industrial consumers.

Manitoba Hydro to the NEB in the 1980s shows that the province's industrial structure changed very little between the mid-1970s and the mid-1980s. In the 1970s, Manitoba Hydro's three mining customers used 34 percent of the industrial load, while its fifteen major industrial customers consumed 65 percent of Manitoba's total industrial electricity (Table 6.3). When Kettle Rapids started supplying new electricity in November 1974, primary resource industries, such as mining, pulp and paper, chemicals, steel and oil refining, were the main industrial consumers, as can be seen in the consumption pattern for 1975 (Table 6.4).[66] After all the plants except Limestone were installed during the 1980s, little changed in the mix of industries: the ten largest customers remained in the mining, pulp and paper, and chemical sectors (Appendix 13). According to manufacturers' reports to Statistics Canada, firms processing primary metals, pulp and paper, food and beverages, and chemical products consumed 82 percent of the electricity purchased in Manitoba in 1975 and 81 percent in 1984 (Table 6.4).

I mentioned previously that one corporation, the International Nickel Company (Inco), one of Manitoba Hydro's largest industrial customers since 1961, used its mineral production to supply the Cold War effort. For Inco, the Kelsey plant (on the Nelson River) may initially have been a major consideration in setting up its mining, smelting, and refining operation in

Table 6.4

Consumption of electricity in manufacturing industries in Manitoba, 1975-84

Year	Total consumption in manufacturing (GWh)	Four major sectors				Sum of four sectors (%)
		Primary metal (%)	Pulp and paper (%)	Food and beverages (%)	Chemical (%)	
1975	2,482	53	15	9	6	82
1976	2,401	48	16	10	7	81
1977	2,555	47	13	10	11	81
1978	2,647	43	15	10	12	80
1979	2,567	41	–	11	10	–
1980	2,541	40	16	11	10	76
1981	2,631	45	15	10	9	79
1982	2,289	42	15	11	11	79
1983	2,320	41	15	10	14	79
1984	2,530	46	15	9	11	81

Source: Statistics Canada, Manufacturing and Primary Industries Division, *Consumption of Purchased Fuel and Electricity by the Manufacturing, Mining, and Electrical Power Industries*, 1975-84, cat. 57-208.

Thompson, Manitoba. Between 1969 and 1972, the Kelsey plant's capacity was augmented by two more generators to 224 MW.[67] As Melissa Clark-Jones has reported, nickel was used for military equipment manufacture and became part of the US program of stockpiling strategic resources during the Cold War.[68] Part of the nickel demand resulted from the military needs of the Vietnam War. In its System Forecast in 1983-4, Manitoba Hydro reflected that 'Inco grew rapidly from 1968 to 1971 and has declined since so that [its] 1982 use [of electricity] was only half that recorded in 1971.'[69] While Manitoba Hydro added more plants, Inco used less electricity in its semi-processing of nickel once the Vietnam War ended.

Had it not been for the expansion of the Hudson Bay Mining and Smelting Company, which became a major customer of Manitoba Hydro in April 1981, electrical consumption by the primary metal sector would have declined further.[70] John F. Funnell, general counsel and secretary of Manitoba Hydro, claimed that industrial electricity rates overall were standard, except for the two mining companies enjoying discounted, 'off-standard contracts' (Inco and Hudson Bay Mining and Smelting). He predicted that these two companies would pay standard rates upon contract renewal.[71]

Inviting Foreign Manufacturers

Unlike Churchill Falls, where Brinco signed power contracts with the largest customers before the hydro project was built, Manitoba Hydro built its hydro dams first, leaving government officials scrambling to create a market for large industrial customers. Manitoba Hydro did not itself solicit industrial customers but left this task to departments and offices of the Manitoba government. Initially, during the 1960s, the Department of Industry and Commerce assessed the strategy of using Nelson River power to enhance industrial development; during the mid-1970s oil crisis, the Premier's Office pursued electricity trade for manufacturing reciprocity; and during the 1980s, officials in the Manitoba Energy Authority actively practised a policy of 'industrialization by invitation.'[72] These strategies fell flat because often the energy-based industrial diversification goals of provincial governments and those of transnational corporations did not coincide.

In June 1966, the Department of Industry's 'Assessment of the Impact of Nelson River Electric Power on Industrial Development in Manitoba' was presented by Arthur D. Little Incorporated, of Cambridge, Massachusetts.[73] The consultants predicted that 'low-cost electric power in Manitoba [would] be attractive to firms in the silicon carbide, chlorine and caustic soda, nickel and pulp and paper industries.'[74] Important opportunities for expansion in the pulp and paper industry, the Cambridge consultants advised, could be created by 'the ability of the Manitoba Government to influence the attractiveness of a Manitoba Location by its policy in relation to stumpage charges, rents and the provision of access roads and other facilities.'[75] In addition, they recommended that the power rate and supply contract be competitive with those of other regions.[76]

Favouring such advice, the Manitoba Development Fund advanced loan money to the developer of the The Pas forestry complex, Dr. Kasser. He had been solicited through an advertisement placed by an industrial development officer:

> For 20 years Rex E. Grose, a senior provincial civil servant, had struggled to recruit some entrepreneurs to design and construct a viable pulp and paper mill that would utilize the spindly conifers of Manitoba's pre-cambrian forest. The problems that faced such a venture were many and obvious but the solutions were not. The full resources of the [Duff] Roblin administration were pledged to assist such a venture if the right person could be found in 1965 in Dr. Alexander Kasser ... [His] determination [was] to obtain not less than 100% financing of the project with public funds without letting the providers of such funds become fully aware of the full extent of the assistance [the government] provided.[77]

Soon afterwards, 'Dr. Kasser was in Austria fighting extradition to Canada for his allegedly criminal conduct,' and the project was taken over by Manitoba Forestry Resources, a Crown corporation.[78] The scandal involved a gross waste of public funds. Not surprisingly, Premier Edward Schreyer's office took more direct control of industrial promotion in the subsequent decade, but this time with an emphasis on getting industrial corporations from Ontario and Québec to extend their investments to Manitoba by using renewable energy leverage.

In the 1970s, Premier Schreyer responded to the oil crisis by advocating reciprocity, by which central Canada would help advance manufacturing plants in Manitoba in return for low-cost renewable hydro energy. As he stated publicly in 1974, 'I believe in reciprocity ... but we haven't had that when you look at the size of Western Canadian manufacturing.'[79] His idea bore some fruit, as was seen when Manitoba captured manufacturing content in its hydro projects from central Canada (although these contracts were based on procurement rather than on energy-for-manufacturing reciprocity). However, overall, the fortune of the reciprocity policy declined with the fall of oil and other energy prices subsequent to the oil crisis.

Following these failures, responsibility for the promotion of economic growth shifted to a new department. In 1981, the Energy Authority Act charged the Manitoba Energy Authority with responsibility for creating an energy-intensive and diversified industrial market. 'Part of the mandate of the Manitoba Energy Authority is to encourage the use of Manitoba's hydro-electric power to strengthen and diversify the province's industrial base. An increased emphasis has been placed on informing the international business community of the advantages of Manitoba's hydro-electric power for energy intensive industries. Attracting energy intensive industries to Manitoba will increasingly become a priority of the Manitoba Energy Authority.'[80] In carrying out its mandate, the Manitoba Energy Authority published glossy information brochures about the already diversified nature of Manitoba's economy, and the officers of the authority travelled abroad promoting Manitoba's energy advantage.[81] Between 1987 and 1989, 'the Authority ... met with corporate representatives from Europe, Asia and ... North America to examine potential developments in these areas.'[82] The energy-intensive industries that Manitoba Hydro officials identified included 'fertilizer manufacturers, chemical processors, pulp, paper and lumber operators, food and beverage companies, foundries and metal casting plants.' Representatives of the authority shared with hydro officials the view that since Manitoba still had the potential to almost triple its generating capacity, the number of energy-intensive industries in the province was bound to increase in the years ahead.[83] By 1988, the authority had signed industrial 'cooperation agreements' with a number of Japanese and Korean financial institutions, trading companies, and manufacturing firms. Manitoba Hydro reported in its *Annual Report* of 1986 that, from time to time, 'several aluminum companies [had] expressed an interest in building multi-million dollar smelters in the province.'[84] Furthermore, 'promotional research and commercial activities' were pursued with Japanese, European, and American silicon firms.[85] Yet according to authority reports, between 1982 and 1989, not a single energy-intensive industry came to Manitoba to take advantage of its low-priced hydroelectric power. It was further agreed, by two industrial development officers of the Manitoba Power Authority, that those operations that had established themselves in northern Manitoba – Inco and two pulp mills – had done so in order to take advantage of forest and mineral resources rather than the electrical resource.[86]

The authority came closest to a major deal in its negotiations with the Aluminum Company of America (Alcoa). On 28 March 1984, Alcoa agreed to undertake with Manitoba a joint study of the engineering, economic, and environmental aspects of an aluminum smelter; Manitoba was 'prepared to participate in the funding of and the results from the project through the Manitoba Energy Authority.'[87] Yet 'in December 1984, Alcoa decided to withdraw from the [$2 million] study, citing as reasons a poor international market in aluminium and changes in its corporate priorities.'[88] As it turned out, Alcoa had used the Manitoba negotiations as leverage to gain a better deal from Québec. In addition, a 'gag order' (agreed upon by both parties) was placed on the Manitoba Alcoa negotiations over the $700 million smelter.[89] This is another case in which the profit interests of multinational corporations did not coincide with the regional development goals of provincial governments.

What I have outlined in this section is Manitoba's continued dependence on foreign industrial firms (Inco, The Pas Forest Complex) and on hydro equipment firms from outside the province (CGE from Ontario, DEW from Québec, Mitsui from Japan, and Leningrad Metal Works from Russia) during the recent decades of hydroelectric development. A similar dependency obtains in engineering: in 1985, Manitoba Hydro awarded 'the $236 million Limestone general contract to a consortium known as Bechtel-Kumagai, headed by Kumagai-Gumi Ltd. of Tokyo and the Toronto-based Canadian subsidiary of the Bechtel Group Inc. of San Francisco.'[90] If provincial reliance on foreign firms and control continues, Manitoba contractors will scarcely develop beyond the subcontractor and temporary component-manufacturing stage.

Planning the 'Unexpected' Surplus

During the early 1970s, controversies erupted over the purpose of damming more Nelson River sites. Some viewed Manitoba's hydroelectricity as a valuable renewable resource to be developed only for the benefit of Manitobans, with any surplus being directed to the Canadian market. Others, especially during the 1970s oil crisis, advocated that an acceleration in building hydroelectric projects could provide the catalyst for a transnational energy regionalism, including the creation of industrial load, trade of electricity with neighbouring provinces, and export to US states. Favouring such transborder (US-Canadian) regionalism, Edward Schreyer, Manitoba's premier, and L.A. Bateman, the chairman of Manitoba Hydro, launched the Limestone hydro project in 1975 – only for it to be 'mothballed' in 1978 because it was surplus to the Manitoba system (Table 6.5). Events leading up to this mistake of pre-building the $1.73 billion project show once again the capriciousness of demand projections based on such indicators as predictions by transnational corporations, the inflated hopes of political leaders that industries will flock to new sources of energy, the false demand created in the mineral sector by exports to the US during the Vietnam War, the expectation of interprovincial coordination in hydroelectric planning, and the optimism about the US export market.

In 1972, the Schreyer government set up the Manitoba Hydro Task Force to examine the economic, financial, and east-west aspects of constructing hydro plants on the Nelson River for long-term export to the US.[91] One view among task force members was that Manitoba Hydro tended to assume that its financial and other problems 'could be solved by external interconnections and export sales.'[92] Manitoba Hydro took its cues from Mitchell Sharp's national power policy of 8 October 1963, which encouraged 'power exports and interconnections between Canadian and United

Table 6.5

Chronology of the controversy about hydro development in Manitoba, 1972-92

1972	Manitoba Premier Ed Schreyer establishes Task Force on hydro matters.
1973	Unreleased report by Manitoba Hydro Task Force recommends cancellation of proposed 1,000 MW deal with NSP, preparation of more detailed economic and financial studies for future long-term exports, coordination with and direct surplus power sales to provinces, and raising concept of national power network with federal government (see details below).
1974	Transnational oil companies and Prime Minister Trudeau predict depletion of all Canadian oil reserves in a dozen years.
1974	Manitoba Premier Ed Schreyer, in his Opening Remarks for the First Ministers Conference on Energy in Ottawa, proposes interconnection of four western provinces.
1975	Inflation of industrial forecasts by hydro officials; Manitoba Hydro applies to the National Energy Board to export power to the US.
1975	Limestone Project construction launched by Premier Schreyer and Manitoba Hydro Chair L.A. Bateman.
1976	Task Force Report tabled in the Manitoba Legislature three years after its initial completion.
1974-7	Major energy-intensive industrial customers have unpredictable loads, yet padded industrial forecasts continue.
1978	Limestone Project 'mothballed.'
1979	Tritschler Commission of Inquiry into Manitoba Hydro releases final report in December 1979.
1985	Limestone Project resumed solely for US export.
1992	Limestone Project completed.

States power systems where such induce early development of Canadian power resources.' In the early 1960s, Manitoba Hydro and the Northern States Power Company (NSP) in the US began studies of the 'economics of increasing transmission interconnections between the utilities.'[93] In March 1973, they completed a study that 'recommended that Manitoba Hydro and NSP should interconnect their respective systems further with a major transmission line and that Manitoba Hydro should accelerate the development of its generating facilities to provide for a firm base load export of 1000 MW to NSP for a fifteen year period commencing in 1979.'[94] Meanwhile, the task force examined the financing alternatives, pricing mechanisms, east-west options, the impact on the construction industry, economic alternatives, and other issues related to accelerating the building of Manitoba's hydro plants dedicated to export (Table 6.5). The members of the Manitoba Hydro Task Force made five recommendations in their 1973, then unreleased, report:

(i) That Northern State Power be told that a 1,000 megawatt sale will not take place;

(ii) That any similar proposal for large scale [sic] export of power be subject to the types of economic analysis outlined in this report, [and] pursued in much greater detail;

(iii) That studies be undertaken to ascertain the financial needs and resources of the province for the next 10 years and beyond, [and] to provide a framework within which major capital projects can be compared and priorized;

(iv) That Manitoba actively pursue with the Governments of Saskatchewan and Ontario the possibilities of closer co-ordination and surplus power sales;

(v) [and] That the Government of Manitoba raise again with the federal government the concept of a national power grid.[95]

With respect to financing, Professor Eric Kierans, of the Department of Economics at McGill, recommended in an addendum to the task force report that the utility should properly price industrial (e.g., Inco) and export sales to build up its equity to at least 20 percent of assets; in 1972, the utility, in effect, had owned only 5 percent of its total assets of $932 million.[96] This situation was akin to owning a house with a 95 percent mortgage and borrowing more money for unnecessary additions. He advocated that excess investment in hydro project construction be curbed (1) because it limits growth in other sectors of the economy and (2) because it was not needed, as rising energy exports in the 1970s indicated that hydroelectric 'capacity exceed[ed] the needs and rate of growth of Manitoba.'[97] He suggested that government should scrutinize all long-term contracts with heavy power users and that Manitoba Hydro should properly price its mining, industrial, and export customers.[98] Similar to the task force recommendations, he suggested that 'Manitoba initiate discussion with Ontario and Saskatchewan, initially on a first Minister or [other] Ministerial level, leading to joint efforts to use most efficiently existing sources of energy.'[99] Instead of the 1,000 MW US export planned by Manitoba Hydro, Kierans recommended 'cancellation of further discussion with N.S.P. on the acceleration proposal.'[100]

To bring the extra northern power southward, Schreyer and Bateman appealed in 1973 and 1974 to Donald S. Macdonald, federal minister of energy, mines and resources. In a letter to Macdonald dated 28 November 1973, Schreyer argued that the two 1,000 MW projects (Kettle and Limestone) should be advanced because 'Canada as a whole' required this 'substitution of energy.' He recommended that Manitoba Hydro and energy ministry officials 'meet as early in December [1973] as possible to consider possible acceleration of Nelson River downstream development and the financing arrangements [$500 million federal funds] related to such an undertaking.'[101] In the meeting on 12 December 1973 with ministry officials in Winnipeg, Batemen, chair of Manitoba Hydro, outlined Manitoba's interest in high-capacity interconnections with the Northern States Power Company of Minnesota, accelerated lower Nelson hydro site development, and additional transmission links between the Nelson area and southern Manitoba.[102] On 7 January 1974, in his response to these proposals, Macdonald did not offer financial assistance, noting the more pressing needs in other provinces 'where security of energy supply is obviously much more critical than it is in the case of Manitoba.'[103]

Before the incomplete task force report was released to the public on 10 May 1976, Manitoba Hydro had already decided not to 'accelerate the development of its generating facilities' on the lower Nelson, as was recommended in its joint study with NSP, because of engineering limitations: 'technical and scheduling constraints alone,' according to Manitoba Hydro, in a later explanation in 1976, 'eliminated the possibility of constructing the generating facilities in the time available.'[104] Instead, after further joint studies with NSP were completed in 1975, Manitoba Hydro pursued a power exchange agreement and international power line interconnection with NSP. 'A letter of intent dated August 22, 1975 was signed by the Executive Officers of Manitoba Hydro and NSP wherein it was agreed to interconnect both utilities with a single 500 kV transmission line and undertake power exchange [involving 300 MW, and the sale of 200 MW and 1.5 billion kWh] between the utilities for a 12½ year period commencing on May 1, 1980. This was approved by Order-in-Council # 1199 of September 18, 1975.'[105]

According to Manitoba Hydro, this agreement with NSP of Minnesota allowed postponement of building the Limestone Generating Station for a few years, and the interconnection with the US was recommended in the 1975 studies which the 1973 task force had not been able to assess two years earlier.

This 1973 task force had also recommended the pursuit of a national power grid and the international coordination of power supply when provinces took the initiative to develop a national power grid (see Chapter 2 above). Furthermore, the suggested reduction in investment in hydro construction by Eric Kierans had little effect, because Premier Schreyer and Manitoba Hydro chair L.A. Bateman proceeded with their plan to construct the next Nelson River project by launching the Limestone project in 1975. Schreyer and Bateman argued for expanding Manitoba's hydro infrastructure to suit a transborder regional energy market interconnected in a north-south direction with the US and in an interprovincial east-west direction within Canada. They planned to invite new manufacturers from around the globe to Manitoba, to export power to utilities in the US, to promote hydroelectricity in markets where electricity is generated from fossil fuels, and to use electricity trades with central Canada as leverage to shift more industrial development westward from the Golden Triangle (Toronto, Ottawa, and Montréal).

In their vision of hydroelectricity as a catalyst for multifaceted development, Schreyer and Bateman were influenced by a misconception that was popular during the 1970s oil crisis. The impending 'oil shortage' was used, probably in good faith, by Manitoba's hydro planners as a rationale for building extra hydro plants for domestic use. In 1974 Prime Minister Pierre Elliott Trudeau gave credibility to the so-called 'oil shortage,' and thereby to oil company forecasts (and oil company price hikes), by predicting that existing Canadian oil reserves would last 'for not more than a dozen years.'[106] Stopping oil exports to the US in order to preserve Canadian reserves seemed impossible because the Liberal government of 1963 had worked out a continental energy compromise, and, as Trudeau reasoned, 'we could not, as a good neighbour, suddenly cut off our exports to the United States.'[107]

In fact, estimates of the extent of oil shortages were exaggerated by oil companies, who were gratified by the resulting price rise. But from the perspective of hydro planners, the moment seemed ripe to begin replacing energy from oil with energy from river power. Manitoba Hydro saw in this 'oil crisis' an opportunity to displace fossil-fuel energy with hydro energy not only in neighbouring provinces, but also in neighbouring US states, through its plan to advance the construction of Limestone.[108] In this plan, the valuable, renewable hydro-generated electricity would simply be integrated into the Mid-Continental Area Power Pool, thus reducing the pool's reliance on coal-based electricity.[109] Convinced that hydroelectricity would replace oil as an energy base, supply the growing US market, and shift industry westward, Manitoba Hydro decided to proceed with the Limestone generating station.

Unpredictable Industries

There is little evidence that Manitoba Hydro achieved any of its reciprocity goals in the Limestone project. Except for temporary component manufacture, a westward shift of industry did not occur. Furthermore, Manitoba Hydro had serious difficulties in forecasting its industrial market requirements. As shown above, Manitoba's forecasting problems and surplus infrastructure were, in part, an outcome of its reliance on the expectation that foreign industry would settle in Manitoba. Its attempts to plan for industrial growth in response to its promotions were also hindered by the volatility of many industrial loads, such as those required by Inco and Interprovincial Pipelines.[110]

Forecasting Demand

A comparison of what Hydro planners predicted commercial and industrial customers would use and the amount of electricity they actually consumed in the years from 1978 to 1986 shows that consumption was 29 to
47 percent below Manitoba Hydro's predictions (Appendix 14). Manitoba
Hydro planners relied on historical industrial growth, based on a few large
customers who had been attracted during the 1960s, but, as the new
plants on the Nelson River neared completion, half of the expected new
commercial and industrial demand did not materialize. For example, in
1969, Manitoba expected a new electricity demand from northern mining,
which was to increase to 1,300 GWh by 1976 (double the existing mining
requirement in Manitoba); this load was never required (Appendix 15,
primary metals column shows consumption decline).[111] Overall, the Commission of Inquiry into Manitoba Hydro found that energy growth rates
dropped drastically from 7.8 percent before the mid-1970s to 1.6 percent
between 1974 and 1977.[112] The poor forecasting and industrial growth
assumptions became evident as the emerging massive surplus contributed
to the mothballing of Limestone. Construction work was well under way
at the Limestone project, including the building of a permanent townsite
at Sundance, when the project was halted in 1978. Nevertheless, seven years
after construction at Limestone had stopped, Manitoba Hydro resumed
work at the site, following renewed optimism about a market for the export
of hydroelectric power.

Legitimation of Exports to the US

Although Manitoba Hydro tended to look south for an export market, in
the 1960s and 1970s some provincial premiers, notably Premier Schreyer,
also advocated strengthening interprovincial and national transmission
connections.

The objectives of the Canada-Manitoba Agreement on the Development
of the Hydro-Electric Potential of the Nelson River (1966) stated that the
'low-cost hydro-electric energy' from the Nelson River would meet 'national
objectives in the field of economic and industrial development' and 'would
be a major potential export to other provinces and to the United States.'[113]
During the 1970s, Premier Schreyer publicly advocated the need for the
national energy grid as part of industrial development reciprocity within
Canada, but he also allowed exports.[114] During the National Energy Conference on 22 January 1974, he asked that social and regional benefits be
part of a national energy policy. For such purposes he proposed a national
energy planning board that would base its judgment on social interests
(including concerns for regional equality, job creation for Native peoples,
energy security, and national integration) rather than commercial considerations (e.g., that the price of electricity exports should constitute the

major criterion of national interest). 'The Board should ensure that goals or ideals of national energy development are continually being described and articulated for the benefit of all Canadians. In this way the planning board would discuss policy with broad social and regional considerations rather than focusing mainly on commercial considerations [as done by the National Energy Board].'[115] Schreyer's public pronouncements for more national integration of provincial hydroelectric networks nevertheless did not exclude his support for exports. In 1976, Manitoba Hydro embarked upon the export process by applying to the NEB for an export licence.

Export Hearings and Interest Groups

From 13 to 17 January 1976, the NEB held hearings in Winnipeg to evaluate Manitoba Hydro's application for a 230 kV international power line and for five licences, effective November 1976, to export electricity to the Minnesota Power and Light Company. As reported by the NEB in its March 1976 report, three of the five export licences were approved, but a second public hearing was held on 13 September 1976 in Winnipeg to resolve outstanding landowners issues concerning the routing of the 230 kV international power line. Manitoba Hydro's first 230 kV transmission line and exports to the US had already been approved by the NEB in 1969. The approval of the second international power line and the granting of the 1976 licences would set a precedent for future Nelson River power exports. The application proceedings were controversial, however, with a number of interest groups opposing the southward integration of Manitoba Hydro's network.

Among the parties supporting the application was the 80,000-member Manitoba Federation of Labour, the Minnesota Power and Light Company, and the Progressive Conservative Party of Manitoba. The representative of the Manitoba Federation of Labour 'suggested that export sales should proceed without delay.'[116] The Progressive Conservative Party of Manitoba 'urged the Board to encourage the development of a national grid. It also supported the export of power and energy that was truly surplus to the needs of Manitoba.'[117] The Conservative Party thought that Manitoba Hydro could supply a national network as well as export to the US. The witness from the Minnesota Power and Light Company (MP&L), the US importer, testified that starting 1 November 1976, his company needed the 100 MW export to supply the Taconite mining load in its Northern Range Division, whose consumption was to increase from 285 MW to 590 MW. He argued that, 'while the total power supply is adequate under normal conditions, the failure of [MP&L's] 350 MW generating unit would result in a marked capacity deficiency, with consequent financial losses to MP&L customers. Hence for security reasons the line from Manitoba is needed as soon as possible.'[118]

Many other groups, however, opposed mid-continental electricity integration. The Association of Major Power Consumers in Ontario (AMPCO) made three points: (1) the Manitoba surplus should be used to substitute for 'fossil fuels in adjacent provinces'; (2) since the Nelson River transmission system is federally financed, 'other Canadians should benefit' from Manitoba's surplus; and (3) in 1974, the federal minister of energy, mines and resources had stated that 'Federal Government assistance to develop stronger regional interconnections is available.'[119] The Manitoba Environmental Council feared that 'adverse planning, environmental, economic, and social ramifications' would result from such an export program and that the application was only the 'first stage of a program' that would 'result in the export of up to 1,000 MW of firm power' and in the installation of additional international power lines.[120] The Northern Flood Committee, whose 'Board of Directors consisted of the Chiefs of five Indian bands of Cross Lake, Nelson House, Norway House, Split Lake and York Landing, all in Manitoba,' contended that 'the generation of the hydroelectric energy surplus which Manitoba Hydro proposes to export is dependent upon the operation of the Churchill River [Manitoba] diversion, which will result in the flooding of Indian reserve lands.'[121] The Interchurch Task Force on Northern Flooding 'urged that the export should not be permitted until all matters pertaining to the acquisition of Indian reserve lands for this project have been resolved.'[122] Furthermore, the minister of energy for Ontario cautioned the NEB 'to ensure that Canada could *repatriate* power upon the termination of an export licence' (my emphasis).[123]

All views opposing the transmission line were subordinated to the NEB's primary consideration that the Canadian national interest would be damaged if Manitoba Hydro's export contract obligations with MP&L were not met. The export licences were approved by R.F. Brooks, presiding member of the NEB, in March 1976, and the required transmission line was approved by Robert A. Steed, the presiding member of the NEB, in September 1976. Both made pro-export-to-the-US decisions typical of the case-by-case nature of the NEB's approval process. Manitoba Hydro had advanced its export strategy in stages by applying in 1969 and again in 1976 to construct international transmission lines and to export power. A third 500 kV line was also planned, and exports (commencing 1 May 1980) were already agreed upon by a letter of intent between Manitoba Hydro and Northern States Power, a US utility.[124]

Making a case against exports in such a tribunal-like hearing process was difficult. Opponents of exports, for example, had to provide evidence before the NEB that the individual export application caused the utility to operate its facilities 'in a different manner in order to meet export commitments as opposed to domestic loads.' Providing such proof was difficult, if not impossible, for opponents because they did not have the same access to, or expertise in, hydroelectric systems operations as did utilities. Thus, according to Brooks, the presiding NEB member, the Northern Flood Committee was 'unable to establish any relationship between the proposed exports and the level of flooding on the [Churchill River] diversion.'[125] In contrast, Manitoba Hydro claimed in the January 1976 hearings that its current export application had no effect on operations, even though its overall export strategy was evident to the NEB. Brooks reported in March 1976 that 'the Applicant's policy witness, in cross examination, stated that the Churchill River diversion [in Manitoba] would not be constructed or operated in any different manner by reason of exports.'[126] The presiding member concluded that 'exports would have negligible incremental environmental impact'; therefore, 'the [environmental] impact would be effectively the same, regardless of whether the export were allowed or not.'[127] Brooks and his board were satisfied that 'the power to be exported' in the three export licences was 'surplus to reasonable foreseeable Canadian requirements and that the price to be charged is just and reasonable in relation to the public interest.'[128]

In September 1976, Robert Steed, the presiding member of the second NEB hearing on the 230 kV international transmission line, approved the line with some modifications. His report pointed out the detrimental penalty clause and breach of contract clause. He noted that if the international power line were not in place by '1 November 1976 ... approximately $200,000 per month' would be charged to Manitoba Hydro, and he felt that honouring the 'transaction agreement between MP&L and Manitoba Hydro' was of key national interest:

> Account must be taken of the substantial economic cost to the Province of Manitoba of delay in completing the international connection and, most importantly, the probable detrimental effects to Canadians generally and to public utilities in particular of failure on the part of Manitoba Hydro, for whatever reason, to meet its contractual obligations with United States utilities. This latter is truly a matter of national interest which could have undesirable ramifications in the future for areas and people in Canada beyond the borders of Manitoba.[129]

Yet such continentalization support weakened coordinated system development and interprovincial trade within Canada: interprovincial sales have in fact declined, while the north-south sales have increased.

Nelson Surplus as Transborder Energy
Manitoba's electricity flowed in increasing volumes from the Nelson River plants into its provincial system and to US power companies through new high-capacity international transmission lines. Since 1976, Manitoba's

exports to the US have steadily increased from 534 GWh in 1977 to 6,153 GWh in 1986 (Appendix 16), more than twice the total reported electricity purchased by all manufacturers in Manitoba, which was 2,530 GWh in 1984 (Table 6.4). Manitoba sales to Saskatchewan and Ontario have declined, whereas sales to the US have increased more than tenfold between 1976 and 1986 (Appendix 16). This export growth continued over two decades into the 1990s. Whereas in 1977 electricity sold from Manitoba to the US amounted to 534 GWh, in 1996 it totalled 9,733 GWh, nearly twenty times the original amount.[130] To transport these additional amounts of electricity southwards, Manitoba Hydro, with the approval of the provincial and federal governments, has strengthened its international transmission lines so that in 1994 they were capable of carrying 1,325 MW (the capacity of the Limestone plant), while the interprovincial power lines have remained at much lower capacities (by 1994, with Ontario, they amounted to 260 MW, and with Saskatchewan to 500 MW).[131]

Manitoba, like Québec, is now extracting power from dammed northern white-water stretches of river strictly for export. In 1992, Manitoba Hydro held eight export licences; those commencing in 1987 were mostly firm exports (whereby generators are dedicated to US export).[132] To supply such US-destined electricity, the 'mothballed' Limestone project was restarted in January 1985, when 'the Premier of Manitoba announced that Manitoba Hydro would call for tenders on the general civil contract for the Limestone Generating Station. The contract was contingent upon Manitoba Hydro's obtaining approval from the National Energy Board to export power to Northern States Power of Minneapolis.'[133] In March 1985, the NEB approved the export for a twelve-year period. Since the Limestone generating station (1,330 MW) was completed in 1992,[134] Manitoba Hydro can export 500 MW/3405 GWh per consecutive twelve months from 1 May 1993 to 30 April 2005.[135]

The overbuilding of Manitoba's hydroelectric system for such exports is substantial. Manitoba Hydro predicted that in 1996-7 its electrical capacity surplus to Manitoba's own needs could reach 1,304 MW.[136] The federal government played a contributing role in building the system beyond provincial requirements. With the Manitoba planners' knowledge that the NEB would support the federal government's continentalized energy policy, Manitoba Hydro and the provincial government were able to exceed the provincial mandate and overbuild the system through pre-arranged or hoped for export contracts. In the 1970s, whenever hydro facilities were built for provincial use in the long term but had sufficient generating capacity for intermittent exports, the NEB considered such exports as 'temporary' and, therefore, capital cost allowances for plants were usually not charged to US consumers.

Also frequently evident in its decisions, the NEB has virtually ignored views opposing the transport of electricity to the US and views favouring Canadian system integration; instead, its decisions (also dependent on Cabinet approval) fostered the tenfold flow of electricity to the US over the period between 1977 and 1986. It did this by permitting Manitoba Hydro's building of international power lines, granting export licences, and allowing the Limestone plant to operate solely for export. What becomes evident is that by the time the NEB got around to examining the applications for export, the dams had been built already (or were under construction), the Churchill River diversion infrastructure was in place and water was flowing (doing more damage than expected), and the bottom had already dropped out of the energy demand.[137] Manitoba and its neighbouring provinces, as discussed in Chapter 2, participated in the provincial initiatives favouring the development of a national power grid and a Western power grid but did not coordinate the development of their electricity supply and transmission systems; instead they followed individual provincial expansion policies and export strategies. Soon after, in the 1990s, provincial electricity policies were continentally harmonized under changed export conditions. Just retaining the export market under new federal energy regulations in Washington after the Free Trade Agreement resulted in such new south-north repercussions as utility restructuring, transmission access reciprocity, competition for wholesale customers, and Manitoba Hydro Act amendments.

1990s Exports and Implications
In the 1990s, in order to retain and expand their export market in the mid-continental US, both the utility and the government have made changes to the way the electrical system is operated and advanced in Manitoba. Manitoba Hydro joined a US power pool and complied with US regulations that require that (1) transmission line access be given to wholesale power suppliers into Manitoba and (2) power supply and power line use be functionally separated into distinct business units. The provincial government amended the Manitoba Hydro Act not only to accommodate these US regulatory requirements, but also to increase the utility's corporate autonomy and to allow the building of hydro projects dedicated for export. At the beginning of the twenty-first century there is no immediate need for expansion. Based on forecasts of Manitoba's requirements for electricity, Manitoba Hydro claimed that 'a new source of generation will not be needed until 2015.'[138] Although an opportunity existed for Manitoba Hydro to supply electricity to Ontario from the Nelson River in the 1990s, little was accomplished.

The significant drops in the Lower Nelson River form a staircase of dam sites before the river flows into Hudson Bay.[139] The uppermost of the last three sites, the Limestone dam, had been completed in 1992; down one level, the Conawapa project (1390 MW) development has been delayed; and down another level and closest to Hudson Bay, the Gillam Island location, a potential site, remains undeveloped. In 1989, Manitoba Hydro signed a deal to sell Ontario Hydro 1,000 MW of power for seventy-two years (beginning in the year 2000) from the planned Conawapa generating station.[140] This deal would have provided the basis for constructing Conawapa, a mega-project costing an estimated $5.7 billion, including the dam, powerhouse, and transmission line work. Ontario Hydro, however, experienced fiscal difficulties and opted out of the deal in 1991. A lawsuit was launched, and when Ontario Hydro started shutdowns of nuclear reactors in 1997, hope in Manitoba briefly rose again for the resumption of this project. However, an ongoing obstacle to replacing Ontario's nuclear power with hydro power from Manitoba is that north of the Great Lakes the transmission capability of the power lines to Ontario remains low.[141] Meanwhile, Manitoba's western trading partner, Saskatchewan, uses coal-based thermal power, which could be replaced by hydroelectric power from Manitoba in order to reduce Canada's contribution to global warming (the concern of the 1997 Kyoto conference). However, the low-cost of coal-based thermal generation, according to the manager of Manitoba Hydro's Power Planning and Export Marketing Division, makes such energy substitution unlikely.[142]

Continuing to permit Manitoba to sell its surplus electricity in the changed electricity market in the US in the 1990s, the NEB issued a certificate to upgrade Manitoba's 500 kV international transmission line (4 May 1994) and authorized several power and energy export permits. In the late 1990s, while the NEB lacked or had not exercised national jurisdiction over electricity trade – in the trans-provincial transport or import of electricity from the US – the US FERC regulations came to be applied in Manitoba, as they had in other provinces.

Manitoba Hydro's decision in 1996 to join the Mid-Continent Area Power Pool (MAPP) as a full partner had regulatory and organizational implications. When Manitoba Hydro signed the MAPP agreement and became part of this electricity pool's US regional transmission group, the utility became subject to FERC's new requirements. FERC regulations require reciprocal access to transmission systems; that is, Manitoba has to open its transmission lines to wholesale marketers of electricity from MAPP if it wants to continue selling electricity that needs to be transported across more than one state to more distant US customers. As a result, since 31 January 1997, Manitoba Hydro has opened its transmission lines (to protect its ability to export electricity) and now specifies tariffs and transmission fees to wheel wholesale electricity transactions for members of the US regional transmission group. Before being able to provide wholesale electricity trade reciprocity, according to FERC rules, utilities that have vertically integrated generation, transmission, and distribution must separate these functions into business units. On 4 April 1996, Manitoba Hydro's president and chief executive officer, R.B. Brennan, announced the formation of, and executive appointments for, the 'distinctly accountable business units: Power Supply, Transmission and Distribution, and Customer Service.'[143] Brennan, at the time the chair of the board of the Canadian Electricity Association, the Montréal-based association of major utilities in Canada, was very supportive of harmonizing Canadian utilities with US regulatory requirements, and he identified the forces behind this initiative:

> It is the demand of industrial and commercial customers for low cost electricity that is leading to re-regulation of the [electrical] industry in the U.S. [allowing US industrial customers access to new supplies of electricity unburdened by charges stemming from de-commissioned nuclear plants and levied on common customers]. In Canada, we are also experiencing the market forces that will create new structures for the electricity industry. In this climate, it is apparent that customers don't care what the energy source is. Manufacturers want to focus on producing whatever the plant is making, not on the source of energy.[144]

Brennan's scenario, in which industrial and commercial customers in Manitoba may not care whether their source of electricity is a Manitoba hydro plant or a generating facility in a neighbouring province or US state, is possible. This group of corporate customers may pressure the Manitoba government to open the transmission system to retail customers, like themselves, so they can buy their electricity at the cheapest rate from either Canadian or US suppliers. In contrast to some areas in the US, where transmission firms offer retail and other services, Manitoba Hydro's administrators serve only wholesale customers such as the city-owned Winnipeg Hydro.

To clarify Manitoba Hydro's mandate to build dams solely for electricity export to the US and reciprocally to integrate Manitoba's transmission network with that of US utilities, amendments to the Manitoba Hydro Act were necessary. Therefore in 1997, the former act, which focused primarily on the electricity needs of the province, was amended as the Hydro Amendment Act (Bill 55), which involved two major changes: (1) that legislation allow the transmission system to become a common carrier for wholesalers who buy and resell electricity in Manitoba (this is similar to provisions in the US Energy Policy Act, 1992) and (2) that it allow the corporation to offer new services and to create subsidiaries, joint ventures, and business alliances.[145] The likely users of these new services (e.g., selling spare capacity on transmission lines) are utilities or marketers who wish to carry electricity through Manitoba to destinations in the US, Ontario, or Saskatchewan.[146] The amended act also allows Manitoba Hydro to wholesale electricity to delivery points inside the US and to form power marketing subsidiaries (subject to Cabinet approval if such investments are larger than $5 million) in the US to sell Manitoba electricity. In addition, Section 16 of the act was intended to remove former legislative ambiguities about building for export and now, 'with approval by Order in Council,' allows Manitoba Hydro 'the right to construct new generation for export should the right opportunity arise.'[147]

In June 1997, in the Manitoba Legislature and before the Standing Committee on Law Amendments, politicians and representatives of labour raised issues related to the amendment of the act – issues such as building for export and privatizing Manitoba Hydro.[148] Members of the Legislature from the New Democratic Party, who had been the government when the Limestone project was restarted in the 1980s but were now the Opposition, took the opportunity to profile themselves as supporters of building the Limestone dam for export. They also accused the Conservative government of scuttling the 1,000 MW deal with Ontario Hydro that the NDP government had initiated in the late 1980s. For example, NDP member and leader of the Opposition, Gary Doer, said on 11 June 1997:

> So, of course, we believe in selling power to the United States and to any other jurisdiction. We negotiated 200 megawatts with Ontario and signed it. It led to another thousand megawatts sale to Ontario that the members opposite [Conservative government] fumbled and blew, and we negotiated interchange agreements under [sic] the North American market, and we negotiated a Northern States Power Sale that basically means Limestone is paid for by the Americans. This was opposed by the Tories. Who opposed Limestone and the Northern States Power sale? The members opposite. Now they are born-again marketers of Hydro. You know what you are doing? You are only clipping the coupons that were negotiated by the New Democratic Party of the past.[149]

After the leader of the NDP emphasized his support for building dams for power export, he argued that acts to reorganize and deregulate public utilities can hide privatization intentions. Doer, on 11 June 1997, argued that a 'Trojan Horse amendment is contained within the Manitoba Hydro Act ... that says that the government will not sell Hydro or any part of Hydro as an amendment to this Act.' He claimed that this amendment was redundant because the Tories had sold the Crown-owned Manitoba Telephone System (MTS), and the government had 'a set procedure to deal with privatization of Crown corporations':

> Reorganize, deregulate, deny sale during an election, break your promise after election, hire brokers, do not tell the minister. Later on, when the

opposition breaks it, tell your own caucus. Get brokers to evaluate the situation. Break your promise and then legislate, again, away from public ownership without any referendum and any desire of the public to have a chance to speak.[150]

He thought that since brokers pocketed about $35 million from the sale of the telephone system, the privatization of Manitoba Hydro would probably bring brokers about $100 million – 'that is a lot of Jaguars,' he added. He argued that the Tories used 'the excuse of deregulation to privatize MTS' and predicted that the 'deregulation' phenomenon – akin to the 'devil made me do it' argument – will be part of the Tory's Manitoba Hydro privatization procedure.

> They will go in a couple of years, if they get back in, and say, well, we brought in this Section 15 [in the Hydro Amendment Act of 1997], but we really did not mean it.
> An Honourable Member: The brokers made them do it.
> Mr. Ashton: Yes, the devil made us do it. Well, the brokers made them do it. Bay Street made them do it. Madam Speaker ...

The NDP called for a postponement of the second reading to allow the public to participate in discussing the Hydro Act amendments; this motion, however, was defeated.

Before the Standing Committee on Law Amendments, such members of the labour movement as Ron McLean (business manager of the International Brotherhood of Electrical Workers, Local 2034, representing about 2,250 employees in Manitoba Hydro), who had regularly discussed the deregulation with Manitoba Hydro president Bob Brennan (who supported the passage of Bill 55), and David Tesarski (Manitoba Council of the Canadian Federation of Labour) were also generally supportive of building for export and liked the fact that Manitoba Hydro would remain the only retailer of electricity in Manitoba, as stipulated in Sections 15.1 and 15.2. They also indicated that Manitoba Hydro 'should not be privatized' or face competition from 'fly-by-night power producers.'[151] Thus both major political parties and some representatives of labour in Manitoba were, overall, supportive of building hydro projects dedicated for US export and, to a degree, of adjusting the organization and function of Manitoba Hydro to meet US regulatory requirements.

Conclusion

Until 1930, the federal government assigned rights to develop and distribute hydroelectric power to a variety of owners in Manitoba – private, municipal, and provincial. A private and municipal utility came to share the lucrative urban Winnipeg market, while as early as 1914 the province of Manitoba proposed the conservation and reappropriation of water power for a province-wide hydroelectric system that was assumed to stimulate manufacturing and population growth. In the urban market of Winnipeg, it was primarily businesspeople who supported municipally owned power, the idea being that competition would level prices and advance manufacturing. Private investors willing to transmit electricity from Winnipeg River plants to southwestern Manitoba towns did not emerge; as some studies show, such transport of power appeared to be insufficiently profitable at the time. Partly to improve rural supply and to attract manufacturers, the provincial government increased its involvement. With the takeover of Winnipeg Electric by the Manitoba Hydro-Electric Board in 1953, the earlier privatization of water power began to be reversed. As the larger southern power sites had been developed by the late 1950s, the provincial government set up Manitoba Hydro and soon launched its premature expansion into the Nelson River region.

By 1966, the federal and Manitoba governments agreed that the province would build plants on the Nelson River and that the federal government would construct and finance transmission. Both governments envisaged several markets for Nelson River electricity: in neighbouring provinces using the national power network, in Minnesota and North Dakota by transmission to the mid-continental grid, and in Manitoba itself. Hydro planners expected that one of the collateral benefits of building hydro-electric projects would be the emergence of mechanical and electrical supply industries in Manitoba.

After initial reliance on imported generating equipment, Manitoba Hydro strengthened local component manufacturing, albeit temporarily. In the early 1970s, Manitoba Hydro imported Japanese and Russian generating equipment for its Nelson River projects, but then Long Spruce and Limestone Hydro officials negotiated with US branch-plant manufacturers to increase the local mechanical and electrical manufacture of hydroelectric equipment. This led to an increase in Manitoba content in hydro projects on the Nelson River.

Several branches of the Manitoba government based their industrial promotions on Manitoba's increasing electricity supply: in the 1960s the Department of Trade and Commerce promoted low-cost power in combination with natural resource allocations to create the conditions for secondary industrial development; in the 1970s the Premier's Office, particularly under the direction of Schreyer, advocated manufacturing reciprocity in Canada, whereby renewable hydroelectricity would be exchanged for more industrial investment from the Golden Triangle; and in the 1980s the Manitoba Energy Authority invited industrial corporations from around the northern globe – Asia, Europe, and the US – to bring about energy-intensive and secondary manufacturing. However, similar to the outcome in other provinces, Manitoba's new supplies of electricity were primarily applied to the semi-processing of export goods in four sectors: primary metals, pulp and paper, food, and beverages. Statistics show that, since the lower Nelson expansion in the mid-1970s (and even earlier), manufacturers in these four sectors have reported that they consumed close to 82 percent of all electricity consumed by manufacturers.

The type and amount of industrial development that did occur was not what politicians and government officials claimed would result. Manitoba Hydro expanded the Kelsey plant capacity to serve the declining Cold War demand from the International Nickel Company. Only a few large energy-intensive industries came to Manitoba between 1962 and 1975, and no major industrial customers (other than the Hudson Bay Mining and Smelting expansion) have been added during the subsequent decade. One invited entrepreneur, Dr. Kasser, left the scandalous legacy of the Churchill Forest Products Complex with the Manitoba government. The aluminum company Alcoa used negotiations with the Manitoba Energy Authority as leverage to obtain better industrial benefits for establishing a smelter in Québec. The continental oil swap, as Trudeau described it in 1974, led to volatile electricity demands by the Interprovincial Pipelines Company. Such unpredictable demands, including the incorrect shortage predictions by oil industry multinationals and the prime minister during the 1970s energy crisis, contributed to the building of too many projects and surplus capacities. The results: the Commission of Inquiry into Manitoba Hydro planning and the 'mothballing' of the Limestone project.

Other factors contributed to an increase in surplus power, such as the failure of the national power network and the building of extra capacity for export – a process legitimized, despite resistance by many Manitobans, through NEB licence approvals. As discussed in Chapter 2, even modest attempts to construct a western grid (linking Alberta, Saskatchewan, and Manitoba) failed in 1980.[152] Whereas interprovincial initiatives weakened after the completion of the Lower Nelson River plants, continentalization of Manitoba's electrical system through integration with the US mid-continental utilities increased. The construction of the Conawapa plant for the Ontario market was put on hold, while the Manitoba Hydro Act, 1997, (1) accommodates US regulatory requirements for allowing power-supplier access to wholesale electricity customers on the interconnected Manitoba and mid-continental grid; (2) increases the authority of Manitoba Hydro to conduct more business activities in the US electricity market; and (3) permits the building of future hydro plants solely for exporting electricity to the US.

7
Peace, Pulp, and Power Hunger (British Columbia)

> British Columbia, traversed by three distinct mountain ranges and with, on the whole, a high rate of precipitation, has many mountain rivers which offer opportunity for power development and consequent secondary industrial growth.[1]
>
> – T.L. Sturgess and R.W. Bonner, Department of Industrial Development, Trade, and Commerce, Bureau of Economics and Statistics, Government of British Columbia, 1956

Introduction

During British Columbia's dam-building rush in the 1960s to the mid-1980s, most of the power development occurred on the Columbia River, flowing from the southern interior of the province across the US border through Washington State and into the Pacific Ocean at Portland (Oregon), and on the Peace River, flowing from the northern interior across the Alberta border and into rivers and lakes that drain into the Beaufort Sea. The dams on the Columbia River were built to generate electricity and to regulate the river flow for optimizing electricity production at generating plants in the US. Building hydroelectric plants on the Peace River, which is the major focus of this chapter, became part of the attempt to open northern British Columbia to industry.

Since the 1940s, and particularly since the 1960s, members of British Columbia's political elite and their appointed utility managers have claimed that hydro development would be the leading force in transforming the BC economy from primary resource-based industry to secondary manufacturing industry. They invited industrial developers and manufacturing-plant investors to British Columbia to use the plentiful hydro power together with forest and mineral resources to develop these anticipated secondary industries, which would provide a new electricity market and employment for British Columbians. Decision makers in government thought they would need only to privatize such valuable resources, in some instances whole watersheds, by allocating them to developers of international stature who would then carry out the necessary industrial transformation. Pictures promoting mega-dams, transmission lines, powerhouse control rooms, and modern electrical appliances gave the appearance of an already existing industry that would naturally produce hydroelectric generating equipment and electrical instrumentation: it seemed patently logical that a secondary consumer-goods-manufacturing industry would soon develop.

At times, political leaders acknowledged that the building of large plants in the interior of British Columbia could lead to some temporary surplus electricity, but such surpluses, they proclaimed, could be sold, either into the national energy network or to 'power-hungry' California. A closer examination of such claims reveals the now familiar development pattern – albeit with different specifics – of something going wrong with the initial privatizations of hydro resources, with the kind of industrial development the addition of new hydroelectric projects was expected to bring, with the timing and size of the provinces' electrical supply expansion, with the national and regional initiatives to link these major projects to a trans-Canada power grid, and with the assumptions about easy exports to the California market. A most blatant example of initial privatization and subsequent buy-back of hydro power rights was the allocation in November 1956 to the Swedish industrial promoter, Axel Wenner-Gren, of hydro development rights and mineral and forestry rights in an area in northern British Columbia that extended over the 'watershed of the Peace River and tributaries above Hudson Hope and the watershed of the Kitcheka River and its tributaries' and the partial watershed of the Parsnip River.[2] Industrial developers like Wenner-Gren promised to establish production facilities and to provide infrastructure. Notwithstanding such promises, particularly during the economic downturn in the early 1980s, some industrial corporations neither kept their industrial development commitments nor their loyalty to British Columbians.

I argue that, because British Columbia's hydro expansions were required to serve the provincial government's 'industrialization by invitation' strategy and the provincial policy favouring continentalism, only temporary dependent industrial development took place. For the most part, semi-processing of natural resources for export continued, and the provincial power surplus became stand-by electricity for transborder use. Five points support this proposition.

First, the nature of government intervention in hydroelectric development was influenced by the economic limitations of a privatized electricity supply. Narrowly profit-oriented utilities in British Columbia, such as BC Electric, which supplied the lucrative Victoria and Vancouver markets, failed to provide the less profitable inputs that the provincial government deemed necessary for regional industrial expansion. Therefore, in 1945, through the involvement of the BC Power Commission in rural industrial electrification and then, in 1961, of BC Hydro in power supply to most of the province, the government reversed its earlier policy of leaving most hydroelectric development, transmission, and distribution to the private sector.

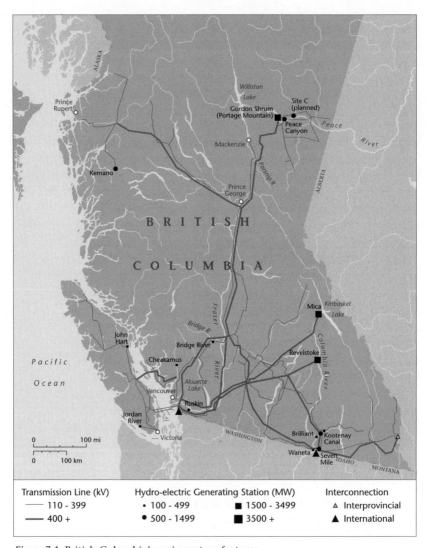

Figure 7.1 British Columbia's main system features

Source: Government of Canada, Department of Energy, Mines and Resources, Energy Sector, *The National Atlas of Canada*, 5th edition, Electricity Generation and Transmission, 1983, MCR4069 and MCR4144; BC Hydro, 1998.

Second, despite the government's taking over hydro utilities in order to advance industrial expansion, the kind of economic growth associated with hydro expansion produced few hydroelectric supply industries, and the use of electricity as an industrial input did not overcome British Columbia's export-oriented, natural-resource-centred growth. In part, this failure occurred because the government's attempt to establish linkages through inviting foreign industry ultimately fostered development that was dependent on imported technology and foreign industrial developers.

Third, because production and planning for industrial expansion was largely determined by forces outside British Columbia, much of the knowledge necessary to plan and build electrical infrastructure for industry at the right time and of the right size was unavailable to the province. Also, in many instances, invited industrialists did not share provincial goals to develop secondary industry, and their lack of regional loyalty contributed substantially to British Columbia's 'unplanned' surplus of electrical capacity. That became obvious when industrial customers requested that BC Hydro build generating facilities that would supply the extra industrial loads they claimed were needed; then these large industrial firms quickly cut back their operations or withdrew their investments in the 1981 recession, which left BC Hydro with substantial surplus capacity.

Fourth, British Columbia's 'unplanned' surplus development was also legitimized by the federal government's national-continental power policy. Through public hearings until the late 1980s, the NEB 'democratically' legitimated this policy: (1) by approving the disposition through export to the United States of British Columbia's surplus industrial electricity and (2) by integrating British Columbia's transmission grid into the US northwest power-dispatching control centres which, until recently, treated British Columbia's electricity primarily as stand-by energy.

Fifth, the 1990s restructuring of BC Hydro is, in part, an outcome of the utility's and the federal and provincial governments' efforts to enhance the marketing of BC electricity in the western US through trade reciprocity involving (1) taking advantage of the special transmission access provision in the 1988 US-Canada FTA, (2) weakening the regulatory power of the NEB, and (3) harmonizing BC Utilities Commission regulations with those of FERC. However, in 1998, anticipating that under conditions of electricity trade reciprocity within a western US regional network US power marketers would 'steal' major customers, BC Hydro began an advertising campaign to retain the loyalty of its industrial clientele.

In many ways, British Columbia failed to learn from the experience of other provinces. By avoiding a national or regional electricity network, and by following its own strategy of building hydro projects for secondary industry and export to the US, British Columbia, in effect, replicated the now familiar development pattern: the privatization of river rights and then its reversal, the dependence on technology and entrepreneurs from abroad, the repetition of industrial load forecasting errors, and the yielding to the allure of access to the US energy market.

Privatization and Its Reversal

Until 1961, with the exception of the BC Power Commission (established in 1945 and supplying electricity to industrial and rural customers), British Columbia's 'free-enterprise' government believed that private-sector firms would diversify economic growth throughout the province. After initially endorsing this policy in the hydroelectric sector in the 1950s, Premier W.A.C. Bennett was the first to reverse it and rapidly brought hydroelectric development under provincial control. An examination of actions by political leaders, public servants, and corporate executives within the Vancouver-based BC Electric Power and Gas Company (BC Electric), the BC Power Commission, and BC Hydro will demonstrate this dramatic shift in hydro policy.

In 1948, Cecil Maiden, the historian for BC Electric, in *Lighted Journey: The Story of the B.C. Electric*, describes how in the first years of hydroelectric development in British Columbia small power companies snapped up rights to the water-power sites near the growing city of Vancouver and how, by 1925, the England-based BC Electric monopolized these sites. The plant at Stave Falls, at the outflow of Stave Lake, had been owned by the Western Canada Power Company; after taking over this firm, BC Electric increased its generating capacity at this site between 1923 and 1925.[3] Water-power rights at Alouette Lake, 140 feet above Stave Lake, had been granted to the Burrard Power Company in 1894; years after its takeover by BC Electric in 1912, the new owners developed the Alouette plant and, in 1928, installed generating equipment.[4] While both Stave Falls and Alouette Lake are about 64 kilometres east of Vancouver, Bridge River is situated near Lillooet, about 250 kilometres north of Vancouver. In 1912, the Bridge River Power Company secured development rights on the Bridge River site; in 1925, BC Electric bought this firm and in 1948 generated the first power from this plant (Appendix 17).[5] 'Because BC Electric was an English Company, interest, or dividends on any of its securities, if sold locally in British Columbia, would be subject to British Income Tax. The difficulty could only be overcome in one way, and on March 4th, 1926, an entirely new holding Company, the British Columbia Electric Power and Gas Company (BC Electric), was incorporated under the British Columbia Companies Act.'[6] From its parent organization, the new company acquired the entire issued share capital of its subsidiaries, including the Bridge River Power Company and several gas companies in Victoria and Vancouver.[7] Following Maiden's convention, I will refer to the new British Columbia-based, private utility simply as BC Electric, although the ownership and control remained British.

BC Electric focused on providing service to large industrial customers, such as the Britannia Mining and Smelting Company near Squamish, and the lucrative markets of Greater Vancouver and lower Vancouver Island; like the other large privately owned power utilities in Ontario, Manitoba, and Québec, it showed less interest in supplying rural and remote industrial customers. Therefore, in 1945, the provincial government established the BC Power Commission and made it responsible for the generation and

Table 7.1

Privatization and its reversal: hydroelectric resources in BC, 1912-97

1912	Bridge River Co. incorporates privatized water-power rights to Bridge River sites.
1925	BC Electric buys Bridge River rights, holds Alouette Lake water-power rights and transfers ownership to its holding company, BC Electric Power & Gas Company.
1926-45	BC Electric serves Vancouver and Victoria.
1945	BC Power Commission, established by the BC government to serve rural and industrial customers, reappropriates smaller hydro resources.
1956	The Wenner-Gren Memorandum of Intent allows the Wenner-Gren BC Development Co. to start exploring and developing the hydro resources in watersheds of the Peace, Kitcheka, and Parsnip Rivers.
1959	BC Energy Board established by provincial government to enforce cooperation between the private BC Electric and the provincial BC Power Commission.
1961	Two-river policy by Gordon Shrum, Chair of the BC Energy Board, recommends constructing hydro dams on the Peace and Columbia Rivers.
1961	The BC provincial government buys back hydro resources from Wenner-Gren's Peace Power Development Co. and BC Electric and forms BC Hydro.
1995-7	The BC Utilities Commission advocates that BC Hydro be vertically deintegrated into three private-sector-like corporate entities responsible for generation, transmission, and distribution. (This type of restructuring has taken place and offers possibilities for privatization.)

---

distribution of electricity to areas unserviced by BC Electric, except for the Kitimat/Kemano area, the South Okanagan, and the East Kootenays (Table 7.1).[8] Throughout the province, the BC Power Commission took over small power utilities in rural and remote areas reliant on resource industries.[9] As a result of the commission's subservience to industry, the costs to industrial customers dropped, for they no longer had to buy their own generating equipment and, instead, received low-cost power, while the operating costs of the BC Power Commission climbed because of its increased capital investments in plants, transmission, and distribution.[10] Overall, privately owned BC Electric had captured the lucrative markets of the major population centres, while the BC Power Commission was left with the expensive and unprofitable task of serving outlying communities.[11] Despite its limited success, the commission was, however, a sign of things to come, for its establishment constituted the first major shift in the production and distribution of hydroelectric power from the private to the public sector.

After August 1959, the Bennett government established the BC Energy Board 'to oversee the production and distribution of all power generated in the province. This measure was really intended to enforce cooperation between the BC Power Commission and BC Electric' in the provision of power.[12] Gordon Shrum, a physicist from the University of British Columbia and a known expediter of major projects, was appointed its first chair.[13] 'The board's first assignment grew out of Bennett's vision of developing the north country and his consequent involvement with Axel Wenner-Gren, a Swedish millionaire industrialist.'[14] Although on a smaller scale, the Bennett government's allocation of northern hydro sites and other natural resources to Wenner-Gren's interests was similar to Newfoundland's concession to Brinco's interests in Labrador.

In November 1956, the government had reserved natural resource rights for the Wenner-Gren group. Watersheds of the Peace River, including its tributaries (above Hudson Hope), the watershed of the Kitcheka River, and the partial watershed of the Parsnip River were identified as proposed areas of development. To secure rights in these areas 'the Wenner-Gren group posted a $500,000 bond and were to be given timber and mineral rights over about 40,000 square miles of northern BC.'[15] Allocation of these rights was conditional upon the Wenner-Gren group making timber surveys, railway surveys, and a 'survey [of] the water resources of the proposed area of development with the object of hydro development.'[16] The public responded angrily to such large transfers; 'controversy increased when Wenner-Gren announced his intention to build a massive dam on the Peace River, and the government announced that it had granted water rights as well as timber and mineral rights in exchange for this construction.'[17]

The Peace River project was stalled until 1959 when a further complicating factor arose: the federal government's preference for proceeding with large power projects on the Columbia River in south-central British Columbia.

> Premier Bennett, who very much wanted the Peace project to proceed, was in a major struggle at the time with the federal government, which wanted to develop the power potential of the Columbia River. This would be an easier project because there were ready markets, not only in B.C. but also south of the border, and financial assistance would be forthcoming from the Bonneville Power Authority in the United States. These points were strongly pressed by Davie Fulton, the federal minister of justice and Prime Minister John Diefenbaker's spokesman on Columbia development ... Bennett objected, however, that if the Columbia went ahead, it would kill the Peace and, with it, his hopes for developing the north.[18]

In 1960, Gordon Shrum was charged with the task of developing policies for the Peace and Columbia Rivers in consultation with the W.A.C. Bennett government. His report was published on 31 July 1961.

Two-River Policy and Takeover

Shrum's *Report on the Columbia and Peace Power Projects* argued that the solution to stagnant development, particularly in the northern part of the province, was to proceed with the Peace River project.[19] He pointed to the following benefits: 'The Peace River project is in a region of low economic development ... The reservoir area will act as an inland waterway which will open up the Trench area for timber removal, [and] mineral exploration ... It is felt by the [BC Energy] Board that an early start on either project [the Columbia or Peace] will provide the Province with a much-needed economic stimulus and contribute to curbing unemployment.'[20] The government expected northern British Columbia's energy infrastructure to provide the appropriate impetus for the diversification of the resource-based economy. Reminiscent of Brinco's abandoning of mineral exploration in Labrador in favour of hydro development, the Wenner-Gren group did not proceed with mineral exploration but, instead, 'concentrated on the hydro-electric power prospects ... They did not find many good sites until they got through the mountains onto the plateau. They located one near Hudson Hope.'[21] The large capacity of potential power from the Peace River could be absorbed only by BC Electric's urban market, but Dal Grauer, chairman of BC Electric, would not cooperate by signing any agreement to purchase power from Wenner-Gren's development in northern British Columbia. In the summer of 1961, Shrum advised Bennett that in order to proceed with both projects he needed to finance the two-river scheme with capital borrowed by the BC government and to take over both private utilities:

> I had to tell him in order to fulfill his northern vision, the government would have to take over both Peace River projects and the B.C. Electric Company. Obviously, if the Peace were to go ahead, there had to be a market for the power: for this to happen, the power must be delivered at the right price, which could not be done by either Wenner-Gren or B.C. Electric because they did not have access to low-cost capital. Only the government could get money at the right interest rate. I told him I was absolutely convinced that there was no other solution, if he wanted the Peace to go ahead. Mr. Bennett reminded me that in the recent election he had opposed a threat by the NDP [New Democratic Party of BC] to take over the B.C. Electric. He did not say much more, but when I walked out of there I had a feeling that he was going to do it.[22]

The appropriation of BC Electric was authorized on 1 August 1961 when the BC Legislature gave unanimous approval to the Power Development Bill,[23] which provided for the acquisition of the BC Electric Company and the Peace River Power Development Company by the province of British

Columbia.[24] After having privatized hydro sites earlier in the twentieth century for the generation of electricity in the private sector, the BC government's establishment of the BC Power Commission was the first direct government intervention in the production of hydroelecticity; the act to acquire the two power companies was the second. In this trend towards public ownership British Columbia followed the pattern of other provinces.

Premier W.A.C. Bennett chose Shrum to become the chairman of the board of BC Electric. An issue of $100 million parity bonds at 5 percent was sold by the government to buy the common shares of the BC Power Corporation (parent company of BC Electric) at $38 each.[25] Major share-holders, 'led by the Power Corporation of Canada,' received $172 million, followed by another $25 million.[26] Since there was no further need for the BC Power Commission as a separate rural utility, it was integrated into the urban BC Electric Company in January 1962 to form BC Hydro.[27] Federal Cabinet records of 9 March 1962 indicate that W.A.C. Bennett planned to sell in the US the extra power generated at US Columbia River plants for storing water in BC and to take 'cash instead of free power ... to get the capital to develop the Peace River Power,' and this he did in September 1964, when his government received $274 million dollars.[28]

The reappropriation of the water-power sites on the Peace River, allo-cated as part of a hydro-linked and resource-based industrial strategy to the Wenner-Gren group, had stirred political controversy. Elmore Philpott, a former Liberal Member of Parliament, in his column in the *Vancouver Sun* (5 August 1961) advocated rising above party politics and supporting this reversal of the earlier privatization mistake: 'By taking over the whole elec-tric works, lock stock and barrel, Premier Bennett's government has done more than reverse its earlier mistake in the Wenner-Gren franchise.'[29] But some sectors of the business community were not so enchanted by the takeover of BC Electric. Howard T. Mitchell, newly elected national vice-president of the Canadian Chamber of Commerce, perceived such a move to be a socialist plot and to have no place in a free-enterprise system. On 21 September 1961 he told the chamber (as recounted by Paddy Sherman in his book, *Bennett*):

> 'You see a still-dazed British Columbian publicly on view trying to collect his thoughts after discovering that his beloved free-enterprise province, without even a vote to favor such a course, has arrived at the position of being the most highly-socialized society in Canada, clearly well in the lead of socialistically backslid Saskatchewan in this respect.' He ticked off the things that disturbed him: the government was the biggest employer, ran its own railway, a giant power system, a transit system, bought and sold natural gas, controlled the operating policies of forestry and mining

companies to a crucial degree, and could punish and reward by patronage on a scale previously unknown in Canada. 'And remember,' he said, 'all this has happened in a decade in the name of free enterprise.'[30]

Mitchell's objection failed to recognize that the government takeover (by paying off shareholders) occurred precisely because the profit-driven private companies did not have the interests of the entire province at heart. BC Electric had not wanted to risk its profits by involving itself in northern development, and so it stayed with its urban markets; and the Wenner-Gren group had wanted to separate profitable hydro development from other resource development in the Peace River region. Now, as a public utility, BC Hydro almost immediately embarked on hydro projects. In this way, the government's hydroelectric equipment needs constituted a new public market, and the establishment of supply industries for hydro projects became possible.

Industrial Linkages

Despite the industrial and technological potential of their substantial population base, British Columbians historically thought they could rely on entrepreneurs from elsewhere to develop their secondary industries.

Dependent Linkages

Dal Grauer, a political economist and former professor of social science at the University of Toronto, was one of the key promoters of the industrial use of electricity within BC Electric.[31] While its executive vice-president, Grauer oversaw the establishment of an industrial development department within the company in 1945, 'its job being to foster industrial growth' and to 'assist in the conversion of wartime industries back to peace time production.'[32] Newspaper advertisements in eastern North America and England stressed the opportunities provided by BC Electric's energy and British Columbia's 'vast natural resources' for the development of 'secondary industries.' By the end of 1945, 594 inquiries had been received as a result of the Industrial Development Department's campaign, and new industrial consumers were planning an annual production of over $12,000,000, with annual payrolls of $3,440,000.[33] Such inquiries by 'new industrial customers' were taken into account when planning the appropriate size and timing for new hydro plants. Such invitations had become a way to import immigrant entrepreneurs or, more often, to solicit the investment of absentee industrial owners and thereby (often with an inadvertent colonial mentality) continue the dependent industrialism which had existed since the time that British Columbia was developed, in Williams's terms, as a lesser economic region of the British Empire.[34] This province also having been a lesser region of the Canadian and US economy, it is not surprising that some British Columbians have even made it a point of pride that most investors in British Columbia are foreign. John Raybould, the author of 'Paper for the World,' published by BC Hydro in 1966, was enthusiastic about the ability of foreign corporations to increase the production capacity of energy-intensive pulp mills:

> The list of major world companies now operating here, or with mills planned or being built, is impressive. Among these companies are the Reed Group of England, Svenska Cellulosa of Sweden, Unso Gutzeit of Finland, Feldmuehl of Germany, East Asiatic of Denmark, Weyerhaeuser and Mead of the U.S., and Mitsubishi and Honshu of Japan. *All the new pulp mills in B.C. will have an initial operating capacity ranging from 250,000-350,000 tons per year. Each mill represents a capital investment from $50 million to $100 million.*[35] [original emphasis]

During the 1980s, solicitations of foreign investment included the availability of travel grants ($2,000 and assistance while in British Columbia) to those firms and associations whose 'products or services contain significant British Columbia content.' The Ministry of Industry and Small Business Development extended special assistance to those who wanted to sell their firms to a company that might want to vertically integrate its operation in British Columbia.[36] However, one reason foreign corporations have been unlikely to develop secondary industries in British Columbia is that they do not want to compete with their own subsidiaries, parent companies, and branches outside the province. The competition from BC-finished products would not be, as BC Hydro's senior economist T.D. Buchanan suggested, in the interest of vertically integrated companies that manufacture their own finished products elsewhere. The combination of foreign ownership in pulp supply and foreign trade policy, according to Buchanan, has retarded the process of diversification: 'Foreign trade policies have hindered the development of a diversified forest products export industry in B.C. Recent announcements of new "captive pulp" capacities are a manifestation of these policies. The pulp customer has every reason to resist B.C. competition in overseas markets for his [sic] finished product. He is in a strong position to do so because he already controls, in part, his B.C. source of supply.'[37] If high-grade finished paper was produced from BC pulp, then it could become a competitive product in the home country of a transnational company sourcing its pulp from British Columbia.

Economic nationalists, when they assumed that the provision of industrial energy would reverse Canada's history of economic dependence, had not foreseen that decisions by utility managers and the political elite to pre-build hydro projects and then canvass industries would only perpetuate Canada's dependence on foreign technology and entrepreneurs. Given the right conditions, utility managers and politicians thought they could continue relying on outside owners to install 'their' manufacturing facilities for them.

Although many hydro projects were built on the Columbia, Peace, Kootenay, and Pend d'Oreille Rivers between the 1930s and 1980s, British Columbians, unlike Quebecers during the Quiet Revolution, have not been associated with highly politicized economic nationalism favouring local industry over imports.[38] Prior to the 1960s, BC utilities bought powerhouse equipment from branch plants in central Canada; subsequently, during the 1960s and 1970s, BC Hydro imported turbines, generators, and other electrical equipment primarily from Japan, Russia, Europe, and the United States (Tables 7.2 and 7.3). The generators and turbines in the two power plants on the Peace River – the G.M. Shrum powerhouse (2,416 MW), and, further downstream, the Peace Canyon powerhouse (500 MW) – were mostly manufactured in Japan and Russia. Japanese manufacturers supplied these units for the G.M. Shrum powerhouse: the turbines were built by Mitsubishi, Toshiba, and Fuji, and five out of the ten generators were built by Toshiba and Fuji (the remaining five were manufactured by Canadian General Electric; Table 7.2). Downstream from the Bennett Dam at the Peace Canyon plant all turbines came from the Leningrad Metal Works, and all generators were manufactured by Mitsubishi (Table 7.2 and Appendix 17).

Since the 1960s, British Columbia's reliance on East Asian technology has increased even more than Manitoba's. Prior to the 1960s, the bulk of generating equipment was purchased either locally, within the province, or – more commonly – from central Canada. The Vancouver Iron Works manufactured six turbines for the Bridge River plants between 1948 and 1957 and two during 1959-60. Installation records show that utilities in

Table 7.2

Turbines and generators in Peace River power stations, 1968-80

Site	Capacity of plank (MW)	Years of turbine installations and manufacturer	Years of generator installations and manufacturer
Gordon M. Shrum	2,416	1968-80 5 MITI, 3 TOBA, 2 FUJI	1968-80 5 CGE, 3 TOBA, 2 FUJI
Peace Canyon	700	1980, 4 LMW	1980, 2 MITI

MITI = Mitsubishi, TOBA = Toshiba, FUJI = Fuji, CGE = Canadian General Electric, LMW = Leningrad Metal Works.
Source: Statistics Canada, *Electric Power Statistics: Inventory of Prime Movers and Electric Generating Equipment*, cat. 57-206, 31 December 1986.

British Columbia from 1930 to 1968 purchased powerhouse equipment, manufactured as import substitutes, in central Canada for plants on rivers throughout British Columbia (Appendix 17). In fact, Canadian General Electric supplied seventeen generators, and its subsidiary, Dominion Engineering Works, manufactured twenty turbines (Table 7.3). The other major equipment manufacturer, American-owned Canadian Westinghouse, manufactured twenty-nine generators (Table 7.3). Since 1968, BC Hydro has become increasingly reliant on technology from East Asia for dams on the Peace River (W.A.C. Bennett, 1968; and Peace Canyon, 1980) and on the Columbia River (Duncan, 1967; Keenleyside, 1968; Mica, 1972; and Revelstoke, 1985), and has purchased the vast majority of core components for its generating stations from Japanese manufacturers. In sum, in British Columbia, '100 percent of ... the turbines and 60 percent of the major generators purchased between 1968 and 1985 were direct imports. Of these suppliers, 90 percent were Japanese firms (Fuji, Hitachi, Toshiba, and Mitsubishi). The only major Canadian[-based] supplier was CGE, which built generators at G.M. Shrum, Kootenay and Mica.'[39]

To encourage local manufacture, BC Hydro allowed a 10 percent preference provision in the 1980s, and possibly earlier, for tenders (under $100,000) from British Columbia, and a 5 percent preference for Canadian tenders.[40] Yet even these premiums for local bids did not substantially increase BC manufacturing content. The hydroelectric industries that were

Table 7.3

Turbines and generators in power stations (over 100 MW) by manufacturer in BC, 1930-84

Manufacturer	Turbines	Generators
Canadian Allis-Chalmers (CAC)	2	
Canadian General Electric (CGE)		17
Canadian Westinghouse (CWES)		29
Dominion Engineering Works (DEW)	20	
English Electric (EE)		2
Fuji (FUJI)	6	6
Hitachi Ltd. (HITA)	2	3
Leningrad Metal Works (LMW)	6	
Mitsubishi (MITI)	12	5
Neyrpic (NEYC)	2	
Pelton Water Wheel (PWW)	3	
Toshiba (TOBA)	3	3
Vancouver Engineering Works (VEW)	2	
Vancouver Iron Works (VIW)	6	

Source: Statistics Canada, *Electric Power Statistics: Inventory of Prime Movers and Electric Generating Equipment*, cat. 57-206, 31 December 1986.

established in British Columbia were minor and temporary; at times, custom fabricating work has been awarded to some Vancouver-area companies, such as Canron, Western Bridge Division, Ebco Industries, and Norelco Industries in Surrey, which built small cranes (less than 25 to 50 tons). Control-panel housings were manufactured in the Lower Mainland by Westinghouse and Federal Pioneer, and printed circuit boards and battery chargers were made locally. Still, when all items manufactured in British Columbia are added up, local inputs received the advantage of only a tiny portion of the overall supply opportunities. Stephen Tang, the senior electrical equipment inspection engineer of BC Hydro's Quality Control and Inspection Department, estimated that 'less than five percent of all mechanical and electrical equipment used at dams in British Columbia originates from British Columbia.'[41] He confirmed that, despite seven decades of building hydroelectric projects in British Columbia, no long-term hydroelectric supply industries have emerged, whereas British Columbia's dependence on Asian technology has increased. The great optimism concerning secondary industrial growth emerging from the development of British Columbia's plentiful natural resources was misplaced. Meanwhile, the decrease of forest resources in the southern coastal regions resulted in the progression of timber extraction inland. The provincial government's preference for large, energy-intensive forest industries for the interior regions of British Columbia accelerated the demise of small sawmills and the increased predominance of pulp mills.

Public Power and the Expansion of Industry
The opening of Crown forests in British Columbia's interior provided the basis of a shift from a sawmill economy to a pulp mill economy. As the profitable varieties of trees were first cut in the most accessible coastal regions and then in the interior, the W.A.C. Bennett government encouraged moving from a diverse small-ownership structure and more integrated forest economy to a transnational corporate structure and pulp and paper economy. T.D. Buchanan, BC Hydro's senior economist in the 1960s, was aware of the extensive removal of coastal forests and recognized that this exploitation would propel expansion to the interior.[42] He predicted that 'the Coast zone will move away from an integrated forest economy producing saw logs, peeler logs and pulp logs towards one in which the production of pulp logs will predominate.'[43] The coastal zone itself, he observed, is moving from 'the age of fir' needed in the sawmill economy to the 'the age of hemlock' used in the pulp economy.[44] This trend became evident from 1946 to 1963 as the pulp industry's consumption of coastal log production increased from 12 percent to 22 percent of the total annual harvest.[45] At the same time, with the encouragement of British Columbia's Social Credit government, large forest companies

'purchased small sawmills to obtain their harvesting rights and reduce competition for the resource.'[46] Thereafter followed the 'construction of lumber and pulp mills in combined operations in central locations,' resulting in a decrease of the number of sawmills 'from over 2000 in the 1950s to 330 in 1978.'[47]

The BC Power Commission assisted these trends by adding extra power plants, such as the John Hart Dam near Nanaimo,[48] and by setting low rates for the pulp and paper industry on Vancouver Island. In fact, many mills gave up generating their own electricity because it was cheaper for them to buy it from the government. By 1956, the power commission's rural electrification program included the development of forest-based industries on Vancouver Island and the interior. In the southern interior of British Columbia the power commission's Whatshan power project, together with the nearby forest resources, was promoted as an industrial magnet. In 1951, the commission's publication, *Progress*, boasted that 'a $75,000,000 industry using Whatshan power and the forest resources of the Arrow Lakes sharply pointed [to] the contention that power, wherever it is developed and made available at reasonable cost, would be utilized by industry for the advantage of the country as a whole.'[49] The commissioners assumed that once a fresh supply of power and resources was made available, industrial use would automatically follow.

As an illustration, C.W. Nash, the BC Power Commission's director of electrical load development, strove to increase the load from pulp and paper companies. In his promotional study, *Pulp and Paper Opportunities in Central British Columbia* (1956), he wrote that the central interior was ready to develop because it 'contained the elements of timber, power, water, and transportation to support the profitable operation of the pulp and paper industry.'[50] He advocated centralization, rather than diversification, of resource ownership and use. In line with the government's preference for large companies, he reasoned that the resources of the smaller mills could be combined and additional forest licences granted for pulp production. The wastes of the 450 small mills operating around Prince George and the 160 small mills near Quesnel would support a 300-ton unbleached kraft pulp mill. And should there be any need for additional timber licences, in Nash's words, 'the Provincial Government [was] anxious to grant a license in this area to pulp and paper interests.' In his solicitations to industry, he clarified how private pulp and paper firms would gain advantages and how the power commission would underwrite the necessary costs and risks for hydro development:

> The construction and maintenance of hydro projects by the Commission is to the advantage of private enterprise. When this is done, the private company will realize, through the Commission, the advantage of low interest

rates, low depreciation rates, and non-liability for corporation income tax. Also the fact that the company's investment is held to a minimum is an added advantage. ... It is suggested that potential power is a tremendous asset that can be developed for industry into low cost electrical energy. This can be done and will be done. When? Depends on industry.[51]

The BC government's power commission increasingly relied on information from industry for its timing of hydroelectric expansions, a practice that, because of industry's basic self-interest, contributed to costly generating-capacity surpluses. Not only low-cost power, but other infrastructural benefits were made available to forest-based industries.

BC Forest Products, Finlay, and the Williston Lake Reservoir

Damming river sites creates reservoirs, such as the Williston Lake reservoir, that have a water surface useful for the transportation of resources by log boom and by barge, or over ice bridges in the winter. To accelerate the development of northern British Columbia, the Bennett government in the 1950s privatized natural resource rights, including hydro-power sites, handing them over to corporate investors. The Wenner-Gren group's assets, for instance, included timber rights received for their intention to build the massive Peace River dam. These holdings were close to the hydro reservoir that Shrum had identified as key for timber transport purposes in his *Report on the Columbia and Peace River Projects* (July 1961).[52] Initially, the Wenner-Gren group incorporated its timber rights as Alexandra Forest Holdings and Alexandra Forest Industries, both of which were soon bought by transnational corporations: 'Alexandra obtained a tree sale harvest licence in the Finlay Public Sustained Yield Unit in 1964 (on the promise of constructing a pulp mill in the region) and further cutting rights the following year in the Peace River drainage system ... In 1967, BCFP [BC Forest Products] bought out Alexandra and its timber holdings. At that time BCFP was dominated by the Argus Corporation, headed by E.P. Taylor, together with the Mead Corporation of the United States, Brunswick of the United States (itself owned 50 percent by Mead and 50 percent by Scott Paper), and Noranda Mines.'[53] The construction of lumber and pulp mills in combined operations, as predicted by Buchanan, was supported by government infrastructure, including rail linkages and the provision of water surface for barge transportation on the Williston Lake reservoir. The Peace River basin became an important transportation infrastructure for both BC Forest Products and Finlay Forest Products (43.7 percent owned by BC Forest Products).[54]

In 1986, BC Forest Products moved its entire cut of 1.7 million cubic metres of timber by water. In the summer, the timber was transported by booming and towing; in the winter, by ice bridge across the reservoir or by ice-breaking barge. BC Forest Products produced kraft pulp, and its three sawmills produced dimensional lumber for sale to the US railway market. The manager of Finlay Forest Industries in Mackenzie (an instant forest company town of 6,500 near Williston Lake) also stressed the importance of towing log booms across the hydro reservoir, claiming that if the reservoir 'was to go dry it would be quite a crisis around here.'[55] But although both companies used the reservoir for transport, only Finlay used BC Hydro electricity. Instead of using the Peace River power as a primary source of energy, BC Forest Products self-generated 80 to 90 percent of its power from hog fuel (wood residue such as sawdust, shavings, and bark generated from processing raw timber), and it aimed for energy self-sufficiency. When asked in an interview whether building the dam and providing the reservoir made any difference to the forest industry, the answer given by a spokesperson in the corporate communications department of BC Forest Products was that it probably 'speeded up the process' of tapping the forest resource in the Peace River area.

Finlay also used the reservoir, but unlike BC Forest Products, Finlay used BC Hydro electricity. Finlay had purchased land from the BC Railway prior to the formation of the reservoir: 'In 1965, the BC government provided a free grant of land to [the] BC Railway for an industrial park on the southeast shores of the proposed Peace reservoir. This was subsequently sold to Alexandra Forest Industries and Finlay Forest Industries at nominal fees for construction of their saw- and pulpmills. The railway completed a spur line into the region in 1966.'[56] Finlay had obtained extensive timber licences in the Trench (in the Peace River region) and used the reservoir's water surface (provided by BC Hydro), rather than expensive logging roads, to transport logs to its mills. The transportation distances from the timber supply to the mills were 75 to 100 miles across the water surface, and, by 1989, timber transport encompassed the full length of the lake. In addition, logging roads were built to truck the higher elevation logs to the reservoir. Finlay Forest Industries operated one sawmill and one energy-intensive groundwood pulp mill; the pulp was shipped to destinations within British Columbia, but also to Japan, India, and Mexico. The company was the largest bulk power customer of BC Hydro in the Central Interior Division (actual consumption 239 GWh in 1981-82, expected consumption 330 GWh in 1982-83).[57] This example demonstrates how the provision of electricity and the related infrastructure (e.g., the reservoir) serves the continuation, if not the intensification, of a staple industry – the forest products industry, which provides semi-processed paper inputs (pulp) for manufacturing into finished paper products elsewhere.

Both companies (BC Forest Products and Finlay) agreed that the increased accessibility to forest resources afforded by the reservoir gave rise to the town of Mackenzie. Its instant development was analyzed by Marchak in *Green Gold* in terms of class structure and population transience. Since both companies were still only 'looking at' the development of a secondary paper production industry, the inhabitants of Mackenzie were employed in the primary production of semi-processed forest products for export.[58] Jobs typical of one-industry towns were created (with their inherent instability and lack of variety) not so much by the electricity supply per se as by the reservoir and privatized rights to forests. More jobs could have been created if secondary, rather than primary, industries had been developed. However, the W.A.C. Bennett government had great confidence in its strategy of inviting development corporations from abroad, believing that they would automatically combine energy with natural resources to build the secondary industrial base for British Columbians.

Peace Power Insufficient to Attract and Transform Industry
Evidence mounted that the existence of a new power plant was not enough to attract industry to a region. In June 1959, the *Vancouver Sun* carried an article indicating that experts were puzzled by Wenner-Gren's Peace River Power Company and the provincial government's claim that 'much of the power will be used at the site by industry "flocking" in.'[59] In contrast to the government's public proclamation, BC Hydro's own Industrial Development Department found that on-site industry was unlikely to develop. In a report entitled *Power Intensive Industries for Peace River at Site Power* (13 August 1963), prepared to assess the possibilities of primary metal industries using power at site, the department discovered what had been known since transmission lines where installed at Niagara Falls in 1896: it is generally cheaper to move electricity from the powerhouse to industrial sites than to move materials to the source of electricity.[60] For example, in a detailed study for the BC Electric Company in 1961, Ivan Bloch and Associates of Portland, Oregon, 'concluded that in order to attract new aluminum producers to BC it would be necessary to offer 2½ mill power at tidewater, and concessions such as tax holiday [sic] and subsidized sites. Power would have to be given away free at Portage Mountain [former name of Peace River site] if a saving in power cost had to be used to offset extra transportation costs resulting from a site at Portage Mountain.'[61]

Even cheap on-site power and subsidized rail transportation would not suffice to attract aluminum and other metal smelters to the Peace River area. The study concluded that 'resource industries such as pulp and paper mills, processors of local minerals, etc., will locate in the area primarily because of the resource and not primarily because of low cost power.'[62] An exception to these findings is offered by Mary Doreen Taylor, who notes that 'the availability of large-scale, low-cost hydro-electric power motivated the establishment of the aluminium smelter at Kitimat' and helped Cominco's development of the large smelting and refining complex at Trail.[63]

In both these cases transportation of minerals was optimal (extracted near Trail and available at tidewater in Kitimat) in relation to the power sites, owned in each case by the smelting companies.

As seen so far, energy-intensive industrial users, whether they generate their own electricity or buy from BC Hydro, show little interest in developing secondary industry. The combination of manufacturing requirements and cheap energy did little to create diversity based on natural resource extraction: the provision that wood be manufactured within British Columbia into finished end-products (such as fine-grade paper, furniture, doors, and window sashes) has existed since 1891; cheap hydroelectricity to transform natural resource products was added during the 1960s; and the New Democratic government attempted to impose manufacturing conditions on mineral extraction during the mid-1970s.[64] In spite of all this legislation, the private sector brought little secondary industrial development based on natural resource growth to British Columbia. As we have already seen in the other provinces, despite the installation of new hydro projects between 1965 and 1985, when Revelstoke, the last dam, was completed, forest-sector enterprises tripled the consumption of this refined form of energy in the pulp and paper industry (Appendix 18) rather than diversifying production within the province. Most semi-processed products of British Columbia's resource-based industries are shipped to other countries as supplies for finished goods production.

For instance, the government's own publication, *British Columbia Facts and Statistics* (1956, 1964, 1983), indicates that wood products are largely shipped to the US, the United Kingdom, Europe, and Japan (in 1956, 71 percent went to the US, 17 percent to the UK; in 1964, 70 percent went to the US, 12 percent to the UK; and, in 1983, 57 percent went to the US, 16.6 percent to Europe, and 14 percent to Japan). State policies favouring large corporations, which direct export-oriented commodity extraction from outside the province, have exacerbated provincial reliance on primary industry.

Out of twenty major manufacturing groups categorized by Statistics Canada, three – wood industries, paper and allied industries, and chemical and chemical products industries – purchase, on average, 85 percent of the total electricity used in manufacturing in British Columbia. A historical review of the purchase statistics of the three major industrial sectors for the hydro expansion period from 1962 to 1983 indicates that the variation of purchases, as a percentage of the total manufacturing, has stayed within 81 to 87 percent (Table 7.4 and Appendix 18). Products from these industries, such as pulp and chemicals, are primarily semi-processed and are used for inter-plant transfer for more sophisticated manufacturing activities elsewhere. Government and corporate policies not only failed to diversify production and ownership, but also contributed to an electricity surplus that was perhaps not as 'unplanned' as BC Hydro administrators claimed it was.

Table 7.4

Consumption of electricity in manufacturing industries in BC, 1962-84

Year	Total consumption in manufacturing (GWh)	Three major sectors			Sum of three sectors (%)
		Wood (%)	Pulp and paper (%)	Chemical (%)	
1962	3,808	12	37	35	84
1963	3,866	13	35	35	83
1964	4,244[a]	12	36	35	84
1965	4,950	12	41	32	85
1966	5,830	11	42	33	86
1967	6,288	12	44	31	87
1968	6,852	11	44	31	86
1969	7,381	12	46	28	86
1970	8,105	12	50	25	87
1971	8,093	13	46	27	86
1972	7,857	16	54	15	85
1973	7,794	19	48	18	85
1974	7,564	19	47	18	84
1975[b]	6,910	20	48	13	81
1976	8,900	19	49	18	86
1977	8,916	21	49	16	86
1978	9,372	21	48	16	85
1979	10,008	21	49	15	85
1980	10,042	21	48	16	85
1981	9,076	20	46	16	82
1982	8,964	20	50	13	83
1983	9,780	20	52	13	85
1984	9,765	21	50	12	85

Source: Statistics Canada, *Consumption of Purchased Fuel and Electricity by the Manufacturing and Primary Industries, 1962-1974*, cat. 57-206; the data for this period include BC, Yukon, and the Northwest Territories.

[a] The totals for the twenty manufacturing categories had to be adjusted for the reporting of high anomalous purchases (1964-76) in the primary metals industry. The anomaly is probably due to Cominco's report of the purchase of power from its own utility, West Kootenay Power & Light. Therefore, the total was adjusted by allowing only 200 GWh for the primary metals industry. Statistics Canada uses a more comprehensive classification of twenty manufacturing industries, which includes the manufacture of consumer goods. To assess the use of electricity in manufacturing, the four major industrial manufacturing categories are examined.

[b] The period 1975-84 originates from cat. 57-208, which is for BC purchases only.

Roots of the 'Unplanned' Surplus

Gordon Shrum was in charge of expansion policies and the construction of the Peace River project during the 1960s, first as chairman of the BC Energy Board and later as chairman of BC Hydro.[65] As early as 1961, he predicted that even if the Peace River plants were phased into service over a number of years, a 'very large market outside the Province' would have to be found for the power. Nearby, the US Pacific Northwest market did not need power for at least a decade. The other market, California, posed transmission access difficulties. The *Report on the Columbia and Peace Power Projects* by the British Columbia Energy Board (1961), a key document for proceeding with the BC Electric takeover, clearly identifies the export-dependent nature of the two-river policy:

Dependence upon Export of Power
Since the minimum efficient development of either the Peace [River] or the Columbia [River] will provide more power than British Columbia can absorb in the initial years of the project, our consultants have pointed out that it is not possible to start the two concurrently or even phase them [in] without finding a very large market outside the Province for power which British Columbia cannot currently absorb. The low cost of thermal power in Alberta makes development of a market there uncertain; therefore, the only possibility lies in the Pacific Northwest or California ... The Pacific Northwest areas of the United States may not require substantial amounts of power from other areas until the early years of 1970 [sic], but that an immediate market for the large blocks of firm power might be arranged in California.[66]

The two-river policy also pointed to precedents for the sale of power across international boundaries and to ways to finance the required surplus capacity with foreign capital. In his two-river report, Shrum cited Trygve Lie, the chairman of Norway's National Energy Council, who thought that more power plants could be financed with loans and paid for through exports: 'With foreign loans more power could be developed and the extra could be exported until the loans have been repaid.'[67] Such a policy, however, can lead to overinvestment of US portfolio capital in electricity infrastructure. Fearing this outcome, not all public utility planners or regulators were in accord with Shrum's recommendations. H.F. Angus, the chairman of the Public Utilities Commission, disassociated himself from the preference for additional public power from the Peace River since 'it takes no account of the contingent risk to the taxpayer attaching to an investment of this character, a risk so clearly evidenced in the history of our publicly owned railways.'[68] Angus was not the only one who questioned the wisdom of starting the Peace River project for export (as will be discussed below); others soon saw the problems of planning projects for expected industry.

The overinvestment that did, in fact, result was partly planned and partly the result of the failure of the 'industrialization-by-invitation' strategy. This latter policy continued during the 1960s, 1970s, and 1980s when extra load allowances were made for the production facilities that were promised by invited industrialists:[69]

[In 1961] a large allowance was added to the base provincial forecast for new power intensive loads coming into the province. There was no attempt made in 1961 to allocate the new load to any particular area.

[In 1970] substantial allowances were made for industrial loads in individual E.S.A.'s [the eighteen Electric Service Areas into which B.C. is divided] over and above those already included in B.C. Hydro, Alcan, and Cominco projections.[70]

By the early 1980s, the 'unplanned surplus' – the equivalent of the entire capacity of the $2 billion Revelstoke dam (1,843 MW) – became evident, and BC Hydro was called to account for its planning decisions. The statement of Robert Bonner, BC Hydro's chairman at the time, was reported as follows: 'Hydro was merely responding to what its corporate customers thought they would need. He said those industries could not have forecast plant closings and the general slump in the economy that resulted in a drop in the demand of electricity.'[71] Similarly, Norman Olson, the president of BC Hydro, acknowledged the lack of demand for power from the $2 billion Revelstoke dam. A *Globe and Mail* article reported his explanation as being that 'the earlier forecast was based at least in part on firm inquiries [i.e., commitments, but not contracts] from companies wishing to set up or expand their businesses and their use of electricity in British Columbia.' He asked rhetorically: '"Where are they now?"'[72]

The lack of promised industrial consumption was offered by both the chairman and the former president of BC Hydro as a cause of the 'unplanned surplus.' In October 1982, this problem of expanding hydro capacity for industrial requirements became a major focus at the BC Utilities Commission hearings that were reviewing the application to construct a third dam at Site C on the Peace River. J.R. Brassington, in charge of BC Hydro's Special Power Contracts Department, which has dealt with bulk inquiries since 1970, presented tables concerning thirty-four companies. All had been planning to expand their existing operations or start new accounts. To verify these tables, R. Overstall, a geologist and researcher on behalf of the Society for the Promotion of Environmental Conservation (SPEC), independently surveyed the inquiring companies. Both Overstall and Brassington concurred on many of their findings, particularly on the sudden economic downturn and sharp increase in 'contract' cancellations that occurred in 1981 and 1982. Within a four-month period (October 1981 to February 1982), the following Canadian and foreign-controlled companies either cancelled or reduced plans for establishing new industrial plants or expansions of existing facilities in major industrial sectors: Ferro-Silicon Smelter: SKW Canada Ltd., Mitsui & Co. (Canada) Ltd.; Coal Mining Expansion: Fording Coal Ltd., B.P. Canada Ltd., Petro-Canada Coal

Division; Lead and Zinc: Cyprus Anvil Mining Corporation; Copper: Teck Corporation, Kennco Exploration; Pulp and Paper: Eurocan Pulp & Paper Ltd., Canadian Cellulose Company Ltd., Doman Industries Ltd; and for pulp mill Chemicals: Canadian Occidental Petroleum.[73]

There is no evidence that these companies were held responsible for not purchasing the electricity for which they had made firm inquiries. Overstall contacted the companies that had made industrial power inquiries (requests) to BC Hydro – totalling 85 percent of the industrial capacity (1,349 MW) requested – but found that 49 percent of them no longer needed the power (equivalent to 667 MW).[74] Upon revising and reconfirming the inquiries made by potential industrial customers, BC Hydro found that its estimates were now less than half of the original forecast.

The preceding discussion indicates the inability of the province to plan and produce the appropriate size of its hydroelectric infrastructure when such planning is based on corporate planning information provided by industry. Large foreign corporations have, as it were, flirted with BC Hydro, the Ministry of Energy, and the government, only to withdraw once those agencies committed themselves to the new hydro projects. Abruptly changing corporate decisions in the major sectors of British Columbia's resource-based economy denies BC Hydro, the Ministry of Energy, and the government the power to plan comprehensively for future industrial consumption. By 1981, the pattern of overinvestment was clear. The BC Utilities Commission came to the following conclusions with respect to British Columbia's expectations of industrial loads that never materialized:

> The evidence showed clearly that a major difficulty Hydro faces in developing load [consumption] growth forecasts and hence system plans is in estimating the future new industrial loads. On the one hand, Hydro feels obliged to make provisions for such loads so that electricity supply does not constrain new economic development. On the other hand, since these loads are not committed, it is not clear to what extent they should be taken into account since new facilities might be built in anticipation of loads that never materialize.
>
> This problem relates directly to provincial industrial policy and the extent to which the province wishes to encourage the development of electricity-intensive industry and to gear Hydro's planning to accommodate such development whenever it occurs.[75]

The expected surplus in the period from 1984 to 1990 was projected to decline from 2,139 MW in 1984 to 1,384 MW in 1990 (Table 7.5). BC Hydro did not add new plants during this period. In other words, two decades of constructing hydroelectric facilities had come to an end as the total capacity of its entire system was reported to be the same: 9,769 MW from 1984 to 1990 (Table 7.5). In 1983, the recommendation by the BC Utilities Commission (established three years earlier) to postpone the Site C dam on the Peace River contributed to halting further large-scale expansions.[76] By 1997, neither the Site C nor any other dam had been built by BC Hydro; thus, its installed hydroelectric capacity of 9,746 MW has remained virtually unchanged since the mid-1980s.[77] Following new initiatives since 1989, BC Hydro has bought some power from independent electricity producers and has continued demand management programs such as the Power Smart program (aimed at reducing consumption through energy efficiency in homes, institutions, and businesses). In addition, BC Hydro's mid-1990s strategy of maximizing production from existing facilities, including generating-station replacement (at Stave Falls) and upgrading (Seven Mile and Revelstoke), was predicted to add 700 MW of capacity.[78] In the period from 1998 to 2003, as part of the Columbia River Treaty, half of the power from US plants on the Columbia River (amounting to 1,400 MW) was scheduled to be received by BC Hydro as payment for BC water storage on the Columbia River.[79]

Not for Export?

Four years before the announcement that the Revelstoke dam had increased the 'unplanned' surplus capacity by 1,843 MW (to 2,139 MW; Table 7.5), British Columbians had became suspicious that BC Hydro was building power projects for export rather than for its provincial mandate. This suspicion is evident in the record of proceedings from the NEB hearings in Vancouver in December 1979 and January 1980. BC Hydro applied to have four licences renewed to export power from 1 October 1979 to 30 September 1984 in increased maximum quantities of 25 GWh, 2,000 GWh, 3,000 GWh, and 10,000 GWh.[80] In opposing the licence during the 1979-80 hearings, 'the NDP contended that BC Hydro has no mandate from the people of British Columbia to construct facilities for export.'[81] The Communist Party of British Columbia urged the board to deny the

Table 7.5

BC Hydro capacity, reserve, and surplus, 1984-90 (MW)

	1984-5	1985-6	1986-7	1987-8	1988-9	1989-90
Total capacity	9,769	9,769	9,769	9,769	9,769	9,769
Reserves and surplus	3,409	2,909	2,699	2,549	2,309	2,109
Surplus	2,139	2,159	1,974	1,874	1,584	1,384

Source: National Energy Board Reasons for Decisions: In the Matter of the Application Under the National Energy Board Act of British Columbia Hydro and Power Authority, July 1984, Appendix 4, p. 25, NEB Library, Ottawa.

application on the basis that it was part of 'a new policy referred to as a "continental energy policy."'[82] The United Fishermen and Allied Workers' Union questioned the utility's ability to plan. It pointed to BC Hydro's poor record in forecasting loads and 'presented its own calculation to show that Revelstoke will not be needed during the licence period ... The Fishermen were opposed to any building of facilities for export, but if any are built the cost should be fully recovered in the export price.'[83] Yet the utility refused to admit that it was exceeding its provincial mandate. For example, on the hearing day of 29 January 1980, the NEB members summarized the utility's argument: 'BC Hydro has clearly stated that its generation planning is based solely on its domestic load [consumption].'[84]

The submissions by opponents had little impact on the NEB's decision, since long-term evidence of planning and construction of facilities for export did not fit into the narrow parameters of individual licence applications. Although noting the company's forecasting difficulties, the NEB supported BC Hydro and its application, accepting that BC Hydro had planned for domestic use only and not for export:

Disposition of the Matter of Construction and Operation for Export
– 29 January 1980

The Board has considered all the evidence and argument presented on the question of whether B.C. Hydro has constructed, is constructing, or will operate its system differently because of exports.

B.C. Hydro has clearly stated that its generation planning is based solely on its domestic [provincial] load. Its method of forecasting that load has been the subject of extensive cross-examination in these proceedings. The Board notes that the forecasting of energy demand has been difficult over the past few years due to events on the domestic and international scenes. It appears to the Board that the load forecasts of B.C. Hydro which were made at the time of decisions were taken to add generation to their system are quite high when compared to actual experience ...

Under critical water flow conditions the evidence presented establishes that all projects including Revelstoke must be on stream during the period under consideration to meet domestic load.

Therefore, it is the decision of the Board that B.C. Hydro has not constructed, is not constructing, and is not operating its system differently because of exports.[85]

Either BC Hydro successfully pulled the wool over the NEB's eyes or the NEB was primarily established to facilitate continental network integration (Table 7.6). Back in 1961, Shrum had advocated building for export, and there is ample evidence that BC Hydro, in fact, has planned for export. For instance, the two-river policy in 1961 planned for an 'immediate market ...

arranged in California.' Two 500 kv transmission lines to the US (4,300 MW) were approved by the NEB in 1964 (the first only three years after Shrum's report) and in 1973 and subsequently installed. The NEB also approved export licences in 1970, 1975, 1980, and 1984 for increasing amounts of electricity. But the NEB in making its decisions focused only on the time period of the terms of the licence. The NEB, once it determined that BC Hydro planned only for domestic need, took the position that 'the National Energy Board Act does not confer jurisdiction upon the Board to regulate the generation planning practices of a provincial utility and the manner in which that utility operates its generating system in order to meet its domestic load.'[86]

The NEB did discuss the possible relevance of evidence that BC Hydro had built or operated plants for export, a case that it deemed to be hypothetical. Evidence that showed the utility had built or operated plants for export does not lead to an export licence denial but, at best, an increase in

Table 7.6

Chronology of building for export, BC, 1961-90s

1961	Two-river policy: *Report on the Columbia and Peace Power Projects* by the BC Energy Board states: (1) 'it is not possible to start the two concurrently or even phase them without finding a very large market outside the Province,' and (2) 'an immediate market for the large blocks of power might be arranged in California' (p. 30).
1961	Political rhetoric that California is 'power hungry.'
1962	W.A.C. Bennett rejects federal ownership of national power grid within British Columbia.
1963	Mitchell Sharp's national power policy advocates interconnection with US systems, exports to US, and pre-building of dams in BC.
1964	National Energy Board approves licence for 500-kV power line to US and BC Hydro installs it.
1974	National Energy Board approves licence for second 500-kV power line to US, and BC Hydro installs it. Design capability for both lines totals 4300 MW and is higher than that of Québec.
1970, 1975, 1980, 1984	National Energy Board approves export licences to transmit power to the United States; the size of exports increases.
1980	Representatives of political interest groups intervene at the National Energy Board hearings, challenging BC Hydro's denial that it is building for export.
1980s to 1990s	Access to California market is restricted by the US federal Bonneville Power Authority – US national utilities obtain preference; BC Hydro power exports have stand-by status; Powerex obtains wholesale marketing licence for US.

export price. The export price was the *key* criterion with regard to determining whether exports were in the Canadian interest. 'The Board stated that it would hear evidence to establish whether the generating system of BC Hydro had been constructed, or was being constructed, or would be operated in a different manner to meet export sales as opposed to domestic load within the period of the licences applied for. Only if the evidence established that such was the case would the Board hear any evidence to quantify additional costs in Canada attributable to the exports.'[87] Despite its assertions to the contrary, a review of NEB hearings shows how such public hearings were used to legitimate this process 'democratically.' The NEB allowed the utility to build for intermittent export and 'eventual' domestic use.

The Allure of California

As we have seen, supporters of the two-river policy, including influential managers and administrators in BC's electrical utilities – Shrum and the BC Power Commissioners among them – looked to California as the land of opportunity for the export of BC power. Yet as early as 1961, at least one sceptic anticipated that, given the potential of a northwestern 'power glut,' the so-called California 'power hunger' might actually be fictitious. In an article entitled 'Power Hunger Claims for U.S. Misleading' (12 August 1961), Gordon Bell, the *Victoria Times* business editor, questioned whether California and the Pacific Northwest really needed to import electricity from British Columbia; and, there being a tendency in British Columbia merely to absorb US rhetoric about hydroelectric development, he suspected that hydro expansion planners in BC adopted, often without enough attention to their economic rationale, the political reasoning of the Kennedy administration, which favoured public power projects. Bell claimed that the notion of 'power hunger' had more to do with politics than economics.[88] He probed further: 'Is it not possible that any power "vacuum" – fictitious or real – will be filled by the U.S. public power [the Bonneville Power Authority], leaving no room for import of foreign power? Certainly it seems doubtful that Interior Secretary Udall and his cohorts have done all their propagandizing on the need for power for the benefit of B.C. ... Mr. Udall is pushing hard for a multi-million dollar western U.S. grid system.'[89] Being similarly endowed with mountain ranges, mountain rivers, and hydro facilities, the US Pacific Northwest had indeed become a competitor with British Columbia for electricity sales to California and hardly needed to import power from elsewhere. Indeed, the federal Bonneville Power Authority (BPA) itself developed an increasing surplus of power, which was revealed in its publication *Issue Alert*. An article entitled 'Selling South: BPA Seeks Ways of Marketing Surplus Power to the Pacific Southwest,' probably published in the early 1980s, indicates that two decades ago the BPA made long-term surplus capacity sales contracts with California, which were due for renewal in 1987 and 1988. Until the mid-1980s, the BPA was trying to negotiate further contracts with California. The US Department of Energy affirmed that the 'BPA would sell up to 2,000 megawatts of surplus capacity for the 20-year term, and would sell firm energy (based on surpluses forecast by BPA) on a rolling five-year basis. Currently it looks as though 1,000 megawatts of surplus firm energy might be available for the first five years.'[90] The British Columbia government and BC Hydro had not anticipated the possibility of US Pacific Northwest surpluses for that length of time.

The development of a 'surplus conflict' put additional strain on BC electricity access to Bonneville transmission lines in Washington and Oregon and inter-linkage with facilities in California. Although the inter-linkage problem had then lasted twenty-five years, BC Hydro's surplus exports continued to occupy a backseat to US states' access to the Western US market throughout the 1980s.[91] In 1986, for instance, BC Hydro was excluded by BPA from the export of firm electricity. As shown in Appendix 19, exports from British Columbia to the US remained volatile in the 1980s, varying from 2 to 21 percent of British Columbia's total generation (1,429 to 12,322 GWh of a total of generation in BC of 60,942 GWh). The BPA allowed only restricted access to the California market for extra-regional utilities, thus relegating BC Hydro's electricity to stand-by vendor status. The US federal authority's access policy states: 'BPA will not provide Assured Delivery to extra-regional utilities' outside of Oregon, Washington, Idaho, a portion of Nevada, Utah, and Wyoming.[92] Not only has access to California's market been uncertain throughout the 1980s, but the BPA has also backtracked on a major agreement with the BC government. In the early 1990s, the province renegotiated the Columbia River Treaty, only to have the BPA cancel the agreement in 1995 because of changing market conditions in the US.

The policy failure of building for export and for anticipated new industry had repercussions for the employees of BC Hydro. In the early 1980s, BC Hydro laid off more than 5,000 engineering and technical employees, and since then has no longer been the premier development company of British Columbia. During the early 1980s, it became an operating company – maintaining existing power plants and electricity distribution – rather than a development company building hydro projects. The BC government has since stopped its capacity pre-building program for 'temporary' export to the US.

1990s Exports and Repercussions

In the 1990s, British Columbia wished to retain or enhance its access to the electricity market in the western US. Decision makers at the federal, provincial, and utility level took several steps to obtain the potential export revenue: federal trade negotiators included a special transmission access provision in the 1988 US-Canada FTA; and provincial officials in the BC Utilities Commission (BCUC) set new terms and conditions for transmitting electricity – terms and conditions that fit the rules of FERC. In the 1990s, utility executives changed BC Hydro's internal structure (by separating administratively supply from transmission of electricity in distinct business units) and its internal mode of operation (by running these divisions for profit and as if they were independently owned) in order to comply with FERC's reciprocity rules for transmission line access and to obtain a US licence for wholesaling electricity in the western US. Soon after, the province's contentious 1997 'win-win' strategy to establish a 'level playing field' for electricity trade reciprocity over soon-to-be-open transmission lines in western North America required a defensive reaction from BC Hydro in 1998, which had to 'reconnect aggressively' with its large BC customers so they would not be 'stolen' by US power marketers.

Since the late 1980s, the Canadian and BC governments had been taking several steps to gain better access to the western US electricity market. To start with, trade negotiators included in the 1988 US-Canada Free Trade Agreement an interstate access provision that the US-government-owned Bonneville Power Administration had to modify its Intertie Policy so that BC Hydro would be treated as favourably as those utilities outside of the US Pacific Northwest with regard to attaining access to the California market. Soon after, Canadian and BC regulatory institutions were required to harmonize their procedures to suit US national electricity policy (see Chapter 2). As already demonstrated in the cases of Manitoba, Québec, and Newfoundland, the regulatory powers of Canada's Calgary-based NEB declined while those of the US's FERC increased. For instance, the NEB reduced export regulation through an Act of Parliament in 1990 (to fit with the electricity trade provisions included in the FTA), while the US Department of Energy Policy Act, 1992, allowed FERC to increase regulation over transmission access and wheeling; this mismatch has serious repercussions for those Canadian transmission-line-owning utilities that wished to export electricity to the US. In contrast to the two preceding decades, in the 1990s the NEB has held no public hearings on British Columbia's electricity exports and appears to remain without enforceable jurisdiction over inter- and intraprovincial electricity transmission or over electricity import from the US. For electricity exports to the US, nonetheless, the NEB granted blanket permits which allowed utilities to engage in electricity exports to the US for up to three years without additional NEB approval.[93] For example, the NEB issued permits effective 1 January 1993 to 30 September 1997 (and later extended them to September 1998) to BC Hydro and Powerex (Columbia Power Exchange Corporation) for firm power export of up to 2,300 MW and firm energy export of up to 6,000 GWh as well as interruptible energy of up to 20,000 GWh.[94] These permits allowed BC Hydro to sell in the US electricity market with less public scrutiny than was the case prior to the 1990s.

As indicated above, in light of the 1980s surplus capacity from the Revelstoke dam, BC Hydro executives restructured the utility turning the province's premier development company into a company primarily concerned with operating its generating plants and its transmission and distribution networks. In addition, it sold surplus electricity to US utilities through the newly established British Columbia Power Exchange Corporation (Powerex).[95] In the 1990s, however, this surplus capacity and the desire to retain and enhance access to the US electricity market contributed to the restructuring of the vertically integrated utility. Coincidentally, the BC Utilities Commission, under Commissioner Dr. Mark Jaccard, conducted a BC electricity market review and, on 11 September 1995, recommended to the lieutenant-governor-in-council that 'all electric utilities in British Columbia owning generation and transmission assets ... establish fully separate operating divisions for these [assets such as generating plants, transmission systems], with elimination of cross-subsidies with other divisions of the utility.'[96] Following a similar administration deintegration scheme, BC Hydro established major business units, such as the BC Hydro Power Supply (which supplies power from the utility's generating facilities at dams throughout the province), the BC Hydro Transmission and Distribution (which transports for a fee electricity within the BC network of transmission lines), and Powerex (which trades and exports electricity).

In other countries, such as England, the US, and Australia, similar neoliberal utility deintegration policies led to privatization and the subsequent transnational ownership of formerly integrated utility assets. Utility analysts such as Marjorie Cohen are concerned that such deintegration into business units and increased reliance on independent power producers can lead to the privatization of generating and transmission facilities. Cohen informs us that the World Bank, as part of its structural adjustment policies, has an interest in 'the incremental approach to privatization of electrical utilities' through a strategy of supporting private power producers – even if no shortages of electricity exist and prices are as low as they are in British Columbia.[97] From such modest beginnings, so the World Bank policy paper argues, independent power producers 'can lay the groundwork for an upheaval ending in private ownership of much of the generation, transmission and distribution sector of utilities.'[98] For example, Cohen examined the ownership changes in privatizations of this kind and finds that in England this phenomenon 'rapidly led to the US ownership of two-thirds of the regional electricity companies,' and that 'most recently in the Australian state of Victoria the transmission system network was bought by the GPU of Parsippany, New Jersey, for $2.72 billion Australian dollars ($2 billion US).'[99] In 1998, the BC government appears to have put such ownership changes on hold; although US electricity policy requires that BC Hydro operate its transmission system to match FERC rules, the restructuring of BC Hydro (except for the sale of its gas division a decade earlier) has remained internal to the corporation. However, the administrative and operational autonomy of the transmission system is of particular importance.

In the mid-1990s, for instance, BC Hydro and Powerex pursued a strategy to attain power-marketing status in the western US. In contrast to western Canada, where western power-grid initiatives failed in the early 1980s, in the western US regional transmission groups formed. For Canadian utilities to attain power-marketing status in the US, several steps must be taken: join a regional transmission group (RTG), provide transport of electricity services on open transmission lines to other power suppliers, have such transmission services approved by the US state and Canadian provincial regulatory agencies, and meet all the requirements for obtaining a licence from FERC. The BC Utilities Commission in Vancouver outlined the steps that BC Hydro and Powerex took to become power marketers in the US.

In 1995, BC Hydro and its non-regulated subsidiary, Powerex, joined two RTGs: the Western Regional Transmission Association and the Northwest Regional Transmission Association. RTGs are US-based voluntary organizations, composed of both utility and non-utility members, that agree to adhere to rules for wholesale transmission access. Under the terms of the RTGs, each utility member is required to allow other utility members to use its transmission system in a manner that is comparable to its own use. Included in the bylaws of each of these associations is a requirement that each member utility must file, with its own regulator, wholesale transmission tariffs within 180 days of the bylaws of the RTG being approved by FERC. Accordingly, BC Hydro was required to file its application with the BCUC by 13 November 1995.[100]

A few days before this deadline, on 10 November 1995, BC Hydro filed an application with the BC Utilities Commission to provide wholesale transmission service on BC Hydro's grid.[101] Its 'unbundled electric transmission service' is FERC's centrepiece for establishing a market in which electricity can become a freely traded commodity and in which wholesale customers can shop for competitively priced power. To a large degree following FERC's requirements, the BC Utilities Commission approved BC Hydro's wholesale transmission service application on an interim basis, commencing on 31 January 1996.[102] The potential beneficiaries of a more open transmission system, according to the BC Utilities Commission, were expected to be: '(1) West Kootenay Power Ltd. and the municipalities it now serves as well as the municipal utility of New Westminster; (2) Independent Power Producers who wish to sell to municipal utilities and to the export market; (3) Alberta electricity producers who wish to transmit power through British Columbia; (4) B.C. Hydro's Power Supply Business Unit; and (5) parties [e.g., power marketers] wishing to import electricity into British Columbia.'[103] One way to verify that a utility's open transmission system actually provides an 'unbundled electric transmission service' is to examine whether the tariff such a utility charges to transport the electricity of other power suppliers matches the tariff it charges its own power supply subsidiary. The terms approved by BCUC in providing transmission service to wholesale customers and power suppliers in British Columbia resulted in difficulties because these terms did not meet FERC's 'equal access' requirements when BC Hydro and Powerex tried to obtain a power-marketing licence from FERC.

In February 1997, FERC rejected the initial application.[104] The *Vancouver Sun* reported that 'FERC [had] ruled Powerex couldn't sell power directly to wholesale customers in the US because it hadn't shown that the wholesale transmission terms in this province were as good as [those] south of the border.'[105] For instance, under the BCUC pricing system, power companies in BC were to pay 'significantly higher prices for sending power long distances over transmission lines than for short hauls' to discourage BC Hydro from building costly long-distance transmission lines to potential northern power sources, and, instead, to encourage new private power production with short transmission paths to urban and other major consumers.[106] After BC Hydro's initial application to FERC containing such pricing strategies was rejected, the utility returned to the BCUC to modify the transmission tariffs proposed for BC to coincide with transmission pricing approved by FERC in Washington. These transmission tariffs revised by the BCUC in 1997 eliminated the differential pricing concept.[107]

Reporters, power suppliers, and major BC Hydro customers raised questions about this quick BCUC turnaround on transmission pricing. For instance, Edward Alden, a *Vancouver Sun* reporter, wondered whether the provincial commission 'has knuckled under to the demands of a foreign regulator.'[108] He noted that it soon became evident that 'the new tariff rate – designed to win FERC approval – was modelled exactly after systems used by U.S. utilities.'[109] Industrial customers, independent power producers, and a major wholesale customer criticized BC Hydro and the BCUC for loss of credibility, loss of provincial autonomy over terms and conditions for transporting electricity in the future, and the potential loss of public investment in an electrical infrastructure devalued by hastily-agreed-to encumbrances. Similarly concerned, the City of New Westminster, one of BC Hydro's wholesale customers, argued that 'the urgency displayed by BC

Hydro trying to ensure export revenue in the immediate term may be short-sighted when considering the billions of dollars of investment made [in the public hydro system] by the citizens of BC which need to be protected in the longer term.'[110] Representatives of the independent power producers of British Columbia argued that 'B.C. Hydro has developed a major credibility problem in matters relating to its transmission system' and that its request for a quick interim ruling was 'an attempt to circumvent' the previous BCUC decision supporting distance-based transmission pricing.[111] BC Hydro's biggest industrial customers, including forest, mining, and chemical industries, warned the BCUC in a 30 June 1997 letter that 'it will be giving up all semblance of authority over the terms and conditions for wholesale, and possibly retail [selling of electricity to industrial, commercial, and residential customers], access [to the transmission system] in the future.'[112]

Such criticism, expressed by some of its largest customers, did not deter Brian Smith, chair of BC Hydro, from adjusting the transmission service tariffs to the US requirements. During the summer of 1997, Smith went to Washington almost weekly to talk 'to FERC Commissioners to determine exactly what was required in the refiled tariff.'[113] Agreeing to Washington's regulatory requirements, Smith said 'we met every test' required by FERC to obtain a power-marketing licence.[114] Soon after, in September 1997, FERC reversed its earlier decision and granted BC Hydro's export and trading subsidiary, Powerex, the authority to sell power to US wholesale customers. Powerex was now allowed to deliver market-priced power from British Columbia directly to US wholesale customers and to purchase electricity from US power suppliers for resale.[115] After FERC announced its decision, Brian Smith suggested that 'the ruling allows us to sell some electricity into the U.S. immediately.'[116] However, as Mark Jaccard cautioned, even though the FERC licence has been granted to Powerex, transmission accessibility also depends on the resolution of 'the question of what individual [regulatory institutions in the] states will do' to resolve such issues.[117] For instance, he warned, 'California regulators may well demand full retail access in BC' as the price for allowing BC Hydro to export to industrial customers in California.[118] While exports to the US constituted only about 12 percent of the electricity generated in British Columbia in 1996, they influenced more than domestic factors the restructuring of BC Hydro.

The Threat of Power Marketers

As a market access strategy, Powerex has become a power marketer in the US, and, as noted, BC Hydro has deintegrated its corporate structure by operating its transmission system to suit the requirements of a changing US electricity market. Making all these changes to retain and enhance access to the US electricity market, BC Hydro has assumed it has struck a balance between ensuring potential export revenues and its internal restructuring, short of corporate divestment, to meet US regulations. That an apparent balance has been struck is evident in the 'win-win' claims made by public officials and US regulators.

In his ruling, James Hoecker, chair of FERC, welcomed BC Hydro to open cross-border trade and declared it 'good for consumers' and a 'win-win' situation for the US and Canada.[119] British Columbia premier Glen Clark praised British Columbia's negotiating skills, saying that 'it also demonstrates that when the parties involved in cross-border issues sit down together, the end result can be a win-win for everyone.'[120] After all, by first obtaining BCUC approval to provide wholesale transmission services, and then US federal approval to trade wholesale electricity in the US, BC Hydro appeared to attain a measure of reciprocity of access in a transborder, regional power grid.

Soon, however, BC Hydro realized that with reciprocity it had exposed itself to a new potential threat: US marketers taking over part of its market in British Columbia. In return for Powerex's having been given access to the western US market through US transmission lines, large US electricity supply companies also have to be given access to wholesale customers in British Columbia, such as the City of New Westminster. This possibility called for a new defensive strategy to secure customer loyalty in British Columbia so that BC Hydro would not face its own demise. Smith explained that US power companies transport BC Hydro electricity for a fee; in return, BC Hydro is required to transport US power suppliers' electricity for a fee. In a CBC radio interview on 25 March 1998 Smith stated: 'The American Federal Energy Regulatory Commission gave us a licence to sell our power directly into California; and, so we use their wires [transmission lines], the wires of other companies. Then, we pay for that. But they can do the same with us. What you get you have to give.'[121] Although Smith reassured the public that BC Hydro is a low-cost electricity supplier in a continentalized market, the 'win-win' confidence from the fall of 1997 changed to a defensive $6 million advertising campaign to retain customer loyalty in the face of competition should the transmission system be opened further to transport electricity for retail sale to industrial, commercial, and domestic consumers.

In its 1998 advertising brochures, Powerex is defined as a marketer of energy products that provides consumer benefits to BC residents: 'B.C. Hydro, through Powerex, its power marketing arm, markets electricity to customers in other parts of western Canada and the United States in ways that make maximum use of the flexibility of our hydro system, which helps keep B.C.'s electricity rates among the lowest in North America and contributes to the annual dividend B.C. Hydro pays to the province.'[122] That, however, shows only one side of the Northwest electricity market.[123]

BC Hydro chairman Brian Smith shows another; he knows and fears that US power marketers can unbalance the 'win-win' situation by eroding BC Hydro's customer base:

> I think our major threats are power marketers. These are companies that don't have to pay for dams, don't have to pay for transmission, but they buy and sell power, and they sign up customers and woo them away from the power company at a lower rate. They make margins on the difference on what productions they bought. And these are very successful companies in the United States. The largest and most successful is Enron. It has about 30 per cent of the business in the United States. They are aggressively moving now into Ontario, and they are looking at British Columbia. If I was to sit and just wait for all this to happen, I would not be doing my job with this utility.[124]

Defending the utility's aggressive advertising campaign to solicit customer loyalty, Smith said that 'everyone is to know that this utility [BC Hydro] is changing its mindset; it's gonna be customer-focused. It's gonna do that in large ways and small ways. It's gonna pay a lot more attention to people's needs. Individual attention to customers, large and small. And that's gonna be the focus the next two years.'[125] The ad campaign has been undertaken, in part, to prevent the poaching of BC Hydro customers: the advertisements are geared to reconnect aggressively with customers.

> Because competition is already on us at the wholesale level. We have wholesale competition. And at the retail level, it looks like it should be just around the corner. So rather than sitting and wrapping ourselves in the security of our monopoly position, as other industries did. Like the gas industry in Eastern Canada, and waking up one morning, finding out that somebody has stolen away our customers. We are going to jump the mark and aggressively reconnect with customers and become a customer-oriented company.[126]

Thus Smith explains, albeit ungrammatically, how BC Hydro could especially 'reconnect' with its potentially disloyal industrial customers. When asked by Rick Cluff, the host of the CBC's Radio One *Early Edition*, 'What's your sales pitch to the major customers, like the mining companies and forest products companies? What's your pitch to them?' Smith replied:

> The pitch to them? We are gonna assign accounts people to deal with you. We are going to give you more customized service. We are going to give you real time pricing. That means, if you take additional load, over and above what you are taking now, a portion of that we'll give you at straight

market trading cost – even below actual cost. We are gonna really work to make them stay with us. Because they would be the first ones to leave.[127]

Could it be that not only Canadian rivers, but also Canadian utilities, have fallen into excessive servitude to both industry and exports?

Conclusion

British Columbia's hydro expansions did not reduce reliance on resource-based growth, although many government officials predicted that power development would lead to secondary industrial growth. The industrial stagnation following the major exploitation of the coastal forest and water-power resources during the 1950s, and the W.A.C. Bennett government's desire to attract industries to the northern Peace River region during the 1960s, contributed to the acceleration of hydro development in the interior of British Columbia, first under private Swedish control and later under the publicly owned BC Hydro.

Directors of BC Hydro and government officials, in their efforts to advance industrial expansion in the province, have fallen short of fostering the local development of mechanical and electrical supply industries in British Columbia. They selected core electricity production components, such as turbines and generators from manufacturers outside the province, and have made little effort to reverse British Columbia's technological dependence. The government and BC Hydro continued to invite foreign industrial corporations into the province, hoping that they would create secondary industries if BC Hydro discounted their industrial energy and supplied them with land and infrastructure. BC Hydro also provided access to natural resources via the Williston Lake reservoir so that forestry multinationals could establish pulp mills. The BC government favoured such large, foreign-owned, energy-intensive pulp and paper mills over the many small BC-owned-and-operated sawmills. The outcome: the hoped-for hydro-related economic diversification did not occur in British Columbia.

During the expansion phase of BC Hydro's electricity network (from 1961 to 1983), three major industries – wood industries (sawmills), paper and allied industries (pulp and low-grade paper), and chemical and chemical products industries (suppliers to pulp mills) – constituted between 81 and 87 percent of the Crown corporation's manufacturing industry sales (Table 7.4). The provision of cheap and plentiful hydroelectricity did not overcome British Columbia's reliance on raw material exports, and the 'industrialization by invitation' strategy used by the government and the utility intensified external control of BC's economic growth.

Planners, experts, politicians, and industrial development officers in BC Hydro, the provincial government, and industry have shaped the hydro-electric infrastructure of British Columbia. Despite stated efforts to plan the hydro system for the province, they created an 'unplanned' surplus. That surplus occurred in part as a result of the unreliability of the industrial customers' commitments to expand their semi-processed, raw-material production; BC Hydro's inability to predict markets for industrial consumption; the treatment of electricity as an exportable energy product; an industrial optimism associated with big projects; and the absence of clear government industrial policy.

NEB hearings legitimated the early construction of BC Hydro dams and powerhouses by allowing 'exports by default' in spite of numerous public counter-positions. In the case of British Columbia, the NEB showed a clear bias for the transnational western US-BC electricity-grid integration, justified by the rationale that exports supplement utility revenues. British Columbia did not make the use of its provincial surplus a national priority; rather, misled by California's 'power hunger' in the 1980s, its surplus became trapped as continental stand-by energy.

In the 1990s, with the NEB further weakened by FTA-related amendments to the National Energy Board Act, the fortunes of electricity exports to the western US market have depended increasingly on how well BCUC decisions concerning transmission services and BC Hydro's corporate mode of operation have coincided with FERC rules. In British Columbia, American rules have gradually penetrated the export-reliant public utility and the BCUC. Although BC Hydro attained a marketing licence to sell electricity to wholesale customers in the western US, its chairman now fears that US power suppliers could sign up and woo away customers from the corporation. The utility has tried to pre-empt this possibility by 'aggressively reconnecting' with its own BC customers through advertising campaigns.

8
Conclusion:
Review and Resistance

In conclusion, I will briefly restate my argument and reintroduce the relevant theoretical concepts and research findings in the subsections that address national and regional grid initiatives and the five provincial hydro-electric expansions. I have argued that theoretical insights from the new Canadian political economists allow us to understand the failure of national and regional power-grid initiatives. Canada's first national policy, using the trans-Canada railway as an infrastructure to bind the region's economies together, resulted in a branch-plant industrial development in the centre, leaving staple-based extraction in the regions. Therefore, subsequent national policy initiatives, even if they were technologically advantageous, were often perceived to benefit central Canada and so have often been resisted by provinces. The ambition to build national or regional power networks is a case in point. The technology to 'wire' large northern generating facilities over long-distance transmission systems into a national power network became available in the early 1960s. However, as Hughes argues, if diverse owners fail to cooperate, then the mere techno-logical and economic benefits of national or regional power grids are in-sufficient grounds to proceed with them. Political conflicts over territorial autonomy or the distribution of benefits across geographic space, Brodie argues, have a long history in Canada and can undermine national plans. As she points out, the '*where*' issues of development policy affecting regions or provinces when mega-projects are involved give rise to regional political conflict in Canada. Federally sponsored national policies such as energy policies can be perceived by some provincial governments as internal colo-nialism; and, as Duquette suggests, the economic 'escape' from such poli-cies tends to direct industrial strategy into the logic of foreign markets. With Canada's close proximity to the US, this logic has been evident in the provincial pursuit of a decentralized electricity policy that favours north-south rather than east-west integration of transmission systems.

The Diefenbaker, Pearson, and provincial initiatives to build national power grids were all export oriented; however, the Pearson Cabinet's national-continental policy had a much more pronounced continentalist bias than Diefenbaker's. Although some engineering companies may have had a self-serving interest, their technical reports invariably supported the technical feasibility and economic benefits of national power grids, but their recommendations were undermined by political conflict over capturing local industrial benefits and were eclipsed by export agendas. The setting up of an extra-provincial organization with authority over the generation and/or transmission of hydroelectricity was never achieved.

This outcome is due, to a large degree, to the decisions taken by economic and political leaders with regard to advancing hydroelectric schemes and related development. Rather than providing the basic infrastructure for a national power grid and serving as a major force for secondary manufacturing, expansions of provincial hydro systems, when required to serve a policy of 'industrialization by invitation' and 'temporary' exports to the US, at best foster dependent industrial development and staple production; at worst, they result in surplus electricity for continental, rather than for national, use.

National Power Grids

For the first time in history, in the early 1960s, as noted, Canada's capabilities in transmission technology would have allowed the integration of provincial hydro transmission networks into a national power grid – an entity separate from and transcending provincial grids. In 1961, according to Walter Dinsdale, Diefenbaker's minister of northern affairs and national resources, because planning had been almost entirely provincial, and because provinces had tended not to look beyond their own borders, the federal government started taking a more active role in planning the use of electric energy on a national basis. In an effort both to respect provincial rights and to pursue the national advantage (while still allowing exports), Prime Ministers Diefenbaker and Pearson attempted to get provincial premiers to support the first proposed national electricity policy.

In 1961 the Diefenbaker government suggested that the remote northern reserves of water power in Newfoundland, Québec, Manitoba, and British Columbia could more than satisfy local industrial development needs; therefore, large blocks of power could be made available at these sites for transmission to the more industrialized areas of Canada. To transport this power, Cass-Beggs envisaged a national power grid as a 'common carrier that would transmit, for a fee, the energy that would be bought and sold in deals between the utilities.'[1] Along similar lines, Diefenbaker's Cabinet discussed plans for a national power grid extending from Vancouver, British Columbia, to Corner Brook, Newfoundland; invited all premiers to participate; set up a federal-provincial working committee (which included most of Canada's electricity elite); hosted a first ministers meeting in 1962; and engaged engineering consultants to assess the benefits of a national grid as opposed to individual provincial hydro networks. Even though a variety of engineering firms with extensive experience in hydroelectric development found the national power network to be technically feasible and economically beneficial, and even though the benefits of a national power grid far outweighed those of individual provincial systems, political decisions undermined the considerable efforts to establish such a network. Whereas the Diefenbaker Conservative government emphasized a promising new electricity strategy based on northern and peripheral development,[2] with electricity surplus also being directed to the industrial centre of Canada (and leaving room for some exporting), the Pearson Liberal government, which gained power in 1963, adopted an electricity policy more supportive of continental electricity network integration and soon proposed increases in the size and duration of electricity exports. The Diefenbaker policy *considered* exports, as *some* provinces demanded, whereas the Pearson policy *emphasized* the early development of large low-cost power sources with an eye to exporting hydroelecticity to US utilities. As a result of this emphasis, the coordination of plant development of a national power grid became secondary (Table 8.1).

In the 1960s federal and provincial politics, as well as private and public utility interests, undermined the possibilities of a national power network. Provinces followed separate, not national, development strategies that militated against the formation of a national power grid: Ontario pursued the nuclear-power option; British Columbia expected its Columbia River benefits to finance the Peace River development[3] and dreamed of exporting to California; Québec planned the joint development of Churchill Falls with Brinco and planned to become the 'Kuwait of the North,' 'l'Alberta de l'Est,' and, in the twenty-first century, aims to be the 'Energy Hub of North America'; and Manitoba hoped to export power to the US from the Nelson River plants (over federally financed power lines). Québec did not participate in the federal-provincial national power-grid discussions because, as Premier Lesage argued in 1962, hydroelectric development was seen as a provincial responsibility. Furthermore, in keeping with Québec's views on provincial autonomy, wheeling of electricity across provincial transmission lines was denied to Labrador by the province of Québec in 1965. As Mitchell Sharp, Pearson's minister of trade and commerce, told his Cabinet colleagues in 1964 (the year after announcing his national-continental electricity policy), 'the various provinces and private interests were proceeding in a completely unco-ordinated manner in developing energy facilities.'[4] In addition, on the matter of a power grid, unresolved issues remained as to who would own transmission lines: the federal government, the provincial government, or both.

Table 8.1

Descent from national to continental grid integration: chronology of national, regional, and transnational grids

1959-67	National Power Grid – federal government initiatives: 1963 Diefenbaker policy revised by Pearson cabinet to allow exports for up to 25 years.
1974-8	National Power Grid – provincial government initiatives: NEB and cabinet approve export licences; net export to the US increases from 12,959 GWh (5 percent of all generation in Canada) in 1974 to 20,267 GWh in 1978 (6 percent of all generation in Canada); Québec passes BC, Manitoba, and New Brunswick to become second largest exporter of electricity behind Ontario.
1975-6	Eastern Power Grid initiatives and exports: Québec decides to sell surplus hydroelectricity in the US, and Newfoundland plans to reopen the 1969 Churchill Falls contract.
1976-9	Maritime Power Grid initiatives and exports: from 1976 to 1979, New Brunswick's net exports to the US constituted 39 percent of all generation (the province imported 3,900 GWh on average per year); other Maritime provinces did not export power.
1978-82	Western Power Grid initiatives and exports: in 1979 Manitoba's net exports to the US constituted 20 percent of all generation and BC's amounted to 5 percent; neither Saskatchewan nor Alberta exported electricity in 1979.
1984-5	Dismantling of Trudeau National Energy Policy; Mulroney and Reagan discuss continental energy reciprocity. In 1985, net exports amounted to only 9 percent of all electricity generated in Canada.
1988, 1994	Electricity included in FTA and NAFTA provisions; in 1988 Canada's net export of all its electricity was 6 percent; in 1994 it was 8 percent.
1992	US Energy Policy mandates FERC to order open transmission systems in the US.
1996	Some major provincial utilities in Canada deintegrate transmission to form separate business units in keeping with FERC regulations. In 1996, Canada's net export to the US was only 7 percent of all its generation.
1996-7	Major provincial utilities open transmission systems to US power suppliers and attain power-marketing status in US regional grids.
1998	Building Canadian hydroelectric projects primarily for exports proposed by Manitoba, Québec, and Newfoundland.

Source: Export percentages estimated from Energy, Mines and Resources as well as Natural Resources Canada, *Electric Power in Canada*, as published for years 1974 to 1996.

In 1967, by the time the Ingledow report favouring a national power network was finished (one of several reports on the first national power-grid proposal that detailed its substantial technical and economic benefits) the Federal-Provincial Working Committee on Long-Distance Transmission had stated its preference for pursuing regional power grids. In fact, by the time the report was released, two major projects that drew their supply of electricity from the Nelson River (Manitoba) and the Churchill River (Labrador) – in part developed for export to the US rather than to supply the national power grid – had already moved ahead as independent initiatives. Integrating northern projects into separate provincial networks and serving provincial industrial development and exports to the US, rather than planning for national needs, became the norm. These northern projects allowed for anticipated long-term exports through interconnection with US utilities (Table 8.1), encouraged the early development of northern water-power resources, and made east-west interconnections secondary to north-south interconnections.

In 1974, during the height of the oil crisis, the provinces, rather than the federal government, took the second initiative to attempt to develop a national power grid. The Interprovincial Advisory Council on Energy, consisting of provincial government officials, initially proposed an extensive new east-west transmission system from Selkirk, British Columbia, to Elk River, New Brunswick. Many benefits of a national power network, such as the replacement of expensive oil generation with less costly hydro generation, the reduction in capacity reserve, the diverse use of generators in different time zones, and the security of supply, were cited as advantages by this council and by engineering reports. However, in 1978, with issues such as Canada's preference for north-south electricity trade over east-west trade unresolved, with jurisdiction (provincial or federal) over a national grid unresolved, and with the provincial share from the benefits of a national grid unclarified, the advisory council suggested that a national power grid was premature and that regional studies appeared to be more promising.

One of the main reasons for the demise of this second attempt at establishing a national power grid was the unwillingness of provinces to delegate at least some authority to an extra-provincial body. Other issues that were not resolved included the issue of provincial veto rights over national grid projects, jurisdictional ambiguities as to interprovincial trade responsibility, the proposed sharing of costs and benefits by provinces, the coordination of international trade, and the type and location of new generating facilities.[5] Twice the federal government supported the initiation of a national power system, then postponed its development and, instead, recommended a more modest regional-grid integrations.

Regional Grids
Three groups of provinces undertook initiatives to develop regional power systems in the 1970s and 1980s: Québec, Newfoundland, and other Atlantic provinces considered an Eastern grid; New Brunswick, Nova Scotia, Prince

Edward Island, and the federal government negotiated a Maritime grid; and British Columbia, Alberta, Saskatchewan, and Manitoba investigated the potential for a Western grid.[6]

In the mid-1970s, at a time of escalating petroleum prices, most utilities in eastern Canada were isolated from major sources of hydroelectric power and were highly dependent on oil-fired generation. In September 1975 the premiers of Québec and the four Atlantic provinces agreed that the 'Committee on Interconnections between Québec and the Atlantic Provinces' should identify energy surpluses in the region and allocate where hydro power, through a regional power grid, might be used to replace oil-fired generating units. Although other longer-term options (such as the power supply from the Gull Island project on the Lower Churchill and additional transmission facilities) were considered, the short-term allocation of surplus electricity from Québec, using the existing and planned transmission systems from various utilities, was suggested to replace the least efficient oil-fired generating units in the Atlantic provinces. In considering the latter, the committee's study determined that there would be economic and technical benefits at little cost to the participating provinces, yet no such sales were made by Québec to the Maritime provinces. Given the political conflict over the distribution of hydro development benefits that remained in 1976, Newfoundland (a strong supporter of the national power grid and open electricity market in Canada) planned to reopen the Churchill Falls power contract, whereas Québec, less interested in the regional coordination of hydroelectric resource developments, preferred to sell surplus power to the US. All provinces participating in the Eastern grid initiative decided to take part in the second initiative to develop a national power grid, but some were also involved in the negotiations to establish a Maritime power grid.

In this regional undertaking the federal government, from 1976 to 1979, began negotiating with New Brunswick, Nova Scotia, and Prince Edward Island to establish the Maritime Energy Corporation, through which they could jointly research, own, plan, and coordinate the entire Maritime bulk-generation and transmission system. However, the Maritime Energy Corporation was not formed because federal-provincial disagreements arose on several fronts: over financing, over whether control of the Point Lepreau nuclear plant in New Brunswick should be federal or provincial, over the priority of developing the Fundy tidal project in Nova Scotia, over ownership of transmission lines interconnecting provinces, and over federal payments for the loss of provincial autonomy in matters of electricity policy. The discussions were further complicated by the election of an anti-nuclear government in Prince Edward Island, by questions of the sharing of risks, and by fears that the region might limit New Brunswick's benefits from its external sales to Québec and the State of Maine.

In 1979, Alberta, Saskatchewan, and Manitoba agreed, with only British Columbia dissenting, to the Western Electric Power Grid study, which considered, among other things, the replacement of coal-fired thermal plants with hydroelectricity from the Limestone project on the Nelson River in northern Manitoba. The study concluded that among other benefits to the utilities in the region were savings amounting to $150 million. The western power-grid initiatives, however, were postponed in 1982 and failed to be realized for several reasons. These included the changing economic conditions in Alberta and Saskatchewan (the 'oil boom' of the 1970s turned into the 'oil bust' in the early 1980s); the election of new provincial governments with reduced commitments to a common power grid; the re-evaluation of differential employment benefits in constructing generating facilities in Manitoba rather than in Saskatchewan or Alberta; the proposed attempt by Manitoba to repatriate its generating capacity (Manitoba would sell 1000 MW to Alberta and 500 MW to Saskatchewan) sooner than had initially been anticipated; the fact that utilities were more sceptical than were provinces about possible regional benefits; and Manitoba Hydro's signing of a contract with Northern States Power of Minneapolis for its Limestone power in 1984.

In the 1970s, the continuing agenda of exporting power in British Columbia from the Peace River, in Manitoba from the Nelson River, and in Québec from Churchill Falls and James Bay; the inability to cooperate on electricity policy by more than two provinces; and the federal government's wish to establish an extra-provincial authority led to the failure to form regional transmission grids. Only when two provinces found electricity trade profitable (not always on equal terms, as the Churchill Falls case shows) did they establish provincial interconnections. Even when Canada experienced an energy crisis in the 1970s (a time when one might have thought it possible to coordinate a domestic electricity policy on a national or regional basis), the federal government, in accordance with its policy of allowing exports to the US while also advocating national and regional power-grid integration, contributed to undermining regional grid formation when several provinces opted for exports (Table 8.1). I have argued that there exists in Canada not only a resistance to perceived 'centralizing' national plans, but also an inability among the provinces to decide cooperatively on the kind of extra-provincial authority that would control a national grid or even a regional power grid involving more than one neighbouring province.

Provincial Hydro Expansions

Thus regional power grids did not form in Canada because provinces preferred to formulate their own hydro-related industrial development and export policies. Federal constitutional arrangements in Canada define

administrative boundaries, and within those boundaries provincial governments are assigned rights to water-power development. Provincial governments, therefore, have the jurisdiction to plan, construct, and operate
hydroelectric systems and jealously guard their authority.

With regard to the major provincial hydro expansions, in this study I
have focused on key water-power developments in Ontario, Labrador,
Québec, Manitoba, and British Columbia, at Niagara Falls, Churchill Falls,
James Bay, Nelson River, and Peace River, respectively. Only select aspects
of these particular hydro resource projects were analyzed in their provincial contexts. Each province followed its own development agenda and
privatized the use of waterfalls, canyons, and rapids for hydroelectric
development. It then reappropriated such rights, promoted industrial
linkages, installed surplus capacities, exported electricity, and integrated,
or planned to integrate, its transmission system regionally with US power
grids.

In Canada, provincial governments, by virtue of Crown ownership,
directly intervened in the development of natural resources, mostly by
allocating resource rights (including water power) to private or public
organizations for economic development. Provinces have had their own
development strategies and have, at different times during the twentieth
century, intervened to produce electricity under government ownership or
have reversed natural resource privatization by buying back such rights (at
times complete with generating and transmission facilities). In several
instances, such reversals of privatization were made to enhance the provision of low-cost and reliable electricity for regional industrial development
within the province and to transform the staple economy. However, dependent industrial development and staple-biased economic growth continued
while the modes and frequency of government interventions changed.

Privatization Reversals
To explain the different modes of government intervention, I have applied
Claus Offe's allocative and productive modes of state intervention to
Canadian provinces. For example, there is a difference between merely
allocating natural resources, including water-power rights, to private
developers (privatization) and directly intervening in the production of
hydroelectricity in order to improve or create private accumulation conditions for a variety of industries.[7] To examine the pattern of reversing the
privatization of allocated water-power rights by purchasing both these
rights and the production and transmission facilities developed by private
utilities, I have employed Offe's 'criterion of non-profitability.'[8] His definition of this concept includes the evaluation criterion used by a potential
entrepreneur to assess whether or not developing an infrastructure is sufficiently profitable. Should, for instance, a regional hydroelectric system

not be profitable to an entrepreneur, yet be necessary for enhancing a region's growth, pressures arise for nationalization so that the government may provide the needed infrastructure. Initially, government intervention to provide 'unprofitable' infrastructure often occurred in sub-provincial regions that the private sector failed to show an interest in developing. To understand why the privatization of hydroelectric development in several provinces ended in renewed government intervention in the form of public ownership, the past practices of provincial governments and private hydro entrepreneurs have to be examined.

During the twentieth century, both private and public firms developed new hydro plants, but the trend has been towards increased government intervention. Since the 1880s, in each province under examination, such changes in ownership of, and control over, hydroelectric dam sites took place after provincial governments had initially leased or sold water power for development by the private sector (Table 8.2). Provincial governments initially relied on private hydro entrepreneurs to promote industrial development, but when private companies failed to provide power for the less lucrative industrial market in smaller towns and rural areas, or failed to take the risk of building dams in advance of industrial demand, the provinces took over the development, production, and transmission of hydroelectric power.

Table 8.2

Privatizations and their reversals: hydroelectric resources in five provinces, 1887-1974

	Privatization of key sites (year)	Reversals of privatization (year)	Name of utility
Ontario	1887 (Niagara Falls)	1906	Ontario Power Commission Ontario Hydro
Québec	1885 (Shawinigan)	1944	Quebec Hydro-Electric Commission
		1963	Hydro-Québec
Labrador	1953 (Churchill)	1974	Churchill Falls (Labrador) Corporation / Newfoundland and Labrador Hydro
Manitoba	1897 (Winnipeg River)	1953 1961	Manitoba Hydro-Electric Board
British Columbia	1956 (Peace River)	1945 1961	BC Power Commission BC Hydro

Source: Privatization reversal information derived from Chapters 3 to 7.

The provincial government in Ontario started providing energy to Ontario's small-town industry in 1906. In Québec and British Columbia, such government interventions started in the 1940s, with rural and remote industrial service being provided by public hydro commissions. Most other provincial takeovers occurred from the 1950s to the 1970s (Table 8.2). At the same time, the federal government encouraged the development of large northern water-power sources in order to draw US portfolio capital into the provinces and to interconnect power networks between provinces and the US.

The first reappropriation of formerly privatized water power took place, as noted, in Ontario. At Niagara Falls, Ontario, two US owners hoped to gain higher profits by monopolizing the water-power rights for fourteen years (1887 to 1901) without constructing hydroelectric plants on the Canadian side of the Falls, thereby slowing down the transition from steam to electrical manufacturing for small-town Ontario manufacturers. When the province had subdivided the power rights, all three private utilities operating at the Falls – the Canadian Niagara Power Company, the Ontario Power Company, and the Electrical Development Company (the only Canadian-owned company) – preferred to export most of the electricity to US industries. Southern Ontario manufacturers pressured the government to set up the public Ontario Power Commission (1906) and demanded the rerouting of Niagara power so that it would flow to Berlin (now Kitchener) instead of Buffalo. Likewise, the Québec government sold or leased hydro sites to private anglophone and US owners from 1887 onwards. Before the 1944 and 1963 interventions, private utilities had not sufficiently equalized prices and had failed to provide industrial energy to strengthen economic growth in many regions of Québec. René Lévesque, minister of natural resources in 1963, emphasized that it was urgent to provide the Abitibi, Gaspé, and Lower Saint Lawrence River regions with low-cost electricity and stressed that, before Hydro-Québec took most of them over in 1963, anglophone utilities employed too few francophones.[9] During 1953, the Newfoundland government granted all remaining hydro rights, including those of the Churchill River, to the private British Newfoundland Corporation (Brinco) in return for an investment guarantee of a minuscule $1.25 million within five-year periods.[10] Brinco sold Churchill Falls power, not to Newfoundland or to new industries based in Labrador, but to Hydro-Québec as part of a long-term contract. In 1974, when Brinco wooed Hydro-Québec once more as a customer for another power deal involving its next project, Lower Churchill (or Gull Island), the Newfoundland government intervened and bought back all water rights from Brinco for $160 million.[11]

Industrial Development

John Dales argues that hydroelectric development is a powerful catalyst in the promotion of diversified manufacturing, particularly if plants are pre-built. Dales, however, overstates his case because building power plants and then promoting low-cost electricity to industrialists in different parts of the world has not proven to be a powerful catalyst for transforming the production of resource-based semi-processed goods into the manufacturing of finished goods. In fact, this strategy, in some instances, has merely accelerated staple production or, at best, contributed to the continuation of dependent industrial development. This pattern has materialized because of past decisions on the part of utilities, provincial governments, and manufacturers. I have focused my inquiry on key aspects of this pattern of development by employing analytical concepts taken from the new political economists. To identify industrial linkage from hydro development, I have used Mel Watkins's notion of linkages (backward, forward, and fiscal) to trace turbine and generator manufacturing (backward linkages); the use of new supplies of electricity in activities such as the energy-intensive, semi-processing of export commodities (incomplete forward linkages); and the use of discounted power in manufacturing ventures (fiscal linkage). A refinement of these linkages was achieved by ascertaining whether or not they were dependently developed (through 'industrialization by invitation' strategies), through import substitution (in branch plants), or through 'nationalistic' procurement practices (by utilities and governments).

The 1960s saw the beginning of massive hydro development in Canada, accompanied by politically propagated myths of technological progress. Glossy annual reports of hydro utilities touted proposed hydroelectric developments as symbols of provincial industrial development; and the dam-building rush from the 1960s to the 1980s created the illusion that secondary manufacturing industries, even if they were not already in place, would soon develop. The publications did not dwell on the fact that this technology was largely imported, as were most of the consumer goods in general, or that even the utilities themselves relied on foreign technology and entrepreneurs. The planners' hopes that hydro development would transform the resource-based economy into a secondary manufacturing economy have not been realized.

The technological dependence on foreign sources in Canada's hydro-mechanical industry had taken root, in fact, as early as 1904. The industrial strategy of early Canadian manufacturers – particularly turbine and generator manufacturers – was an early form of 'import-substitution industrialization' (ISI). It was characterized by unchallenged dependence on foreign production technology and a lack of interest in manufacturing for anywhere beyond the domestic market.[12] Many 'Canadian manufacturers, seeking a cheap and effective shortcut, licensed US industrial processes rather than developing their own.'[13] Utilities would then buy turbines, generators, and other equipment in Canada from branch plants, whose initial choices over such technology for Canadian plants were made in the US, not in Canada.

In Ontario, for instance, three private power companies, as well as the public Ontario Power Commission, relied upon US (General Electric, Westinghouse) and European generating technology at Niagara Falls. In Québec, by 1905, the US immigrant owners of the Shawinigan Power Company had installed US and European generating equipment. Later, the US-owned Dominion Engineering Works in Montréal manufactured the turbines, and Canadian Westinghouse (a branch plant) supplied the majority of generators for the next sixty years (Appendix 20). Even with the 1960s 'buy-Québec' procurement policy, most heavy power-plant equipment was manufactured in Québec branch plants of foreign firms. Thus, though hidden by the philosophy of 'maîtres chez nous,' even here a dependent industrialism continued. Furthermore, Québec's nationalistic manufacturing requirements became part of the Churchill Falls power contract and thereby hindered Newfoundland from benefiting from the growth of supply industries for Churchill Falls. Western Canada's strong reliance on Asian power-plant equipment became evident during the 1970s, as Manitoba Hydro and BC Hydro relied increasingly on generating equipment from Japan and Russia. In the west, only Manitoba Hydro insisted upon local manufacture, but here too Canadian General Electric and other companies in Winnipeg and in The Pas were US-owned (Table 5.1 above). Nevertheless, Manitoba did benefit somewhat more than did British Columbia from the local manufacture of hydro-mechanical components.

Between 1900 and 1986, the US-owned Canadian Allis-Chalmers and Dominion Engineering Works, the British-owned English Electric, and the French and Québec government-owned Marine Industries (which used French technology) supplied the majority of turbines in Canada (Appendix 20). Marine Industries and the US branch plants of Canadian General Electric and Canadian Westinghouse supplied most of the generators in Canada. Since the 1960s, British Columbia and Manitoba, as noted, have also purchased Japanese and Russian generating equipment. Thus, in most of these instances, provincial politicians and utility executives relied on foreign firms and technology to advance the centralized formation of industrial linkages located in Ontario and Québec.

In addition, with increased provincial investment in electric power projects and the quadrupling of generating capacity since the 1960s (Table 8.3) industrialists used the fresh loads of electricity mainly to start or continue primary industries, not secondary industries. In Québec, Newfoundland, Manitoba, and British Columbia, this failure is evident from the fact that after power became available from the new plants, about 80 percent of industrial power in four of these provinces was used to produce semi-manufactured products for export. For example, in Newfoundland, despite the installation of the largest hydro plant in Canada, Churchill Falls, by 1974 (Table 8.4), between 1975 and 1978 three sectors – wood, pulp and

Table 8.3

Generating capacity in five provinces, 1960-96 (MW)

	1960	1970	1980	1985	1990	1995	1996
Ontario	7,109	13,700	25,796	29,932	32,733	36,996	36,996
Québec	8,920	14,047	20,531	26,991	28,873	34,422	34,731
Newfoundland	314	1,248	7,195	7,316	7,462	7,415	7,415
Manitoba	1,043	1,794	4,142	4,142	4,414	5,293	5,293
BC	2,963	5,473	10,525	12,451	12,497	13,058	13,058

Sources: *Electric Power Statistics*, vol. II, Statistics Canada, cat. 57-202, cited by Energy, Mines and Resources, *Electric Power in Canada 1998*, p. 30; National Resources Canada, *Electric Power in Canada 1996*, Table 7.3, p. 72.

Table 8.4

Manufacturing sectors that constitute about 80 percent of the total electricity consumption in five provinces, 1962-84

Province	Time period	Manufacture type	Average % of total manufacturing
Ontario	1963-82	Pulp and paper	81
		Primary metals	
		Transportation equipment	
		Chemical products	
		Food and beverage	
		Non-met. minerals	
		Petroleum and coal products	
Québec	1962-84	Pulp and paper	78
		Chemical products	
		Primary metals	
Newfoundland	1975-8	Pulp and paper	89
		Chemical products	
		Wood industries	
Manitoba	1975-84	Primary metal	80
		Pulp and paper	
		Food and beverage	
		Chemical products	
British Columbia	1962-83	Pulp and paper	85
		Chemical products	
		Wood industries	

Source: Compilation from tables containing electricity consumption by manufacturers, Statistics Canada cat. 57-506, 57-208, and from Chapters 3 to 7 of this book.

paper, and chemical industries – still made up 89 percent of manufacturing consumption. Similarly, in Québec, Manitoba, and British Columbia, the industrial energy produced by the large new hydro stations was used mainly in industries that semi-process pulp and paper, chemicals, primary metals, and wood products. Virtually no diversification in manufacturing use, except for food-and-beverage production, is reported by industrial firms in these provinces. Nevertheless, politicians and utility executives continued to overbuild in the expectation that provincial infrastructure was essential for future industries and for export to the US.

Overbuilding Infrastructure
While offering useful concepts for specifying industrial linkages, the new political economists do not probe deeply enough into the overdevelopment of infrastructure. Some explain overbuilding either as a costly 'mania' (Gordon Laxer) or as representative of the overinvestment of foreign portfolio capital (Naylor).[14] In contrast, I argue that state planning becomes uncoordinated when it tries to match the size and timing of infrastructure with industrial growth that is split into domestic manufacturing, branch-plant manufacturing, and staple production (often by foreign-controlled firms), each following its own trajectory. Furthermore, predicting the infrastructure requirements for industries that may arrive from different parts of the globe is more difficult than estimating industrial growth from within the country. In addition, there were at least three aspects to hydro-related development ideology that contributed to the overbuilding of hydro infrastructure: What Watkins identifies as the 'export mentality' of provincial and federal members of the political elite, the outdated belief that the availability of new hydro energy attracts new industry, and the alleged 'power-hunger' of US utilities, which proved to be much less than assumed, resulting in an uncertain market for power export.

For example, since BC Hydro found itself shut out of any regular access to the California market and had the output of an entire dam in the mid-1980s (Revelstoke, $2 billion) as surplus capacity, it changed from being a plant (and industrial) 'development' company to a plant 'operating' company, incurring massive layoffs in the process. Since the building of dams stopped in British Columbia, the number of employees at BC Hydro has been reduced by 6,376 (from 12,195 in 1980 to 5,819 in 1996).[15] As electricity-generating projects were stopped in central Canada, the careers of 9,868 employees at Ontario Hydro were terminated (their number shrinking from 34,839 in 1992 to 24,971 in 1993), and Hydro Québec has continued to reduce its permanent and temporary workforce (from 27,234 in 1992 to an estimated 19,500 by 2000).[16] Furthermore, overbuilding hydro facilities has caused extra social disruption among Aboriginal peoples who, in consequence, have vigorously opposed this process.

Provincial utilities became increasingly unable to plan and coordinate infrastructural and industrial development, because, when hydroelectricity is developed in advance of a manufacturing market, and manufacturing activities are actively encouraged by seeking foreign investors, the task of estimating the size of plant needed and the timing of the new electricity supplies is uncertain.[17] As long as capital is available, provincial utilities tend to overbuild their hydroelectric infrastructure and, as a result, must rely increasingly on short-term exports. Thus, two factors have contributed to the acceleration of the building of hydro projects: (1) the strategy of soliciting industries to locate within the province (with largely unpredictable outcomes) and (2) the strategy of putting Canadian hydro-power sources into service as part of energy continentalism.

The worst aspect of relying on foreign industrial firms turned out to be that, when provincial utilities built plants in anticipation of firm contracts, mostly foreign-owned companies could, and did, cancel their commitments and, thus, contribute to the increase of provincial surplus capacities. For example, BC Hydro built the Revelstoke dam based on inquiries by industrial customers; when these customers no longer needed the power, the entire capacity of the plant was surplus.[18] Likewise, the oil companies' misleading predictions of fossil-fuel shortages during the 1970s 'oil crisis' contributed to the acceleration of the building of Nelson River plants in Manitoba and the subsequent 'mothballing' of the Limestone project in 1978. The 'industrialization by invitation' policy meant (as the Brinco example shows) that key corporate strategies, including expansions, became integrated into other economies and were decided outside the province and, often, outside the country. As noted, following the attempt of one US branch-plant manager to have 'power at cost' turned off in small Ontario towns and redirected to Goodyear's tire plant, two months later, after having signed a deal following a directive from its US parent, his branch plant cancelled the contract.

The inability to rely on continental energy axes for exports, likely affected by the loss of production clout within the US since the 1970s (in part because of US industry relocating to southern states or shifting production abroad, and of goods from East Asia and Europe [formerly produced in the US] making major inroads in the US market), which contributed to a slowing of industrial demand in the northern US states, led to the accumulation of surplus capacities from prematurely built power plants. The underused capacity becomes obvious in the decline of Canada's average load factor (e.g., the ratio of average demand to peak demand in any given year), which indicates the underuse of its generating equipment. The Department of Natural Resources Canada reported that 'in 1960, the [electric power] industry load factor was 72.3 percent, but by 1980 it had gradually reduced to 65.6 percent. Since then the load factor has varied around 65 percent.'[19] During 1994, the load factor in Canada as a whole was 62.2 percent and varied substantially by province, from 58.7 percent in Ontario, to 58.2 percent in Québec, to 68.7 percent in Newfoundland, to 66.0 percent in Manitoba, to 66.2 percent in British Columbia.[20] Showing a drastic descent in generator capacity use, Québec's load factor of 85.6 percent in 1960 decreased to 58.2 percent in 1994.[21] Overall, the capacity at which generating equipment is used, on average, is about one-third below the highest demand during the year, leaving substantial surplus capacity unused.

Provincial utility officials during the 1970s and 1980s fiercely denied accusations of deliberate early building and oversizing of plants with the purpose of 'encouraging exports and interconnection' with the 'United States Power System,' as Mitchell Sharp's federal policy of 1963 encouraged.[22] In 1980, when asked whether BC Hydro was building for export, Bill Best, vice-president of electrical operations at BC Hydro, responded: 'The construction and scheduling of additional generating facilities is based entirely on servicing forecast requirements in this province ... the National Energy Board of Canada ruled in support of our contention that we have planned and constructed our system only for domestic needs.'[23] However, BC Hydro substantially exceeded domestic needs and built capacity for export, as Shrum's two-river policy had already advocated in the early 1960s. Although since 1984 the utility has built no new generating facility, the British Columbia Utility Commission (a provincial regulatory agency) reported that for the period from 1992 to 1997 BC Hydro, under optimal water conditions, still had a surplus capacity of 2,300 MW available for export to the US.[24]

Until the mid-1980s (when some provinces changed utility mandates), in order to conform to their mandates to build only for provincial requirements and not for export, provincial hydro utilities needed to claim that surplus capacity would eventually be needed domestically.[25] A similar strategy was evident in Québec. At the time of signing the Churchill Falls contract, Hydro-Québec predicted industrial power shortages; then, after the plant was connected to its system, half the plant was available to produce for export. In a similar manner, Bourassa predicted a shortfall if the James Bay plants were not developed, yet by 1984 Hydro-Québec reported that half the James Bay plants were not needed during the winter and that all plants were available for electricity export during the summer. Bourassa's dream that Québec would become 'l'Alberta de l'Est' by strengthening its New England energy axis collapsed when eastern US utilities cancelled $32 billion worth of future contracts.[26] Similarly, surplus capacities of US utilities in the Pacific Northwest, and the restriction by the US federal Bonneville Power Authority on BC power imports, added to BC Hydro's capacity surplus in the 1980s and limited its exports.

I have argued that permitting the export of electricity to the US has contributed to undermining the development of a national power grid and that subsequent initiatives to secure access to the US electricity market have increasingly turned electricity into a continentalized commodity. As has become evident, in order to retain and enhance access to the US electricity market in the mid-1990s, Canada has included electricity in the provisions of the free trade agreements and has restructured major utilities to remain in keeping with the demands of these agreements and US regulatory requirements.

Continental Integration

The politics and policies of foreign countries, particularly those of Britain and the US, have influenced the formation of Canadian policy and the well-being of Canada's economy. For instance, some of the 'free-market' orientations of Britain and the US have been adopted by Canada and have culminated in the Canada-US FTA. Increasingly, according to Brodie, as such economic trading blocks are formed, the relative fortunes of regions depend on how they integrate as zones within those larger economies.

In 1988, Canada's Department of Energy, Mines and Resources argued for including electricity within the provisions of the FTA for several reasons: because electricity was not covered in international trade agreements, because of fears that US protectionists might impose restrictions on electricity imports from Canada,[27] because of the economic benefits that replacing non-renewable coal- and oil-based electricity in the US with renewable Canadian hydroelectricity would bring,[28] and because a number of provinces (Québec and British Columbia in particular) would be assisted in their efforts to export electricity.[29] In its rush to sign the FTA, Canada appears not have been as careful as was Mexico when it signed NAFTA. Canadian and Mexican electricity provisions in NAFTA, according to a study done for the C.D. Howe Institute by André Plourde, differ significantly. As stated in Chapter 2, Canada allows foreign ownership in the transmission, distribution, and sales of electricity to the public, allows individual provincial utilities to make NEB pre-approved export arrangements with US utilities, and allows US importers to obtain accrual of proportional access rights to Canadian electricity supply. Presumably 'promoting more egalitarian forms of trading arrangements'[30] and learning from its struggle with foreign ownership in the energy sector, Mexico has exempted its electricity sector from such provisions.

Orders to Open Transmission Lines

As stated in Chapter 1, Grinspun and Kreklewich inform us that the FTA and NAFTA 'can serve as "conditioning frameworks" to promote continental integration and the consolidation of neo-liberal restructuring.' Thus the weaker nation in a free trade arrangement may readily accept the stronger nation's policies, administrative procedures, and rules. For example in Canada, Manitoba, British Columbia, and Québec have adopted the deintegration and transmission regulation policy of the US for their own provincial hydroelectric systems in order to be able to market electricity in the US.

This process began in October 1992 when the US Congress supported the proposal that wholesale electricity markets in the US be deregulated. It did this by ordering vertically integrated electricity monopolies (owning generation, transmission, and distribution facilities) to deintegrate and to open their transmission systems to other power suppliers. Although less than 10 percent of the total electricity generated in Canada is exported to the US, provincial utilities have become very concerned over the *threats and opportunities* posed by US deregulation of wholesale markets and, thus, for Canadian electricity exports.[31] Therefore, in 1996 and 1997, in order to retain and increase their access to the US market, several provincial utilities restructured their formerly integrated administrative organization and opened up their transmission systems.

The US energy policy has also influenced the regulation of utilities in Canada.[32] For instance, as shown in Chapter 2, FERC's orders prohibit 'owners and operators of monopoly transmission facilities from denying transmission access, or offering only inferior access, to other power suppliers in order to favour the monopolists' own generation.'[33] This required 'unbundling' electric transmission service from generation. In compliance with such orders, public utilities in British Columbia, Québec, and Manitoba have restructured their utilities and set up distinct business units for the transmission and generation of electricity. Canadian utilities, unlike their counterparts in the US, are owned primarily by the public and have provided their consumers with relatively low-priced electricity. Nevertheless, the same FERC rules meant for vertically integrated private monopolies that sell overpriced electricity, and the same deintegration procedures, are now being applied to publicly owned transmission services in Canada. The NEB has only weak or no jurisdiction over electricity imports from the US or the wheeling of electricity through provincial transmission lines, and in this area several provinces, including Québec, have adopted FERC's regulatory orders.

Whereas in the 1990s Canada decreased its regulation over Canadian utilities that export electricity to the US, the US has increased its influence over them. For example, besides eliminating export price tests, the amended National Energy Board Act (legislated 1 June 1990) reduced export regulations so that public hearings, formerly part of most export applications, were effectively eliminated. And, in addition, long-term blanket permits were granted so that utilities could reduce the number of export applications reviewed by the NEB. Such blanket permits, granted for up to sixteen years, allowed utilities such as Hydro-Québec and BC Hydro to sign their own short-term (three- to five-year) contracts with US customers without prior NEB approval. In the 1990s compliance with US regulations, applied even to the corporate structure of utilities in Canada, would become increasingly important as Canadian utilities continued to join US regional transmission groups in order to maintain their access to the US electricity market.

Whereas in the 1970s and 1980s provincial utilities, during their failed attempts to form regional or national power grids in Canada, were unable to agree on the establishment of an extra-provincial authority, in the 1990s Manitoba Hydro, Hydro-Québec, and BC Hydro have joined US regional transmission groups and power pools which require them to follow a foreign authority's (FERC's) orders to deintegrate and open their transmission systems to wholesale competition, not only from Canadian, but also from US power suppliers. Similar to paying tolls on private or public roads, electricity marketers pay tariffs (regulated by FERC) on open private or public transmission systems as their electricity is transported to wholesale customers in neighbouring provinces or states. Thus, several provincial transmission systems become small northern attachments to large US transmission regions.

Since 1999, Ontario Hydro, through its successor companies Ontario Power Generation (OPG) and Ontario Hydro Services Company (OHSC) plans to 'provide delivery and energy sales and services' in the North American market.[34] Ronald Osborne, Ontario Hydro's president and chief executive officer, proposes that the supply of electricity from OPG-owned generating facilities be severely curtailed to 35 percent of the Ontario market by 2010, while placing no limits on exports over upgraded OHSC-owned transmission lines to regions outside the province and to the $300 billion per year US market.[35] Also since 1998, Hydro-Québec, through two US subsidiaries – Hydro-Québec Energy Services (US) in Pittsburgh and TransÉnergie (US) in Boston – offers energy and transmission services in the North American electricity market.[36] H-Q Energy Services (US) applied to FERC and, since November 1997, 'has been licensed to sell electricity at market based prices in the United States,' and in 1998 became a full-fledged member in the New England Power Pool. Resembling the practices of the British Columbia Utilities Commission, whose regulation of BC Hydro conforms to FERC's deintegration requirements, Québec's Régie de l'énergie, established as part of Québec's new 1996 energy policy, now similarly stipulates 'that the generation, transmission, and distribution of electricity are regulated activities.'[37] Like other provinces following FERC requirements, in May 1997 Québec opened its wholesale market and transmission system to third parties, presumably including Labrador utilities, which until then had been blocked from transporting electricity through Québec to Ontario, the Maritime provinces, and the US.

The fact that Québec would not allow the wheeling of Labrador electricity across its territory, as Chuck Furey, Newfoundland's minister of energy, argued in 1998, constitutes a thirty-year-old injustice 'created at least in part because Ottawa would not grant rights to sell electricity through another province.'[38] What Newfoundland could not achieve in the early 1980s (concerning wheeling opportunities through Hydro-Québec's system), Washington regulators inadvertently achieved in 1997.[39] With wholesale customers in Québec and New England being able to buy power from Labrador utilities (which could now, theoretically, compete with Hydro-Québec), the premier of Québec, Lucien Bouchard, and Premier Tobin started to plan the joint $12 billion project on the Lower Churchill River in Labrador. Plans for a $2 billion transmission line to St. John's, Newfoundland, would connect Newfoundland and Labrador to the Québec and eastern US grid.

In the Prairies, Manitoba's installed capacity in the late 1990s was sufficient to supply Manitobans with electricity until 2015. Therefore, Manitoba Hydro, in its effort to retain and expand its export market in the US, joined the Mid-continent Area Power Pool. In the process of continuing its transborder regional integration, and following the deintegration trend, by 1996 Manitoba Hydro had formed the 'distinctly accountable business units: Power Supply, Transmission and Distribution, and Customer Service.' In 1997, Manitoba Hydro opened its transmission lines and offered to wheel wholesale electricity transactions through its open grid at specified tariffs for US and Canadian sellers and buyers. In the same year, Manitoba's legislature passed the Hydro Amendment Act to allow Manitoba Hydro's transmission system to become a common carrier for power wholesalers, to expand the utility's wholesale and business activities in the US, and to permit the building of hydroelectric projects solely dedicated to export.

BC Hydro also followed the deintegration scheme and established both BC Power Supply and BC Hydro Transmission and Distribution as distinct and separately accountable business units. Communication between these units now occurs as though they were separately owned, so that US or other suppliers of electricity have the same access to transmission or export services as does the in-house power supplier. In its effort to obtain an electricity marketing licence in the western US, BC Hydro's subsidiary, BC Power Exchange Corporation (Powerex), first joined two regional transmission groups (the Western Regional Transmission Association and the Northwest Regional Transmission Association) and then applied to FERC for a licence. That action resulted in FERC's initial rejection of Powerex's power-marketing status in the US – a rejection that was overturned in September 1997 only after BC Hydro chair, Brian Smith, went to Washington almost weekly in the summer of 1997 to make concessions so as to meet every FERC requirement, guaranteeing that British Columbia's open transmission services and tariffs were equal to those south of the border. Surprisingly, British Columbia's premier, Glen Clark, saw the arrangement as 'good for consumers.'

Threats of Reciprocity

With market access reciprocity in electricity trade being part of the free trade provisions and FERC regulations, Canadian provinces can become both low-cost electricity hinterlands and potential wholesale markets for electricity supplied by US power marketers. As the chair of BC Hydro, Brian Smith, informs us, a continental integration strategy that concedes the opening of provincial transmission lines to wholesale marketing of electricity from US states entails the threat of US power-marketing firms gaining access to Canadian wholesale markets. The increased short-term nature of electricity exports, utilities anticipating additional growth in supplying the US market, and the availability of some low-cost electricity to industry in the US have resulted in different responses by provinces that are regionally integrated with US power grids. Canadian utilities have used several approaches to position themselves in the wholesale electricity market in the US, and one of these has been to start defensive strategies to fend off encroachment by US power marketers in the domestic market.

Hydro-Québec and Manitoba Hydro claim that US power marketers are unlikely to woo away their customers because their electricity rates are much lower than those south of the border. BC Hydro, however, also a low-cost supplier of electricity, saw this as a 'win-win' situation for only about six months. Smith now anticipates that US power marketers, such as Enron, which has 30 percent of the US market share, could woo away or 'steal' BC Hydro's customers. In light of this perceived threat, BC Hydro has started a $6 million advertising campaign to retain and build customer loyalty at home. For example, in March 1998 Smith also offered mining and forest products companies straight market trading prices – 'even below actual cost' – because, as he said, they 'would be the first ones to leave' when wooed by US power marketers.[40]

In concert with the Parti Québécois' counter-paradigmatic and state interventionist approach to integration with the new global markets, as Daniel Salée and William Coleman point out, Québec plans to be an 'energy hub' in the North American electricity market, through the multibillion-dollar expansion of its generating facilities, without privatizing Hydro-Québec. Similarly, Manitoba legislated that Manitoba Hydro would not be privatized and is proposing to increase its building of new hydroelectric plants solely for US export. Furthermore, Newfoundland, with transmission lines across Québec opened in 1997, is in the process of meeting US regulatory requirements and intends to sell future new supplies of electricity, developed jointly with Québec, in the US market. British Columbia, on the other hand, has opted for strengthening the emergence of independent power producers, and the BC Utilities Commission opposes further hydro mega-project expansions by BC Hydro. With the increasingly transnational regional electricity market overlapping the US-Canada border, some regulatory sovereignty has effectively shifted to Washington, forcing provincial utility commissions and legislatures to harmonize utility restructuring requirements and transmission services to comply with FERC rather than with the NEB (which, in the late 1990s, held no such regulatory powers either in Canada or in the transborder regions). Though opponents to power-marketing licences or export licences can state their views in public hearings, such participation is subject to constraints.

Resistance

Anticipated by many opponents of unnecessary expansions and increasing exports, the repercussions of overbuilding provincial hydroelectric systems (rather than building fewer plants but integrating them within regional or national grids) are not only social, financial, and environmental, but, because of international trade reciprocity, also potentially dangerous to the domestic provincial electricity market and have the potential to undermine and disintegrate provincial control.

The limitations on participants in public hearings to resist the building of power plants for export are often due to the pre-selection of imperatives (e.g., the predominance of the mandate to export). However, with sovereignty over sectors in the economy and its development shifting to new public administrative and corporate levels, as Mahon argues, opponents can consider a multiplicity of sites for resistance. With sovereignty over electricity policy fragmenting, transmission policy affecting Canada is increasingly directed from Washington, and with the NEB hearings virtually eliminated in the 1990s, the sites of resistance to building power plants for export include non-governmental locations both inside and outside Canada.

Some institutional and academic responses to tensions between hydro development and oppositional forces (e.g., Native peoples and the environmental movement) were in the form of environmental and social-impact assessments. Ignatius La Rusic, in his social-impact assessment, documents the impact of the river regime changes and the subsequent James Bay Agreement.[41] In addition to wildlife harvesting losses, he describes how negotiating from a subordinate position produces a consultant-dependent, legalistic, and confrontational response to government-sponsored northern development.[42] La Rusic's social-impact assessment is a post-impact evaluation, whereas Boyce Richardson clearly opposes water-development projects that disregard the rights of Native peoples and that insufficiently study the ecology *before* proceeding.[43]

The Cree, in the case of the James Bay hydroelectric development project, were initially treated as part of the environment – as something that simply needed to be rearranged. That Bourassa was more concerned with the Québec and New England power markets than with the Aboriginal population is shown by his merely passing acknowledgment that Aboriginal people already occupied the James Bay area.[44] Overall, despite transnationalized resistance by Aboriginal peoples, such disregard for, and wasteful expansion beyond, Canadian needs make hydro overdevelopment, in Glen Williams's words, 'part of the eco-hostile basis of continental capitalism.'[45]

The interconnection of provincial networks into a regional or national grid would have required fewer generating stations than currently exist and would, therefore, have substantially reduced environmental damage and possibly averted the need for nuclear plants (with their disposal problems) for electricity generation. The implications of this failure to coordinate provincial hydroelectric development are also financial, social, and environmental. Provincial overdevelopment not only had economic repercussions, such as the $90 billion Canadian utility debt in 1996, but also caused wildlife harvesting losses for Aboriginal peoples, had a destructive impact on land and rivers, and broke the careers of at least 25,000 utility and engineering employees. The environmental impact from dam projects has been extensive: 'These include the effect on the local climate, vegetation, fish and wild life [sic] caused by the creation or expansion of a reservoir and the construction of a dam and generating station. Water levels and flows are affected above and below the dam, as are the nutrient content and temperatures of water bodies.'[46]

First Nations, environmental groups, some labour unions, and other groups have made known their resistance to hydro developments in protests near the proposed mega-project sites, in provincial courts, at provincial public hearings, and, because of the virtual elimination of NEB hearings, at FERC hearings in Washington. The northern locations of the large hydroelectric projects are also the homes of First Nations peoples. These people bear most of the environmental impact of hydro projects built for electricity export. They have challenged such developments in the provincial courts, made their case until the late 1980s at NEB hearings (formerly in Ottawa and now in Calgary), and in the 1990s at FERC hearings in Washington. In November 1997, after the Cree had intervened in Washington to no avail, Grand Chief Matthew Coon Come objected to Hydro-Québec's being granted a licence to sell electricity through its US subsidiary in the uncertain short-term US northeast electricity market, protesting that they did not enter into the James Bay and Northern Québec Agreement so that Hydro-Québec could 'export Great Whale River electricity to the United States and play on the energy spot market.'[47] Claiming it has never been compensated for the past injustice of flooding the land for the Upper Churchill, in March 1998 the Innu Nation in Labrador protested Premier Bouchard's and Premier Tobin's exclusion of them from considerations concerning the building of the $12 billion Lower Churchill Falls project.

Labour, on the other hand, is divided between office and construction workers, the latter favouring building hydro projects for exports because it leads to job creation, and the former being against it because it leads to restructuring, potential privatization, and job loss. Further, in Manitoba the New Democratic Party, although fully supporting the building of electricity facilities for export and expanding electricity trade with the US, warned, as mentioned in Chapter 6, of the dangers of privatization and estimated that $100 million could be gained by stockbrokers and friends of the Conservative Party should Manitoba Hydro be privatized. In British Columbia, the representative for BC Hydro's biggest industrial customers, including forest, mining, and chemical industries, in a letter dated 30 June 1997, warned the BC Utilities Commission, that by modifying 'its regulation of transmission service to coincide with FERC orders, it will be giving up all semblance of authority over the terms and conditions for wholesale, and possibly retail access [to the transmission system] in the future.'[48]

What these findings show is that, because Canadian governments allowed electricity exports, included them in free trade agreements, and increased their efforts to integrate provincial transmission systems with US utilities before multi-provincial or national transmission agreements were in place, provincial electricity surpluses have been put to continental rather than national use. The all-Canadian visions of the 1960s and 1970s – visions of interconnecting northern, regional, and provincial electrical systems into a national whole – have been replaced by utilities that have been fragmented into business units that generate and transmit electricity. It appears the lessons taught at Niagara Falls nearly 100 years ago have still not been learned. At the beginning of the twenty-first century, the threat of the re-privatization of generating stations, transmission lines, and urban distribution systems is real. The owners of hydro power are the people of Canada; only by their being informed about their own political and economic history, and by their resistance to repeating the mistakes of history, can this threat be reduced.

Appendices

Turbines and generators in Ontario power stations (over 100 MW), 1905-30

	Capacity of plant (MW)	Years of turbine[a] installations and manufacturer	Years of generator[b] installations and manufacturer
Ontario Hydro			
Ontario Power	101	1905-13 (JMV, WSM)	1905-13 (WE, CGE)
Sir Adam Beck #1	528	1921-4 (WSM, DEW)	1922-86 (CWES, CGE, CRMP)

[a] Turbine manufacturers: J.M. Voith (JMV), Welman Seaver Morgan (WSM), Dominion Engineering Works (DEW), W.M. Cramp (CRMP).

[b] Generator manufacturers: Western Electric (WE), Canadian General Electric (CGE), Canadian Westinghouse (CWES).

Source: Statistics Canada, 'Electric Power Statistics,' cat. 57-206, 31 December 1986.

Appendix 2

Excerpts from the survey of steam and electricity use in small industries by the Hydro-Electric Power Commission of Ontario (Ontario Hydro), 1911

The chief engineer of the Hydro-Electric Power Commission prepared a survey, submitted to Adam Beck [on] 29 August 1911, of industrial and other uses of electrical power in order to consider the extension of transmission lines from Guelph to municipalities further north.

MOUNT FOREST. Population 1500 to 1800
They have a municipal electric light and water works plant ... Coal consumption [is] 35 tons per month, three quarter lump, costs $4.34 per ton, f.o.b. town. Town have approximately 1000 lights connected and sixty transformers in service ranging from 20 to 200 lights. The lighting plant operates from dark until 1 a.m. Could use 75 H.P. Attached is Municipal Report for 1910.

Ernst Bros. (Threshing Machine Manufacturers).
Output 18 to 24 machines per year. Have 25 H.P. engine; burn wood altogether, costs $2,00 per cord; use 200 cords per year; operate 10 hours a day. Could use 20 H.P. motor.

Woolen & Grist Mill.
Owned and operated by Mr. Murphy, utilizing two water powers on the South Saugeen River, owning all water rights. At one development is a 14 ft. head at which two 48″ wheels and one 40″ wheel is [sic] installed; the other development having 6 ft. head and one 48″ wheel. The above wheels are all Vulcan. As a standby for low water he has a 80 H.P. engine, which he states he uses about one month to six weeks in the year. No demand.

Mount Forest Casket Company.
Have a 60 H.P. Engine belt driven to shafting; burn refuse altogether. Operate about three days in the week; ten hours a day. No demand.

Mount Forest Carriage Company.
Have an 80 H.P. Engine; use refuse and coal. Could possibly use 40 H.P. motor capacity. Mount Forest has had set backs from the industries settled there, for which bad management was advanced as a cause. However, up to the present no attempt has been made to improve the lighting or power situation in the town. TOTAL OF MOUNT FOREST POWER – 100 H.[P.]

HANOVER. Population 1500 to 1800
The Town have [sic] a Municipal Water Work system driven by one of 30 H.P. wheel operating from a water power development on the Saugeen. The majority of the industries here are wood-working and furniture factories, using waste or refuse and under boilers in various quantities. The largest plant (Knechtels) state they have sufficient refuse for all purposes.

The other factories can use power to advantage, the greatest power users being the Hanover Portland Cement Company, Mr. Knechtel, the manager, stating they could use 300 to 400 H.P. as their water power development is inadequate and unsatisfactory from which they suffer for two months or more by low water and for which they have a steam plant installed.

The Ball Furniture Company.

State that they could use from 50 to 75 H.P. The present lighting of the Town is had through Herman Greitzner; operating a water power on the South Saugeen about two and a half miles from town. There are four wheels, three Barber and one Kennedy totaling 300 H.P. approximately. The head is 9 ft. Wheels are connected to counter-shaft from which is driven a 150 kW three phase 2300 volts 60 cycle C.G.E. Generator. There are about 2300 16 c.p. lamps connected 19 arc lights from the town, deriving an income for the Town of $600.00 per year. The load is from 36 to 38 amperes per phase in the winter time and about 20 amperes in the summer. This plant also supplies Carlrue & Neustadt with lighting. About 27 H.P. supplied in small motor capacity in the town; rates for lighting and power vary. Town lighting is from dark to 12 p.m. Considerable dissatisfaction was expressed by business men and citizens of price charged and unsatisfactory service given.

J.J. Spiesz. Furniture Factory.

Have a 125 H.P Goldie McCulloch Engine; burn refuse and coal. Coal Bill for 1910 was $500.00. Could use 50 to 60 H.P. The amount of power Hanover could use has been based on 300 H.P.

MEAFORD.

The Georgian Bay Milling & Power Company operate water power on the Big Head River about a mile and a half from town. This Company have [sic] two power sites, another one about a mile below the above mentioned, and the oldest one at which they operate a flour mill.

The Georgian Bay Milling & Power Company have [sic] a perpetual franchise for the sale of [hydro] electrical energy in the town, given them in 1905, and have an agreement with the Town for street lighting which expires in October 1913. They have no agreement for house lighting or any rate agreement covering same, operating with passive consent of the Town Council.

At present there is a by-law before the people to be voted on in a couple of weeks, information of which is briefly as follows:

A 500,000 bushel elevator was built in Meaford some years ago by a Detroit firm. This has been leased and operated to Moore Brothers, or the Georgian Bay Milling & Power Company, who have operated [it] the past two or three years. The Detroit firm who hold the Bonds in Trust have offered the Bonds to the Georgian Bay Milling & Power Company at a certain figure. They, in order to finance the scheme, are asking the town to loan them $16,500.00.

The Georgian Bay Light & Power Company already owe [sic] the Town $6000.00, money advanced them for the present plant, which would mean a total indebtedness to the town of $22,500.00.

This loan ($16,500.00) is to cover a period of 15 years, to be paid back principal and interest. As security for the above Company offered the Town the electrical light and power plant. Another clause in the agreement is that the Town, at the expiration of the present agreement between the Company and the Town renew the lighting contract from year to year for a period of three years: furthermore stating that if Hydro-Electric or any other Company

▶

Appendix 2

desire to operate in the Town they may do so by buying out the Georgian Bay Power plant; and that if in the three years following the expiration of their agreement with the Town in 1913 the Hydro-Electric come into town the Georgian Milling & Power Company agree to reduce their rates to meet theirs. Comment on the above clauses is unnecessary.

My interpretation of the by-law as given to the Mayor and Town Clerk was that the Town receives nothing for this loan—the securities offered by the Georgian Bay Milling & Power Company would be only good as long as the Town further contracted with them for light and power in the town, and they were practically giving them a fifteen year extension of their franchise without any desirable conditions. The present rates were too high and in this agreement there was no reference at all to the service, the rate, or in fact any protective clause whatsoever for the Town, and that this matter should receive their careful attention.

Mr. Albery, Town Clerk, was to mail us copy of the franchise and agreements etc; between the Georgian Bay Milling & Power Company and the Town.

The following are the industries of Meaford, giving the approximate demand if Hydro-Electric power were available:

The Meaford Canning Company.
Operate six or seven months in the year, 10 hours a day; burn three quarter lump, costs $4.20 per ton. Could use 20 H.P. motor capacity.

The Meaford Wheelbarrow Company.
Have two 150 H.P. Boilers, one 200 H.P. McEwan Engine: burn from 25 to 30 ton [sic] of coal per month, coal costs $4.10 per ton. As they burn part refuse, and steam is required for kilns, benders etc.; they could use 75 H.P. motor capacity.

The Meaford Manufacturing Company (Parlor & Dining Room Suites).
Have two 175 H.P. boilers, one 175 H.P. Engine (Wheelock) compound condensing. They use almost altogether edging slabs, etc. However, there is a market for this material, and they could under certain circumstances use 100 H.P.

Seaman, Kent & Company (Hardwood Flooring).
They have three coilers [sic] size 18 x 72 pounds steam pressure, and one 500 H.P. engine and two 50 H.P. engines. They operate waste or refuse altogether and for driving certain other machines could use from 30 to 40 H.P. only.

Charles Barber (Water Wheels).
They have a 25 H.P. engine; coal costs approximately $400 per year. Could use 25 H.P. motor capacity.

James Sparling (Planing Mill).
The have a 25 H.P. engine, burn about three cars of coal per year at a cost of 4.25 per ton; burn in addition refuse. Could use 20 H.P. motor capacity.

Randall Woolen MIlls.
Operating from water power on the Big Head River; Dam timber construction 11 ft. head, 50 to 75 H.P. ten months in the year. No demand.

W.A. Moore (Mantle Makers).
Have a 75 H.P. boiler and a 40 H.P. engine: burn coal and refuse. Could use 25 H.P.

...

Georgian Bay Milling & Power Company.
Have two 24" Barber Water Wheels [possibly made by Barber above] operating under a 48 ft. head direct [sic] connected to 200 KW 2300 volt, 50 amp. A.T.B., C.G.E. generator, form B.

They have no power or motor load in town, supplying street and other lighting only. From information received they have approximately 5500 lamps connected, receiving a gross income of $10,000.00. This includes the town lighting.

I understand from Mr. Moore their maximum load somewhere near Christmas time was 250 to 300 H.P. while at the present season it is between 150 to 200 H.P.

Meaford's estimate for power has, therefore, been made on a 250 H.P. basis.

Source: Ontario Hydro Archives, Archival Folder 3, Sir Adam Beck, General (1911); power survey of towns and their industries for Adam Beck, 29 August 1911.

Appendix 3

Excerpts from the survey of industrial applicants in southern Ontario municipalities by the Hydro-Electric Power Commission of Ontario (Ontario Hydro), 1919-20

		Date of application	Product or business	H.P. applied for
London	Ford Motor Co.	Feb. 1920		1,000
	C.S. Hyman	May 1919		200
	Murray Shoe Co.	do.		75
	London Box Co.	Nov. 1919		100
	H.C. Gillis	Jan. 1920		75
	Lawson & Jones	do.		30
	Canadian Electric Co.	do		3
	W.E. Saunders	do.		3
	D.S. Perin & Co.	Sept. 1919		50
	Hall-Dent	Jan. 1920		6
	Middlesex Mills Co.	do.		25
	Canadian Overall	Jan. 1920		3
	Holeproof Hosiery Co.	Jan. 1919		150
	Ross Ltd.	Jan. 1920		12
	Advertiser Job	do		10
	Webster Constr.	Nov. 1919		25
	Lochlin Brick Y.	Jan. 1920		25
	Walkers Brick Yard	do		30
	Cawrse Brick Yard	do		25
	McOrenere Brick Yard	do		25
	Victor Amusement Co.	do		30
	London & Port Stanley Rlwy	Nov. 1919		500
	Waterworks Pumping	do		100
	Raleigh Co.	Jan. 1920		50
	Republic Truck Co.	do		500
	Canada Oil Co.	Nov. 1919		20
	Loew's Theatre	Aug. 1919		25
	Allen's Theatre	Aug. 1919		25
	Hyatt Bros.	Feb. 1920		10
	Beatty Bros.	do		50
	Carling Brewing & Malting Co.	Feb. 1920		10
	London Paper Box	Feb. 1920		3
	London Shoddy Mills	do		25
	Henry & Colerick	do		5
	I X L Spice	do		10
	Dextie & Willetts	do		7
	W. Peltet	do		10
	L'Air Liquide Soc.	Aug.		150

	The Nut Krust Co.	Feb. 1920		100
Guelph	F.E. Partridge Rubber Co.	Aug. 28/19	Rubbers	100
	Northern Rubber Co.	do	Rubbers	250
	...			
	Sterling Rubber Co.	do	Rubbers	25
	Guelph Carpet & Worsted Spinning	Nov. 3/19	Carpets	500
	International Malleable Iron Co.	do	Castings	200
	Taylor Forbes	Nov. 26/19	Machinery	125
Stratford	Nut Crust Co.	Jan. 1920	Bread	80
	Can. Edison Appl.	Oct.	Elec. Appl.	300
	Grosch Felt Works			35
	Schellenberger			25
Woodstock	Woodstock Wagon & Mfg. Co.	Nov. 5/19		125
	Hay and Co. Ltd.	Dec. 6/19		150
	Jas. Stewart Mfg.	Jan. 1/20		25
	Thomas Organ Co.	do		15
	Canada Furniture Ltd.	Jan. 20/20		100
	Peerless Cereal Mills	Nov. 1919		20
	Bain Wagon Co.	July 1919		75
	Warren & Sons	Nov. 1919		30
	Brunswich Products	Aug. 1919		200
	Karn Piano Co.	Jan. 1920		100
	Breithaupt Leather Co.	Nov. 1919		10
	Bean and Westlake	Dec. 1919		10
	Worsted Spinning Co.	Apr. 1919		50
	Canadian Linderman Co.	Jan. 12/20		50
Brantford	Robins & Meyers	Feb. 18/20	Motors	300
	Robins & Meyers	Aug. 15/19	Motors	300
	Hamm & Nott	Jan. 2/20	Kitchen	200
	Brantford Tin Smelting Co.	Sept. 1/19		300
	Brantford Oven & Rack Company	Oct. 1/19	Ovens	45
	Watrous Engine Works	Jan. 2/20	Engines	400

Source: Ontario Hydro Archives, file OR-510, Corporate Relations, Municipal Customers. General. September 1922. An inventory of municipal and industrial power users, their application date, the product or business constituted in September 1920.

Appendix 4

Consumption of electricity in manufacturing industries in Québec, 1962-84

Year	Total consumption in manufacturing (GWh)	Three major sectors			Sum of three sectors (%)
		Pulp and paper (GWh)	Chemical industries (GWh)	Primary metals (GWh)	
1962	17,656	9,219	1,698	3,677	83
1963	16,979	8,105	1,740	3,864	81
1964	18,264	8,301	1,884	4,501	80
1965	18,712	8,299	1,999	4,530	79
1966	19,679	8,440	2,115	4,881	78
1967	19,872	7,978	2,200	5,201	77
1968	21,110	8,379	2,196	5,751	77
1969	23,343	9,092	2,326	6,933	79
1970	23,322	9,264	2,228	6,381	77
1971	23,074	9,301	2,369	5,848	76
1972	24,504	10,276	2,312	6,054	76
1973	26,549	10,720	2,590	6,694	75
1974	28,475	11,588	2,647	7,830	78
1975	–	–	–	–	–
1976	27,967	12,168	2,796	6,295	76
1977	28,693	12,107	2,654	7,121	76
1978	31,976	13,413	2,886	8,027	76
1979	33,261	13,450	3,037	8,578	75
1980	36,126	14,148	3,451	10,184	77
1981	37,532	15,237	–	10,277	–
1982	34,580	14,468	3,591	8,787	78
1983	36,510	16,211	3,498	8,862	78
1984*	42,277	20,946	3,732	8,996	80

*Survey of consumption by manufacturers has been discontinued by Statistics Canada in 1984, except for a small number of select industries.
Source: Statistics Canada, cat. 57-206, *Consumption of Purchased Fuel and Electricity by Manufacturing, Mining and Electrical Power Industries, 1962-1974*. Cat. 57-208, 1975-84.

Appendix 5

Hydro-Québec, les grands secteurs industriels, 1973 et 1976

Secteur industriel	(Contrat de plus de 3 MW) 1973			(Contrat de plus de 5 MW) dec. 1976		
	N	MW	GWh	N	MW	GWh
Pâtes et papiers	30	1,053	7,144	28	1,071	7,956
Electrométallurgie	16	951	6,567	15	851	6,012
Electrochimie	11	166	1,263	8	191	1,474
Divers	40	270	1,615	21 [sic]	21	1,406
Fer et acier	16	292	1,096	16	330	1,158
Pétrole	7	116	861	7	132	961
Ciment	8	85	538	6	76	427
Textiles	11	57	359	6	46	251
Manuf. générales	36	177	1,018	14	111	629

Note: The categorization of the contracts was changed from 3 MW to 5 MW, therefore, some of the smaller contracts are not included in the 1976 data.
Sources: NEB Applications, Hydro-Québec; October 1974, exhibit 1, p. 39; and October 1977, exhibit 1, p. 42.

Appendix 6

**Hydro-Québec, les grands secteurs industriels, 1981 et 1984
(contrat de plus de 5MW)**

Secteur industriel	1981			dec. 1984		
	N	MW	GWh	N	MW	GWh
Industries manufact.						
Aliments et boissons	6	30	170	6	30	178
Chimie	16	347	2,484	17	370	2,743
Ciment	4	51	272	5	70	395
Divers	11	87	457	14	115	604
Equipment de transp.	9	74	371	9	62	351
Métaux						
Fonte d'affinage	4	522	4,077	5	569	4,059
Sidérurgie	14	579	3,147	14	477	2,724
Autres	4	27	141	2	16	91
Minéraux non métall.	6	94	805	4	78	600
Pâtes et papiers	31	1,215	8,830	34	1,424	9,825
Pétrole et charbon	7	139	1,056	5	126	953
Textiles	8	53	320	9	57	339
Total ind. manufact.	–	3,218	22,130	–	3,394	22,862

Source: (NEB) Demande a ONE Hydro-Québec, exhibit B-2, vol. 1.1, décembre 1982, p. 3 de 4, exhibit B-3, juillet 1985, p. 3 de 3.

Appendix 7

Purchases in Newfoundland by the Churchill Falls (Labrador) Corporation by 1972

Company	Material	Total quantities		Total$
Nesco	rebar	14,350	tons	2,296,000
N.S. Cement Co.	cement	97,686	tons	2,410,560
Standard Mfg. Co.	paint	19,700	gallons	157,702
Nfld. Margarine Co	margarine			138,625
Nfld. Brewers	beer	358,428	cases	1,376,363
Alcan (Cdn.) Prod	alum. cond.	4,724	miles	10,880,386
Total				17,259,636
CFLCo purchases through manufacturers Newfoundland agents				1,801,358
Total				19,060,994

Source: 'Brinco in Newfoundland: A Summary,' n.d., (approx. date late 1971), stamped Confidential, Churchill Falls (Labrador) Corporation, National Archives, Ottawa, MG 28 III, vol. 1, file 7.

Appendix 8

Costs of the Churchill Falls Project, 1971

The Churchill Falls (Labrador) Corporation, in its report entitled 'Churchill Falls Power Project: Project Management Concepts,' issued in Montréal, February 1971, breaks down the $946 million total cost for the project in the following way:

Funds required	Millions ($)
Storage and forebay	115
Power plant and generators	168
Switchyard and transmission	100
Permanent support facilities	25
Temporary facilities and services	83
Management and engineering	31
	522
Escalation	102
Contingency	41
Direct construction costs	665
Interest during construction	189
Administration, overhead, and miscellaneous	92
Total	946

Source: Churchill Falls (Labrador) Corporation Limited, 'Churchill Falls Power Project: Project Management Concepts,' Montréal, February 1971, F1. National Archives, Ottawa, MG28 III 73, vol. 1.

Appendix 9

Turbine and generator contract, press release of the Churchill Falls (Labrador) Power Corporation, 1967

For Release A.M., Friday, June 30, 1967
MONTREAL: Churchill Falls (Labrador) Corporation Limited today announced a contract award with an estimated value of $50 million for the supply of 10 turbines and generators, with related auxiliary equipment, for the hydro-electric project under way in Labrador.

The announcement was made by Donald J. McParland, president and chief executive officer of Churchill Falls, who stated that the machines would be about twice the size of any hydro equipment presently in operation or under manufacture on this continent, and that over 90 per cent of the equipment would be manufactured in Canada.

Manufacture and installation of the equipment will be carried out by a group of three Canadian companies comprising Canadian General Electric at Montreal, Dominion Engineering Works at Lachine, Que., and Marine Industries at Sorel, Que.

More than 75 per cent of the equipment will be manufactured directly in the Province of Quebec, with certain items such as excitation equipment and control boards being manufactured in Ontario. Procurement outside of Canada will be limited to main shaft forgings, certain specialty components for thrust bearings, and similar items not manufactured in Canada.

The 10 turbines, each designed to develop over 600,000 horsepower from the 1060 foot net head available, will be coupled to ten 200 RPM generators rated at 4,500,000 kilowatts output. The capacity of any one of the generators would be sufficient to supply the entire power requirements of a modern city with a population of about 300,000 persons.

The height from the top of each generator to the heel of the turbine draft tube will be 90 feet, the height of a nine-story building. Combined weight of the rotating elements, consisting of the turbine runner, the connecting shaft and the generator rotor, will be in the order of 850 tons.

It is estimated that this contract will provide approximately seven million manhours of work for employees of the three companies and their suppliers. This is roughly equivalent to full time employment for between 500 and 700 men for a period of six years, and is typical of the impact which the Churchill Falls project will have on the economy of eastern Canada ...

Source: David D. Willcock, Director Public Relations, 1980, Sherbrooke Street West, Montréal 25, P.Q. Churchill Falls (Labrador) Corporation News Service, National Archives of Canada, MG28 III 73, vol. 59, file 7.

Appendix 10

Consumption of electricity in manufacturing industries, Newfoundland, 1975-8

| Year | Total consumption in manufacturing (GWh) | Three major sectors | | | Sum of three sectors (%) |
		Wood industries (GWh)	Pulp and paper (GWh)	Chemical industries (GWh)	
1975	1,130	6	816*	149	86
1976	1,468	6	956	370	91
1977	1,639	4	–	529	–
1978	1,883	6	1,050	648	90

*Statistics from category 2710.
Source: Statistics Canada, 'Consumption of purchased fuel and Electricity by the Manufacturing, Mining, and Electrical Power Industries,' cat. 57-208, 1975-84. Only three complete years were reported during this period.

Appendix 11

Electricity as a percentage of the cost of production

These percentages show how little the link between industrialization and hydro development is understood, or how, in the 1960s, the industrial optimism allowed the ignoring of such facts. While electricity may not be the main factor in attracting industry to a specific location, a sufficient and dependable supply is essential. For most industries, the cost of electricity is a very small item in the cost of producing. This is illustrated by the following tabulation, which gives, for a few industries, the proportion of the cost of electricity to the total cost of production in Canada.

Agricultural implements	0.29 %
Automobiles	0.09
Boots and shoes	0.19
Cotton yarn and cloth	0.94
Clothing	0.15
Electric apparatus and supplies	0.34
Machinery	0.32
Petroleum products	0.21
Rubber goods	0.56
Slaughtering and meat packing	0.12

For the following industries, the cost of electricity represents a higher proportion of the total cost of production.

Abrasive products	5.56 %
Acids, alkalis, and salts	6.80
Artificial ice	5.25
Cement	3.60
Compressed gas	3.20
Fertilizer	2.03
Non-ferrous smelting and refining (Québec)	7.65
Primary iron and steel	4.04
Pulp and paper (Québec)	3.83*
Woodenware	2.02

*In the newsprint industry in eastern Canada, this figure is higher, amounting to 5-6%.
Source: *Report of the Royal Commission on Electric Energy*, Government of Newfoundland and Labrador, February 1966, pp. 61-2 (source in Ottawa: Library of the National Energy Board); cited from Huet Massue, 'Highlights of Electric Power in Canada' (Montreal: Shawinigan Water and Power Company, 1954).

Appendix 12

Turbines and generators by manufacturers, Manitoba, 1900-86

Manufacturers	Main turbines installed	Main generators installed
Allis-Chalmers (AC)	1	
Canadian Allis-Chalmers (CAC)	1	
Canadian General Electric (CGE)		33
Dominion Engineering Works (DEW)	38	
John Inglis (JI)	3	
Leningrad Metal Works (LMW)	6	6
Mitsui (MITS)		12
S. Morgan Smith (SMS)	2	

Note: This inventory includes major hydro plants, such as Grand Rapids, Grand Falls, Jenpeg, Kelsey, Kettle Rapids, Long Spruce, and Seven Sisters.
Source: Statistics Canada, 'Electric Power Statistics,' cat. 57-206, 31 December 1986.

Appendix 13

Ten largest customers of industrial electricity in Manitoba, 1989

(1) International Nickel Co.
(2) Hudson Bay Mining and Smelting
(3) Interprovincial Pipeline Ltd.
(4) Abitibi Manitoba Paper
(5) Manitoba Rolling Mills
(6) Simplot Chemical Co.
(7) REPAP
(8) Hudson Bay Mining and Smelting, Ruttan Lake
(9) Can. Occ. Petroleum Inc.
(10) University of Manitoba

Source: Manitoba Hydro-Manitoba Industrial Customers, ranked 31 March 1989; list received from John F. Funnel, general counsel and secretary of Manitoba Hydro, during my interview with him in Winnipeg on 23 August 1989.

Appendix 14

Estimated and actual firm energy supplied to industrial and commercial customers from the Manitoba Hydro and Winnipeg Hydro systems, 1977-86 (GWh)

	Estimated			Actual	Difference
	(1)	(2)	(3)	(4) Power*	Diff. (3)
Year	Commercial	Industrial	Ind. + Comm. (1) + (2)	(Ind.+ Comm. Serv.)	(4) %
1978-9	1,671	5,855	7,526	5,328	-29
1979-80	1,847	6,311	8,158	5,585	-32
1980-1	2,020	6,739	8,759	5,587	-36
1981-2	2,223	7,224	9,447	6,068	-36
1982-3	2,441	7,740	10,181	5,726	-44
1983-4	2,704	8,313	11,017	6,263	-43
1984-5	2,880	8,853	11,733	6,590	-44
1985-6	3,067	9,429	12,496	6,677	-47
1986-7	3,266	10,042	13,308	7,133	-46

* 'Power, General Service and Misc.' is the category used in Manitoba Hydro annual reports as a summary of energy sold in areas other than the Residential and Farm category (statistics for 'energy sole' category are not adjusted for fiscal year).
Source: Manitoba Hydro, Application to the National Energy Board, 1975, vol. 1, exhibit 3, submitted 13 Jan. 1976, Table 7, p. 9; and Manitoba Hydro, *Annual Reports*, 1986, p. F15, and 1989, p. 41.

Appendix 15

Consumption of electricity in manufacturing industries in Manitoba by year, 1975-84

	Total Consumption in manufacturing	Four major sectors				Sum of four
Year	(GWh)	Primary metal (GWh)	Pulp and paper (GWh)	Food and beverages (GWh)	Chemical products (GWh)	sectors (%)
1975	2,482	1,308	364	224	145	82
1976	2,401	1,143	378	250	162	81
1977	2,555	1,192	342	256	274	81
1978	2,647	1,129	394	267	329	80
1979	2,567	1,059	–	285	252	–
1980	2,541	1,022	404	271	246	76
1981	2,631	1,181	400	260	241	79
1982	2,289	956	340	257	247	79
1983	2,320	940	341	228	324	79
1984*	2,530	1,162	371	233	288	81

* Survey of consumption by manufacturers has been discontinued by Statistics Canada in 1984, except for a small number of select industries.
Source: Statistics Canada, 'Consumption of Purchased Fuel and Electricity by the Manufacturing, Mining and Electrical Power Industries,' 1975-84, cat. 57-208.

Appendix 16

Energy purchased by and energy sold from the integrated system of Manitoba Hydro and Winnipeg Hydro, 1977-86 (GWh)

	1986	1985	1984	1983	1982	1981	1980	1979	1978	1977
Energy purchased										
Ontario	0	0	0	0	0	0	0	0	21	0
Saskatchewan	93	61	64	71	179	15	1	1	5	3
USA	17	20	7	43	297	203	0	4	721	528
Energy sold										
Ontario	917	795	1,153	850	1,089	1,616	1,787	1,439	773	1,582
Saskatchewan	222	233	269	231	342	968	1,378	530	364	408
USA	6,153	4,739	5,409	6,533	3,150	3,299	3,966	3,718	1,004	534
Extraprovincial* revenues ($millions)	113	103	106	105	72	82	96	82	35	17

*Sum of Ontario, Saskatchewan, and USA.
Source: Manitoba Hydro, *Manitoba Hydro-Electric Board: 38th Annual Report for the Year Ended March 31, 1986*, pp. F14-F15.

Appendix 17

Turbines and generators by power station (over 100 MW) in British Columbia, 1930-84

	Capacity of plant (MW)	Years of turbine installations and manufacturer	Years of generator installations and manufacturer
BC Hydro			
Ruskin	106	1930-50 (DEW)	1930-50 (CWES)
John Hart	120	1948-53 (DEW)	1948-53 (CWES)
Bridge R.#1	180	1948-54 (VIW)	1948-54 (CWES)
Cheakamus	140	1957 (VIW)	1957 (CWES)
Bridge R.#2	248	1959-60 (VEW, NEYC)	1959-60 (CWES)
G. Shrum	2,416	1968-80 (MITI, TOBA, FUJI)	1968-80 (CGE, TOBA, FUJI)
Jordan R.	150	1971 –	1971 (MITI)
Mica	1,736	1976-7 (HITA, LMW)	1976-7 (CGE)
Seven Mile	608	1979-80 (MITI)	1979-80 (HITA)
Kootenay C.	529	1975-6 (MITI)	1975-6 (CGE)
Peace Canyon	700	1980 (LMW)	1980 (MITI)
Revelstoke	1,843	1984 (FUJI)	1984 (FUJI)
Alcan Smelters and Chemical (BC)			
Kemano	813	1954-67 (CAC, PWW, DEW)	1954-67 (CGE, CWES, EE)
Cominco (BC)			
Brilliant	109	1944-68 (DEW)	1944-68 CWES)
Waneta	293	1954-66 (DEW, CAC)	1944-66 (CWES, CGE)

Note: For full corporate names see Appendix 20, this book.
Source: Statistics Canada, cat. 57-206, 31 December 1986.

Appendix 18

Consumption of electricity in manufacturing industries by year in BC, Yukon, and the Northwest Territories, 1962-74, and in BC, 1975-84

Year	Total consumption in manufacturing (GWh)	Three major sectors			Sum of three sectors (%)
		Wood Industries (GWh)	Pulp and paper (GWh)	Chemical and chemical products (GWh)	
1962	3,808	469	1,411	1,315	84
1963	3,866	521	1,362	1,336	83
1964	4,244 [a]	520	1,536	1,497	84
1965	4,950	596	2,023	1,601	85
1966	5,830	667	2,435	1,908	86
1967	6,288	734	2,743	1,919	86
1968	6,852	773	3,020	2,100	86
1969	7,381	875	3,395	2,080	86
1970	8,105	936	4,093	2,013	87
1971	8,093	1,091	3,702	2,158	86
1972	7,857	1,245	4,248	1,188	85
1973	7,794	1,466	3,705	1,374	84
1974	7,564	1,410	3,561	1,362	84
1975 [b]	6,910	1,404	3,319	884	81
1976	8,900	1,648	4,317	1,634	85
1977	8,916 [a]	1,868	4,339	1,416	85
1978	9,372	1,982	4,492	1,464	85
1979	10,008	2,059	4,947	1,497	85
1980	10,042	2,061	4,801	1,575	84
1981	9,076	1,840	4,216	1,420	82
1982	8,964	1,776	4,453	1,202	83
1983	9,780	1,979	5,056	1,270	85
1984	9,765	2,060	4,909	1,182	83

[a] The totals for the twenty manufacturing categories had to be adjusted for the reporting of high anomalous purchases (1964-76, 1977) in the Primary Metals Industry. The anomaly is likely due to Cominco's reporting the purchase of power from its own utility, West Kootenay Power and Light. Therefore, the total was adjusted by allowing only 200 (GWh) for the primary metals industry. Statistics Canada uses a more comprehensive classification of twenty manufacturing industries, which includes the manufacture of consumer goods. To assess the use of electricity in manufacturing, the four major industrial manufacturing categories are examined.
[b] The period 1975-84 originates from cat. 57-208, which is for BC purchases only.
Source: Statistics Canada, 'Consumption of Purchased Fuel and Electricity by Manufacturing, Mining and Electric Power Industries, 1962-1974,' cat. 57-206; the data for this period include BC, Yukon, and the Northwest Territories.

Appendix 19

Electricity imports and exports by year from British Columbia, 1984-8

Year	Interprovincial trade net exports (GWh)	United States Trade net exports (GWh)	Exporter revenue[a] ($ million)
1988	-855	7,719	111[b]
1987	-189	12,322	
1986	-64	1,429	
1985	32	10,119	
1984	36	6,721	

[a] Revenue from exports includes public and private utilities.
[b] $10 million is revenue from exports by Cominco Ltd.; the remaining $101 million accrued to BC Hydro.
Note: In 1988 BC's capacity was 12,497 MW and it generated 12,497 GWh of energy.
Source: Canada, Energy, Mines and Resources, *Electric Power in Canada 1988* (Ottawa: Minister of Supply and Services, 1989), pp. 77, 78, 20, 30, 80.

Appendix 20

Turbines and generators installed in power plants (over 100 MW) in five provinces, 1900-86

Equipment manufacturers	Turbines installed					Generators installed				
	NF	QC	ON	MB	BC	NF	QC	ON	MB	BC
Canadian Allis-Chalmersmm (CAC)	6	32	22	1	2					
General Electric (GE)							2	18		
Canadian General Electric (CGE)						14	124	60	33	17
Canadian Westinghouse (CWES)							82	55		29
Dominion Engineering Works (DEW)	6	134	49	38	20		1			
English Electric (EE)			19	28		6				2
Fuji (FUJI)										6
Hitachi Ltd. (HITA)						2				3
Leningrad Metal Works (LMW)				6	6					
Marine Industries Ltd. (MIL)	6	28				6	38		6	
Mitsubishi (MITI)						12				
Mitsui (MITS)									12	5
Metropolitain-Vickers (MVIC)							4			
Neypric (NEYC)			8			2		3		
Toshiba (TOBA)						3				
Vancouver Iron Works (VIW)					6					

Source: Statistics Canada, 'Electric Power Statistics,' cat. 57-206, 31 December 1986.

Notes

Chapter 1: Introduction

1 Canadian Broadcasting Corporation, 'The Falls,' from the radio program *Ideas* (no date).
2 Ibid.
3 Rivers have also been dammed for domestic, commercial, but mostly industrial use, with building for export to the US constituting a highly political undertaking. For instance, in 1996, three sectors consumed the largest share of the total electricity consumption (510 116 GWh) in Canada: the industrial sector used 39 percent, the residential sector 26 percent, and the commercial sector 22 percent. Line losses in 1996 amounted to 6 percent. Not all the electricity generated in Canada (547 781 GWh) is consumed in Canada, yet, of all the electricity generated in Canada in 1996, only 7 percent was exported to the United States. Between 1985 and 1996 the percentage of Canada's exports of electricity to the US fluctuated from 3.5 to 9.4 percent of the total electricity generated in Canada. Canada, Natural Resources Canada, *Electric Power in Canada, 1996* (Montreal: Canadian Electricity Association), Table 5.2, Table 5.3, Table 8.1, pp. 50, 51, 82, 84.
4 Electricity is included in the provisions of the Canada-US Free Trade Agreement (1988) and the North American Free Trade Agreement (1994). Because the Energy Policy Act passed by Congress in 1992 gives the Federal Energy Regulatory Commission (FERC) power to order transmission-owning utilities to provide access to suppliers/traders of electricity, Canadian sellers who wish to export to the US market need to conform with such access requirements. Canada's National Energy Board, however, has no regulatory jurisdiction over electricity import from the US (with the exception of approving power exchange contracts that include imports) or over wheeling (transmission of electricity for a fee) inter- or intraprovincially.
5 Canada, Natural Resources Canada, *Electric Power in Canada, 1996* (Montreal: Canadian Electricity Association and National Resources Canada), Table 8.1, 'Canada-U.S. Electricity Trade, 1960-1996,' p. 81. For some utilities the revenues from exports amount to hundreds of millions of dollars, although capital costs often have not been added to export prices.
6 Canada, Natural Resources Canada, *Electric Power in Canada, 1996* (Montreal: Canadian Electricity Association), Tables 5.2, 5.3, pp. 50, 51.
7 British Columbia, BC Utilities Commission, *Site C Report: Report and Recommendations to the Lieutenant Governor-in-Council* (Vancouver: BC Utilities Commission, May 1983), p. 300. British Columbia, Ministry of Employment and Investment, News Release, 'Final Electricity Market Reform Report Released,' 30 January 1998. Also see Mark Jaccard, *Reforming British Columbia's Electricity Market: A Way Forward,* Vancouver: British Columbia Task Force on Electricity Market Reform: Final Report, January 1998, p. 5.
8 'Québec forced out of Great Whale "swamp" Parizeau says,' *Vancouver Sun*, 22 November 1994. Jack Aubry, 'Québec cancels $13 billion project,' *Vancouver Sun*, 19 November 1994, front page.
9 Natural Resources Canada, *Electric Power in Canada, 1996*, p. 100.

10 *Power Grab: The Future of BC Hydro,* a conference on Deregulation and Competition in the Electrical Industry, co-sponsored by the Canadian Centre for Policy Alternatives (BC office) and the Institute for Governance Studies, Vancouver, Simon Fraser University, Harbour Centre Campus, 10 January 1998. Many of the critics focused on the report by Dr. Mark Jaccard, *Reforming British Columbia's Electricity Market: A Way Forward.* Two key critiques are by Alex Netherton, 'International Electricity Trade and Paradigm Shift,' and by Marjorie Griffin Cohen, 'Public Power and the Political Economy of Electricity Competition'; both papers were presented at the Power Grab Conference, 10 January 1998.

11 Ronald Osborne, President and Chief Executive Officer, Ontario Hydro, 'Industry Review: Ontario Hydro,' *Connections: 1999 Electricity Industry Review* (Montreal: Canadian Electricity Association, January 1999), p. 26.

12 Ibid.

13 Load is the amount of electric power or energy consumed by a particular customer or group of customers. For a discussion of loads of varied characteristics see Thomas P. Hughes, 'Technology as a Force for Change in History: The Effort to Form a Unified Electrical Power System in Weimar Germany,' *Industrielles System und Politische Entwicklung in der Weimarer Republik: Verhandlungen des Internationalen Symposiums in Bochum vom 12.-17. Juni 1973* (Düsseldorf: Droste Verlag, 1974), p. 154.

14 Wallace Clement, 'Introduction: Whither the New Canadian Political Economy?' *Understanding Canada: Building on the New Canadian Political Economy,* edited by Wallace Clement (Montreal: McGill-Queen's University Press, 1997), p. 7.

15 Staples are semi-processed resource products – pulp, metal ingots, minerals, frozen fish, pelts, oil – exported to become manufacturing inputs in an industrially more advanced country.

16 Wallace Clement and Glen Williams, 'Resources and Manufacturing in Canada's Political Economy,' *Understanding Canada,* p. 51.

17 Janine Brodie, 'The New Political Economy of Regions,' *Understanding Canada,* p. 252.

18 Hugh Aitken, 'Defensive Expansionism: The State and Economic Growth in Canada,' *Approaches to Canadian Economic History* (Toronto: McClelland and Stewart, 1969), p. 220.

19 Thomas P. Hughes, 'Technology as a Force for Change in History.'

20 Thomas P. Hughes, *Networks of Power: Electrification in Western Society 1880-1930* (Baltimore and London: Johns Hopkins University Press, 1983), p. 5, n. 4.

21 Ibid., p. 17.

22 Christopher Armstrong, *Monopoly's Moment: The Organization and Regulation of Canadian Utilities, 1830-1930* (Toronto: University of Toronto Press, 1986), p. 310.

23 Janine Brodie, 'The New Political Economy of Regions,' p. 240 (emphasis is mine).

24 Ibid., pp. 240-1.

25 Michel Duquette, 'Conflicting Trends in Canadian Federalism: The Case of Energy Policy,' *New Trends in Canadian Federalism,* edited by François Rocher and Miriam Smith (Peterborough: Broadview Press, 1995), p. 409.

26 Ibid.

27 Garth Stevenson, 'Federalism and the Political Economy of the Canadian State,' *The Canadian State,* edited by Leo Panitch (Toronto: University of Toronto Press, 1983); 'Canadian Regionalism in Continental Perspective,' *Journal of Canadian Studies* 15, 2 (1980).

28 Mel Watkins, 'Canadian Capitalism in Transition,' *Understanding Canada,* p. 33.

29 Wallace Clement and Glen Williams, 'Resources and Manufacturing in Canada's Political Economy,' p. 50.

30 Canada, Energy, Mines and Resources, *The Canada-U.S. Free Trade Agreement and Energy: An Assessment,* c. 1988, p. 7.

31 Mel Watkins, 'Canadian Capitalism in Transition,' p. 32.

32 Laura Macdonald, 'Going Global: The Politics of Canada's Foreign Economic Relations,' *Understanding Canada: Building on the New Canadian Political Economy* (Montreal: McGill-Queen's University Press, 1997), p. 183.

33 Ibid.

34 Cited by Laura Macdonald, 'Going Global: The Politics of Canada's Foreign Economic Relations,' p. 183.

35 An idea developed from Leo Panitch's argument that the FTA and NAFTA reflect the roles adopted by their respective states, which represent the interests of their bourgeoisies and bureaucracies that are penetrated by American capital and administration. Cited by Laura Macdonald, 'Going Global: The Politics of Canada's Foreign Economic Relations,' p. 183.

36 Janine Brodie, 'The New Political Economy of Regions,' p. 257.

37 Claus Offe, *Strukturprobleme des kapitalistischen Staates: Aufsätze zur Poltischen Sociology* (Frankfurt: Suhrkamp, 1972), p. 54. Descriptions of the historic and economic conditions that structure the profitability criterion are detailed at the beginning of each provincial case chapter.

38 Rights to extract the power from water flows were obtained by buying title to or by leasing such rights. For example, until 1907 Québec disposed of its water powers by outright sale, either through private contract or by public auction; subsequently, during the first half of the twentieth century in Québec, concessions have been made by 'lease, usually of 50, 75, or 99 years duration.' John H. Dales, *Hydro-Electricity and Industrial Development* (Cambridge: Harvard University Press, 1957), p. 30.

39 John Dales, *Hydro-Electricity*, p. 183.

40 T.C. Keefer, 'Canadian Water Power and Its Electrical Product in Relation to the Undeveloped Resources of the Dominion,' cited by H.V. Nelles, *Politics of Development*, pp. 215-6.

41 Manitoba Hydro, Marc Eliesen, 'Chairperson's Message,' *Annual Report*, 1987, p. 6.

42 Justine Hunter, 'Power plan pushes Hydro toward market rates: Government scheme to lure industries part of transition in electricity pricing,' *Vancouver Sun*, 27 June 1997, p. E4. Vaughn Palmer, 'Clark pulls in an Alcan competitor to, ah, improve negotiations,' *Vancouver Sun*, 27 June 1997, p. A18.

43 Harold Innis, *Problems of Staple Production in Canada* (Toronto: Ryerson, 1933), p. 73.

44 Ibid., p. 75.

45 Ibid.

46 Dales, *Hydro-Electricity*, pp. 182, 184.

47 Ibid., p. 182.

48 Two examples of how energy-intensive semi-processing of natural resources contrast with the capacity requirements of a university: a pulp mill requires about 40 MW (MW= 1,000 kW), an aluminum smelter about 100 MW; this compares to universities such as McGill in Montréal needing an electrical capacity of 12 MW.

49 Based on his analysis of the relationship between hydro development and industrial development in Québec before 1940, Dales argues that 'historically, Canadian staple industries have been characterized by dependency on export markets, and in part the Québec hydroelectric industry fits this pattern: small amounts of electricity are exported directly to the United States.' Dales, *Hydro-Electricity*, p. 183.

50 Canada, Natural Resources Canada, *Electric Power in Canada, 1996* (Montreal: Canadian Electricity Association, no date), Table 8.1, p. 81; *Electric Power in Canada, 1994* (Montreal: Canadian Electricity Association, no date), Table 5.2, p. 50.

51 Mel Watkins, 'The Political Economy of Growth,' *New Canadian Political Economy*, p. 18. Albert Hirschman adds the notion that such developments are, at times, speeded up: 'development is accelerated through investment in projects and industries with strong forward linkage effects.' Albert Hirschman, 'A Generalized Linkage Approach to Development, with Special Reference to Staples,' *Essays in Trespassing: Economics to Politics and Beyond* (Cambridge: Cambridge University Press, 1981), p. 63.

52 Ibid., Hirschman, p. 66; Watkins, p. 18.

53 Whereas Watkins stresses the political aspects of reinvesting rent in industrial ventures to create fiscal linkages, Hirschman points out that reinvestment elsewhere in the economy (not physically linked to local resources) is an aspect of fiscal linkage. (Ibid., Watkins, pp. 18-19; Hirschman, p. 88.) For example, the Newfoundland government discounted the electricity for industry on the Island of Newfoundland from the rent anticipated for electricity from Churchill Falls. In this case electricity could not be transmitted to the Island; this constituted a fiscal linkage without the transmission of the physical input (electricity) to industries on the Island.

54 Tom Naylor, *The History of Canadian Business, 1867-1914*, Vol. 2, *Industrial Development* (Toronto: James Lorimer, 1975), p. 276.
55 Philippe Faucher and Kevin Fitzgibbins, 'The Political Economy of Electrical Power Generation: Procurement Policy and Technological Development in Québec, Ontario and British Columbia' (Montréal: Université de Montréal and Centre for Research of the Development of Industry and Technology, 1990), p. 3. The American connection to technology is discussed by Mel Watkins in 'The Political Economy of Growth,' p. 29.
56 Neil Bradford and Glen Williams, 'What Went Wrong? Explaining Canadian Industrialization,' *The New Canadian Political Economy*, edited by Wallace Clement and Glen Williams (Kingston: McGill-Queen's University Press, 1989), p. 67.
57 Nelles states that 'the principle of crown ownership of the forest survived intact and was used by the landlord state to bind its industrial tenants to contractual manufacturing commitments' (p. 382). For example, 'the sawlog and pulpwood manufacturing clauses were made conditions of tenancy. The government merely wrote these conditions into the crown timber licences and the courts upheld this as a legitimate use of the provincial proprietary right,' p. 106. See H.V. Nelles, *The Politics of Development*, pp. 106, 382.
58 For 'industrialization by invitation' as a conscious development strategy, see Tom Naylor, *The History of Canadian Business*, Vol. 2, p. 276; and for the non-innovative path of industrialization, see Vol. 1, p. 38.
59 Offe, *Strukturprobleme*, p. 55.
60 Adapted from Wallace Clement, 'Debates and Directions: A Political Economy of Resources,' in *The New Canadian Political Economy*, p. 44.
61 Rianne Mahon, 'Canadian Public Policy: Unequal Structure of Representation,' *The Canadian State*, edited by Leo Panitch (Toronto: University of Toronto Press, 1977), p. 173.
62 Adapted from Offe, 'Demokratische Legitimation der Planung,' in *Strukturprobleme*, pp. 139, 146.
63 Ibid., p. 139.
64 Rianne Mahon's adaptation is cited by Laura Macdonald, 'Going Global: The Politics of Canada's Foreign Economic Relations,' pp. 187-8.
65 Ibid.
66 Claus Offe, 'Tauschverhältnis und Politische Steuerung: Zur Aktualität des Legitimationsproblems,' in *Strukturprobleme*, p. 54.
67 For 'Das Kriterium der Nicht-Profitabilität,' see Offe, *Strukturprobleme*, p. 54, n. 36.
68 See Neil Bradford and Glen Williams, 'What Went Wrong?' pp. 60, 67.
69 Armstrong and Nelles focus on how 'organizational style,' business structures, and technology are 'diffused' from the United States to Canada by synthesizing historical descriptions of gas, transit, hydro and other utilities. They summarize: 'Canadian electric lighting utilities were propagated by equipment manufacturers operating under licence, or by branch plants using the same patents and marketing techniques as their US parents. Canada thus bore the stamp of both US technology and US industrial organization.' Armstrong and Nelles, in *Monopoly's Moment*, pp. 90-91.

Chapter 2: Avoiding National Power

1 Walter Dinsdale, Minister, Department of Northern Affairs and National Resources, 'Memorandum to the Cabinet: Long-Distance Power Transmission,' Ottawa, 6 December 1961, Confidential Cabinet Document No. 454/61 (source in Ottawa: National Archives, RG2, B2, vol. 6180, File 454-61), p. 1.
2 Electric power systems usually consist of three interrelated functions: the generating system which produces power, the transmission network which conducts the flow of power from the point of generation to the point of distribution, and the distribution system which delivers the power to consumers. Natural Resources Canada, *Electric Power in Canada, 1993*, p. 93. Much of the national power network initiative was concerned with the installation of a transmission network from the west coast to the east coast of Canada.
3 Cass-Beggs was former chair of Saskatchewan Power Corp., 1955-64, Manitoba Hydro-Electric Board, 1970-3, and BC Hydro, 1973-5. 'Former B.C. Hydro chief lauded,' *Vancouver Sun*, 20 February 1986.

4 BC Hydro, 'Biography of David Cass-Beggs,' approved by Cass-Beggs on 23 December 1974 (New Westminster: BC Hydro Information Centre, Retrieval System No. AK. 138), p. 4.
5 This paper was first presented in October 1959 to the Western Technical Conference of the Engineering Institute of Canada. David Cass-Beggs, 'Economic Feasibility of Trans-Canada Electrical Interconnection.' This paper was also presented to the Canadian Electrical Association, Western Zone Meeting, Edmonton, Alberta, 21-23 March 1960 (New Westminster: BC Hydro Information Centre, Retrieval System No. AK. 138).
6 Cass-Beggs, 'Economic Feasibility,' p. 2.
7 Ibid., p. 16.
8 Ibid.
9 Hughes indicates that after the Second World War, officials in the Soviet Union, the United States, and Great Britain advocated national power networks. Hughes, 'Technology as a Force,' p. 159.
10 Ibid.
11 The law for the socialization of electricity was enacted on 31 December 1919. Hughes, 'Technology as a Force,' pp. 160-1.
12 Ibid., p. 156; Cass-Beggs, 'Economic Feasibility,' pp. 3-4.
13 Natural Resources Canada, *Electric Power in Canada, 1994*, p. 15.
14 Ibid., p. 81.
15 Ibid., p. 15.
16 Ibid., p. 71.
17 Cass-Beggs, 'Economic Feasibility,' p. 9.
18 Ibid., p. 11.
19 Ibid., pp. 3, 12, 8.
20 H.V. Nelles, *The Politics of Development: Forests, Mines and Hydro-Electric Power in Ontario, 1849-1941* (Toronto: Macmillan, 1974), p. 247.
21 Ibid., p. 263.
22 Ibid., p. 301.
23 Sir Henry Drayton was the Power Comptroller of Canada in 1917. See Dal Grauer, 'The Export of Electricity from Canada,' *Canadian Issues: Essays in Honour of Henry F. Angus* (Toronto: University of Toronto Press, 1961), p. 260.
24 Rt. Hon. W.L. Mackenzie King, House of Commons, Official Report of Debates, 1929, p. 415. Cited by Grauer, 'The Export of Electricity,' p. 262.
25 Christopher Armstrong, 'Water Power and the Constitution,' *Politics of Federalism: Ontario's Relations with the Federal Government, 1867-1942* (Toronto: University of Toronto Press, 1981), pp. 166-7.
26 Voltages such as 345-kV, 500-kV, 735-kV and ±450 (DC) were carried by these transmission lines. Natural Resources Canada, *Electric Power in Canada, 1994*, p. 92.
27 Ibid., p. 93.
28 Ibid., pp. 13, 47.
29 Hughes, 'Technology as a Force,' p. 156.
30 Michel Duquette, 'Conflicting Trends in Canadian Federalism: The Case of Energy Policy,' *New Trends in Canadian Federalism*, edited by François Rocher and Miriam Smith (Peterborough: Broadview Press, 1995), p. 409.
31 Ibid.
32 Hughes, 'Technology as a Force,' p. 156.
33 Letter from Walter Dinsdale to Mr. David Cass-Beggs, 17 January 1961. Canada, Department of the Secretary of State, Sessional Paper 199; 'A copy of all correspondence, telegrams, and other documents exchanged between the government and each province since January 1, 1961 [to 16 February 1962], regarding the establishment of a national power grid system,' 35 pp. (Vancouver: UBC Library, Unpublished Sessional Papers, 24th Parliament, Session 5, vols. 912-19 #81-238, 1962, Reel 22/22). The Department of the Secretary of State had received this correspondence from the Department of Northern Affairs and National Resources and from the Prime Minister's Office in Canada.
34 Ibid.
35 Walter Dinsdale, Minister, Department of Northern Affairs and National Resources,

'Memorandum to the Cabinet: Long-Distance Power Transmission,' Ottawa, 6 December 1961, Confidential Cabinet Document No. 454/61 (source in Ottawa: National Archives, RG2, B2, vol. 6180, File 454-61).

36 Ibid., p. 1.

37 Ibid.

38 Ibid.

39 Ibid.

40 Canada, Cabinet Conclusions, Meeting Date 19 December 1961, 'Long-Distance Power Transmission,' Cabinet Document 454-61 (source in Ottawa: National Archives of Canada, Cabinet Conclusions, RG2 A5A, vol. 6177).

41 Canada, House of Commons Debates, Official Report, 18 January 1962, p. 2.

42 A letter from Diefenbaker to all provincial premiers, dated Ottawa, 17 January 1962, inviting them to a conference in Ottawa where, accompanied by their technical experts, they could 'discuss with Federal Ministers the basic problems associated with the development and transmission of electric power within Canada.' Canada, Department of the Secretary of State, Sessional Paper 199.

43 Cabinet Conclusions, Meeting Date 16 February 1962, 'Correspondence with Provinces on Proposed Power Grid,' National Archives of Canada, RG2 A5A, vol. 6192.

44 Province de Québec, Cabinet du Premier Ministre, letter from Jean Lesage to John Diefenbaker, 24 January 1962. Canada, Department of the Secretary of State, Sessional Paper No. 199.

45 Letter from John G. Diefenbaker to Jean Lesage, 22 February 1962; this document bears the word 'copie.' Canada, Department of the Secretary of State, Sessional Paper No. 199A, 20 March 1962.

46 Cabinet du Premier Ministre, Province de Québec, Letter from Jean Lesage to John Diefenbaker, 2 March 1962. Canada, Department of the Secretary of State, Sessional Paper No. 199A, 20 March 1962.

47 Canada, Cabinet Conclusions, Meeting Date 15 March 1962, 'Federal-Provincial Meeting on Long-Distance Transmission of Electricity.' R.B. Bryce signed the minutes for the prime minister. National Archives of Canada, RG2 A5A, Cabinet Conclusions, vol. 6192.

48 Walter Dinsdale, Minister of Northern Affairs and National Resources, Chairman of ad hoc Committee, 'Memorandum for the Cabinet: Federal-Provincial Meeting on Long-Distance Transmission of Electricity,' 14 March 1962, Cabinet Document, No. 100/62 (source in Ottawa: National Archives of Canada, Interim Box # 243, N-1-3(d)).

49 Ibid., p. 3.

50 Notes from the opening remarks by the Rt. Honourable John G. Diefenbaker, Prime Minister, at the Federal-Provincial Conference on Long-Distance Transmission, Ottawa, 19 March 1962, p. 7. Cited by Canada, National Energy Board, 'Inter-Utility Trade Review: Inter-Utility Cooperation' (Calgary: National Energy Board, 1992), p. 2-3.

51 Notes from the 19 March 1962 conference were not available, only references from recollection of the Cabinet minister in the Cabinet meeting of September 1962. Canada, Cabinet Conclusions, Meeting Date 12 September 1962, 'Long-Distance Power Transmission,' and Meeting Date 19 September 1962, 'Long-Distance Power Transmission' (source in Ottawa: National Archives of Canada, RG2 A5A, Cabinet Conclusions, vol. 6192).

52 Third Provincial Premiers' Conference, Victoria, BC, 6-7 August 1962. Proceedings, 6 August 1962, Legislative Library, Victoria, BC, pp. 161-8, CAN ZC, 6 August 1962.

53 Ibid.

54 Ibid.

55 Ibid., pp. 163-4.

56 Ibid., p. 162.

57 Ibid., p. 167.

58 The Quiet Revolution policies, e.g., hospital insurance, secular education, provincial control over pension funds, and francophonization of Québec's civil service, were designed also to foster Québec's social and cultural life.

59 René Durocher, 'Quiet Revolution,' *The Canadian Encyclopedia*, vol. 3 (Edmonton: Hurtig Publishers, 1988), p. 1813.

60 André Bolduc and Daniel Larouche, *Québec: Un Siècle d'Électricité* (no place: Libre Expression, 1979), pp. 260, 268.
61 Canada, Cabinet Conclusions, Meeting Date 26 September 1963, 'Policy respecting the export fo [sic] power.' An explanatory memorandum was circulated and considered by the Committee on Energy and Coal; Minister's memorandum, 19 July [1963], and circulated as an appendix to Cab. Doc. 270-63 of 24 September 1963 (source in Ottawa: National Archives of Canada, RG2 A5A, Cabinet Conclusions, vol. 6254).
62 Ibid.
63 Mitchell Sharp in this policy announcement redefined such exports as being in the national Canadian interest as long as they met the price criterion that 'the export price of the power [set by each utility and its US customer] is just and reasonable in relation to the public interest.' He added a vague commitment that 'a national power system might be developed by a succession of stages.' Canada, House of Commons Debates, First Session: Twenty Sixth Parliament, vol. IV, 1963, 8 October 1963, p. 3301.
64 Canada, Cabinet Conclusions, Meeting Date 23 April 1964, 'Québec legislation to control the export of electric power,' National Archives of Canada, RG2 A5A, Cabinet Conclusions, vol. 6264.
65 Canada, Cabinet Conclusions, Meeting Date 22 August 1964, 'National Energy Policy' (source in Ottawa: National Archives of Canada, RG2 A5A, vol. 6264).
66 Ibid.
67 Ibid.
68 Ibid.
69 Canada, Cabinet Conclusions, Meeting Date 25 March 1965, 'Ad Hoc Committee for the Coordination of Power Policy.' A memo by Sharp recommending an ad hoc Committee of Ministers to report on achieving coordination of power policy was presented before the Cabinet that day, Cab. Doc. 127E65, dated 17 March 1965 (source in Ottawa: National Archives of Canada, RG2, vol. 6271).
70 Ibid.
71 Canada, Cabinet Conclusions, Meeting Day 6 May 1965, 'Hamilton Falls Power Project: Discussion with Premier Lesage and Premier Smallwood' (source in Ottawa: National Archives of Canada, RG2, vol. 6271).
72 British Newfoundland Corporation (Brinco), a consortium of British firms to which the government of Newfoundland had franchised resource exploration rights during the mid-1950s, owned Churchill Falls water-power rights and was planning to develop electricity at this site for possible export to Ontario, Québec, and the US. During 1965, when Lesage, Québec's premier, was 'asked if he would give Brinco a right-of-way for the power he replied, "very emphatically," that he would never agree to that.' Philip Smith, *Brinco: The Story of Churchill Falls* (Toronto: McClelland and Stewart, 1975), p. 204. See also *Le Devoir*, 5 May 1965, front page.
73 Canada, Cabinet Conclusions, Meeting Day 6 May 1965.
74 Power Contract Between the Québec Hydro-Electric Commission and Churchill Falls (Labrador) Corporation, 12 May 1969, p. 16. This copy of the power contract was submitted as part of the Hydro-Québec licence application entitled 'l'exportation d'électricité à Citizens Utilities Company, Contrats et Conventions, vol. 1 [originally dated 'juillet 1985']' during NEB hearings in Montréal, 24-25 September 1985, Ordonnance EH-1-85, exhibit no. B-4 (Ottawa: National Energy Board Library). In 1969, Hydro-Québec bought the electricity from Newfoundland, one of the poorest provinces of Canada, at the extremely cheap rate of 2.5 mills per kWh for sixty-five years, in part for resale to neighbouring utilities in Canada and the US.
75 Federal Members: Dr. C.M. Isbister, deputy minister, Department of Energy, Mines and Resources; Mr. H. Lee Briggs, member, National Energy Board; Mr. T.M. Patterson, acting director, Water Resources Branch, Department of Energy, Mines and Resources; chairs, chief engineers, managers, representing all provincial hydro utilities, including Dr. Gordon Shrum, chair of BC Hydro, and G.F. Pushie, director-general of Economic Development of Newfoundland. T. Ingledow and Associates, Consulting Engineers, 'National Power Network Stage II Assessment,' vol. I (Vancouver: T. Ingledow, February 1967), p. 16.

An interim report with the same title, dated March 1966, was also prepared. Vancouver: BC Hydro Library.

76 Ingledow, 'National Power Network, ... 1967,' p. 16.
77 Ingledow, 'National Power Network, ... 1967,' pp. 16-17.
78 H.G. Acres and Company, 'National Power Network Stage I Assessment,' 24 January 1964, p. ii. Cited by Canada, National Energy Board, 'Inter-Utility Trade Review,' p. 2-4.
79 Ibid.
80 Ingledow, 'National Power Network, ... 1967.' Cited by Canada, National Energy Board, 'Inter-Utility Review,' p. 2-4.
81 This technical and economic assessment report was prepared for the confidential use of the Federal Working Committee on Long Distance Transmission with staff from the National Energy Board, the Department of Energy, Mines and Resources, and various utility managers, engineers and technical advisors on long-distance transmission from most provinces. Ingledow, 'National Power Network, ... 1967.'
82 Ingledow, 'National Power Network Stage II Assessment, Interim Report,' March 1966, p. 1.
83 Ingledow, 'National Power Network, ... 1967,' p. 21.
84 Ingledow study cited by Canada, National Energy Board, 'Inter-Utility Trade Review,' pp. 2-4 to 2-5.
85 Canada, National Energy Board, 'Inter-Utility Trade Review,' p. 2-5.
86 Ibid.
87 The Cabinet discussed the premier of BC's intent to sell the extra power generated at US plants for storing water in Columbia River reservoirs to finance the Peace River scheme. Canada, Cabinet Conclusions, 9 March 1962 (source in Ottawa: National Archives, RG2 A5a, vol. 6192).
88 Victor Mackie, 'National Power Grid Nearer,' *Winnipeg Free Press*, 5 February 1974. During his public speeches, Macdonald referred to the energy conference of first ministers in Ottawa on 22-3 January 1974, at which the federal government offered financial assistance to the provinces to enhance the interconnection between provincial utilities. Donald S. Macdonald, Minister of Energy, Mines and Resources, 'Notes for Statement of Other Elements of Energy Policy,' First Ministers' Conference on Energy, 22-3 January 1974, Document No. FP-4133, p. 3.
89 David Cass-Beggs, 'Energy in Transition,' *Engineering Journal*, February 1973. Leighton and Kidd Limited, 'A National Electric Power Policy for Canada,' March 1973. Canada, National Energy Board, 'Inter-Utility Trade Review,' p. 2-6.
90 Canada, National Energy Board, 'Inter-Utility Trade Review,' p. 2-7. The original sources are footnoted on the same page as EMR (Department of Energy, Mines and Resources) file 6760-0, vol. 1 and vol. 2.
91 Ibid., p. 2-7.
92 Ibid.
93 Interprovincial Advisory Council on Energy (IPACE), Networks Study Group, 'An Evaluation of Strengthened Interprovincial Interconnections of Electric Power Systems,' October 1978, vol. 1. Cited by Canada, National Energy Board, 'Inter-Utility Trade Review,' pp. 2-7 to 2-11.
94 Ibid.
95 Ibid., p. 2-9.
96 Ibid., p. 2-9.
97 Ibid., p. 2-11.
98 'News Release, IPACE Report Evaluates Feasibility of Improved Electrical Inter-Connections between Provinces, 12 December 1978, see EMR (Energy, Mines and Resources) file 1135-J35, vol. 4.' Cited by Canada, National Energy Board, 'Inter-Utility Trade Review,' n. 28, p. 2-11.
99 Canada, National Energy Board, 'Inter-Utility Trade Review,' p. 2-12.
100 Ibid., pp. 3-6 to 3-12.
101 Ibid., pp. 3-12 to 3-24.
102 Ibid., p. 3-12.
103 The differences extended to federal financing of the Lepreau nuclear plant in New

Brunswick, the lower priority assigned to the Fundy tidal project in Nova Scotia, owner-ship over transmission and generating facilities, compensation for loss of provincial hydroelectric development jurisdiction and increased federal authority, and electricity exports from New Brunswick to Québec and Maine. Ibid., pp. 3-20 to 3-23.

104 Ibid., p. 3-23.
105 Ibid., p. 3-1.
106 Ibid., p. 3-1.
107 Ibid., p. 3-2.
108 Ibid., p. 3-3.
109 Ibid., pp. 3-5 to 3-6.
110 Ibid., p. 3-5.
111 Ibid., p. 3-5.
112 Steering Committee, Western Electric Power Grid Study Agreement, *Western Electric Power Grid: Volume I Executive Summary*, prepared by the Study Directorate, December 1980, p. 29. Winnipeg, Manitoba: Energy Policy Branch of Energy Mines and Resources. I was made aware of this 'job issue' related to the western grid by Ron W. Pritchard, director of the Energy Policy Branch of Energy Mines and Resources, Manitoba, during an interview in his Winnipeg office, 24 August 1989.
113 Canada, National Energy Board, 'Inter-Utility Trade Review,' p. 3-6.
114 Canada, Natural Resources Canada, *Electric Power in Canada, 1996*, p. 90.
115 Canada, National Energy Board, 'Inter-Utility Trade Review,' p. 3-39.
116 Canada, National Energy Board, 'Review of Inter-Utility Trade in Electricity: Analyses of Submissions' (Ottawa: Minister of Public Works and Government Services, 1994), p. 5.
117 Ibid., p. 6
118 Laura Macdonald, 'Going Global: The Politics of Canada's Foreign Economic Relations,' *Understanding Canada: Building on the New Political Economy* (Montreal: McGill-Queen's University Press, 1997), p. 183.
119 Mel Watkins, 'Canadian Capitalism in Transition,' *Understanding Canada: Building on the New Canadian Political Economy*, edited by Wallace Clement (Montreal: McGill-Queen's University Press, 1997).
120 Canada, Energy, Mines and Resources, *The Canada-U.S. Free Trade Agreement and Energy: An Assessment*, c. 1988, p. 7.
121 Canada, Natural Resources Canada, *Electric Power in Canada, 1996* (Montreal: Canadian Electricity Association, 1997), Table 8.1, p. 81.
122 Mel Watkins, 'Canadian Capitalism in Transition,' p. 32.
123 Letter from the Minister of Energy, Mines and Resources to Mr. R. Priddle, Chairman, National Energy Board, Ottawa, 19 September 1988. This letter is included as Appendix 1 in the report by Canada's National Energy Board, *Review of Inter-Utility Trade in Electricity* (Ottawa: Minister of Public Works and Government Services Canada, January 1994), p. 27.
124 Ibid.
125 Canada, National Energy Board, *Regulatory Agenda*. Calgary: National Energy Board, Reg-ulatory Support Office, Issue No. 45, 1 June 1993, p. 16.
126 Canada, Energy, Mines and Resources, *The Canada-U.S. Free Trade Agreement and Energy: An Assessment*, c. 1988, p. 44.
127 Ibid., p. 25.
128 Ibid., p. 44. See also chapter nine of the Canada-U.S. Free Trade Agreement that refers to the Bonneville Power Transmission issue.
129 R. Priddle, Chairman, National Energy Board, 'Regulation of Canadian Energy Exports in the Free Trade Era,' notes from a speech presented to the Twenty-First Annual Conference of the Institute of Public Utilities, Michigan State University, 'Emerging Markets and Reg-ulatory Reform: An Agenda for the 1990s,' Williamsburg, Virginia, 11 December 1989 (Calgary: National Energy Board, Library), p. 1.
130 Ibid., p. 2.
131 Bill C-23, An Act to Amend the National Energy Board Act, First Reading 7 June 1989. R. Priddle, 'Regulation of Canadian Energy Exports in the Free Trade Era,' p. 5.
132 Ibid., Priddle, p. 5.

133 Ibid. The new policy of 6 September 1988 and the amendment to the NEB Act (Bill C-23, 1 June 1990) reduce the possibility that exports and international power lines may be 'designated' as requiring public hearing; such hearings are now a rare event; see section 'National Energy Board,' in *Electric Power in Canada, 1996*, pp. 26-7.

134 Ibid., Priddle, p. 1, and see energy trade in Articles 902 through 905 in chapter nine of the FTA. 'Chapter Nine: Trade in Energy,' The Canada-United States Free Trade Agreement, The Department of External Affairs, Ottawa: The Minister of Supply and Services, 1988.

135 Ibid., Priddle, p. 3.

136 See chapter nine, Article 902 of the FTA. Priddle, 'Regulation,' p. 3.

137 André Plourde, 'Energy and the NAFTA,' Commentary, No. 46, Ottawa: C.D. Howe Institute, 1993 [May 1993], p. 2.

138 Canada, *North American Free Trade Agreement between the Government of Canada, the Government of the United Mexican States, and the Government of the United States of America* (Ottawa: Minister of Supply and Services Canada, 1992), Article 601, paragraph 1.

139 André Plourde, 'Energy and the NAFTA,' p. 3.

140 Ibid., p. 6.

141 Ibid.

142 Ibid., p. 6.

143 NAFTA, Annex 602.3, pp. 6-7.

144 André Plourde, 'Energy and the NAFTA,' p. 2.

145 Laura Macdonald, '*Going Global: The Politics of Canada's Foreign Economic* Relations,' p. 183.

146 For a review of the U.S. legislative and regulatory electricity initiatives, see the National Energy Board, *Review of Inter-Utility Trade in Electricity* (Ottawa: Minister of Public Works and Government Services Canada, 1994), pp. 17-20.

147 Ibid., p. 17.

148 Ibid., pp. 17-18.

149 Ibid., p. 18.

150 United States, Federal Energy Regulatory Commission, 'Promoting Wholesale Competition Though Open Access: Non-discriminatory Transmission Service by Public Utilities,' Docket No. RM95-8-001 (Washington: Federal Energy Regulatory Commission, Order No. 888-A, Issued 3 March 1997), p. 1.

151 Ibid., p. 3.

152 Ibid., p. 5.

153 Ibid., p. 31.

154 British Columbia Utilities Commission (BCUC), Decision in the Matter of 'British Columbia Hydro and Power Authority: Wholesale Transmission Services Application' (Vancouver: BCUC, 25 June 1996), p. 45.

155 National Energy Board, *Review of Inter-Utility Trade in Electricity*, p. 18.

156 Ibid.

157 Ibid., p. 25.

158 Canada, Natural Resources Canada, *Electric Power in Canada, 1996*, p. 26.

159 National Energy Board, *Review of Inter-Utility Trade in Electricity*, p. 19.

160 Ibid.

161 Ibid.

162 Garth Stevenson, 'Federalism and the Political Economy of the Canadian State,' *The Canadian State,* edited by Leo Panitch (Toronto: University of Toronto Press, 1983); 'Canadian Regionalism in Continental Perspective,' *Journal of Canadian Studies* 15, 2 (1980).

163 Canada, Natural Resources Canada, *Electric Power in Canada, 1996*, pp. 25-6. Canada, The Constitution Act, 1982, amended by the Constitution Amendment Proclamation, 1983, Section 92A (Ottawa: Minister of Supply and Services, 1986), pp. 16-18. Canada, National Energy Board, 'Regulation of Electricity Exports: Report of an Inquiry by a Panel of the National Energy Board Following a Hearing in November and December 1986,' National Energy Board, June 1987 (Ottawa: Minister of Supply and Services, 1987).

164 Christopher Armstrong, 'Water-Power and the Constitution,' *Politics of Federalism: Ontario's Relations with the Federal Government, 1987-1942* (Toronto: University of Toronto Press, 1981), pp. 166-77.

165 Michel Duquette. 'Conflicting Trends in Canadian Federalism: The Case of Energy Policy,' *New Trends in Canadian Federalism*, edited by François Rocher and Miriam Smith (Peterborough: Broadview Press, 1995), p. 408.
166 Ibid., pp. 408-10.
167 Natural Resources Canada, *Electric Power in Canada, 1996*, p. 1.
168 Ibid., p. 25. See also the *Revised Statutes of Canada, 1985*, The Constitution Act 1867, Section 92A (1) (c) (Ottawa: Queen's Printer for Canada), p. 28.
169 Ibid., pp. 25-6, Energy Resource Branch of Natural Resources Canada derives this interpretation from Section 92 (10) and 91 (29) of the Constitution Act.
170 Ibid., p. 26.
171 Ibid.
172 Walter Dinsdale, Minister, Department of Northern Affairs and National Resources, 'Memorandum to the Cabinet: Long-Distance Power Transmission,' Ottawa, 6 December 1961, Confidential Cabinet Document No. 454/61 (source in Ottawa: National Archives of Canada, RG2, B2, vol. 6180, File 454-61).
173 Janine Brodie, 'The New Political Economy of Regions,' *Understanding Canada: Building on the New Canadian Political Economy* (Montreal: McGill-Queen's University Press, 1997), pp. 240-1.

Chapter 3: Niagara Power Repatriation

1 James Mavor, professor of political economy at the University of Toronto at the time, argued also for electricity exports to the US if local Ontario manufacturers were not willing to match the price that electricity would fetch in the US. James Mavor, *Niagara in Politics: A Critical Account of the Ontario Hydro-Electric Commission* (New York: E.P. Dutton, 1925), p. 17.
2 Merril Denison, *Niagara's Pioneers* (no place, publisher, or date), pp. 17-20.
3 *Official Report of the Ontario Power Commission*, 28 March 1906, p. 16.
4 Adam Beck, *Conservation*, pp. 19-20; cited by K.C. Dewar, 'State Ownership in Canada: Origins of Ontario Hydro,' PhD dissertation, University of Toronto, 1975, p. 252.
5 Cited by Dewar, 'State Ownership in Canada: Origins of Ontario Hydro,' p. 228; from the *Snider Papers*, Petition to the Hon. G.W. Ross regarding Niagara Power (undated).
6 For H.V. Nelles, industrial backwardness – an idea from Alexander Gerschenkron – was a key determinant of the state's direct intervention in the production of hydroelectricity at Niagara Falls. Gerschenkron's industrial change dynamic, assumed to occur in physical terms of tension between pre-development and the promise of industrial growth, is helpful in Nelles's account of how 'backwardness' propelled Ontario's small manufacturers to become supporters of state intervention. It falls short, however, in failing to emphasize the private profit interests of electric power developers and that foreign direct investment can contribute to conditions of backwardness by speculatively holding up power projects, by export of electricity to the US, and by ignoring less profitable transmission gaps in Ontario. In addition, Nelles's concepts of prior Crown ownership and metropolitan versus hinterland tensions (Toronto utility interests versus small-town manufacturers) as determinants of state intervention are transcended in my analysis of several provincial cases by conceptualizing such specific patterns of provincial intervention as 'privatization reversals' and relating them to the profit motives of power developers in supplying electricity to metropolitan, export, industrial, and regional customers. H.V. Nelles, *The Politics of Development: Forest, Mines, Hydro-Electric Power in Ontario, 1849-1941* (Toronto: Archon Books, 1974), pp. 305, 492.
7 Ibid., pp. 32-3.
8 Ibid., pp. 33-4.
9 Ibid., p. 7.
10 Ibid., p. 38.
11 'The Canadian Niagara Power Company was incorporated by an Act of the Legislature of the Province of Ontario in the year 1892'; this act confirmed the 100-year agreement, dated 7 April 1892, between the Canadian Niagara Power Company and the Commissioners for the Queen Victoria Niagara Falls Park. George W. Davenport and the American

Institute of Electrical Engineers, *The Niagara Falls Electrical Handbook* (Syracuse: The Mason Press, 1904), p. 163.

12 Nelles, *The Politics of Development*, p. 34.
13 Davenport, *The Niagara Falls Electrical Handbook*, p. 163.
14 Ibid., p. 163.
15 Ibid., pp. 81-6.
16 Gordon Laxer, in *Open for Business*, employs Gerschenkron's concept of 'late industrialization' to explain why industrialization in Canada was delayed. I argue here, in contrast, that US speculators held up development of hydro-generated electricity, a process which delayed the progress of southern Ontario manufacturers who fell behind in installing electric motor drives for their factory machinery.
17 Davenport, *The Niagara Falls Electrical Handbook*, pp. 76-7.
18 Merril Denison, *Niagara's Pioneers* (printed in the USA, no publisher, no date), p. 28. Foreword by Paul Schoellkopf, president of the Niagara Falls Power Company.
19 Nelles, *The Politics of Development*, p. 223.
20 Ibid., p. 225
21 Ibid.
22 Ibid.
23 Dal Grauer, 'The Export of Electricity from Canada,' *Canadian Issues: Essays in Honour of Henry F. Angus*, edited by Robert Clark (Toronto: University of Toronto Press, 1961), p. 250, n. 3.
24 Blake R. Belfield, 'The Niagara Frontier: The Evolution of Electric Power Systems in New York and Ontario, 1880-1935,' PhD dissertation, University of Pennsylvania, 1981; source: National Research Council, Ottawa, CISTI Library), p. 88.
25 Grauer, 'The Export of Electricity from Canada,' p. 250.
26 Belfield, 'The Niagara Frontier,' p. 94. Wallace Clement argues that Canada's ruling economic interests assumed the US branch plants would Canadianize, as many earlier entrepreneur immigrants had fully integrated their businesses within the Canadian economy. However, such branch-plant firms were vertically and often horizontally linked to their US parent companies. Clement, *The Canadian Corporate Elite* (Toronto: McClelland and Stewart, 1975), p. 79.
27 Belfield found that Canadian Niagara had hopes to supply the Toronto market, but faced competition from the Electric Development Company. Belfield, 'The Niagara Frontier,' pp. 91, 111.
28 Davenport, *The Niagara Falls Electrical Handbook*, pp. 81-6.
29 Nelles, *The Politics of Development*, p. 227; Grauer, 'The Export of Electricity from Canada,' p. 250, n. 3.
30 Grauer, 'The Export of Electricity from Canada,' p. 250.
31 Belfield, 'The Niagara Frontier,' p. 110.
32 Davenport, *The Niagara Falls Electrical Handbook*, pp. 169-70.
33 Nelles, *The Politics of Development*, pp. 227-8; Grauer, 'The Export of Electricity from Canada,' p. 250.
34 Nelles gives an extensive account of the syndicate's stock-watering habits [diluting assets] of floating South American and Caribbean utilities and the public animosity towards its electrical operations. Nelles, *The Politics of Development*, pp. 228-37.
35 Belfield, 'The Niagara Frontier,' pp. 118-9. For a map of the Electric Development Company's manufacturing sites, see Davenport, *The Niagara Falls Electrical Handbook*, p. 171.
36 Ibid., Belfield, p. 119.
37 Ibid., pp. 115, 121.
38 Nelles, *The Politics of Development*, pp. 285-6, 288.
39 Ibid., p. 325.
40 At that time, 'the Ontario Power Company had the right to develop 180,000 h.p. Of this, it had installed 52,000 h.p. and exported 35,000 h.p. to the United States. The Electrical Development Company had the right to develop 125,000 h.p. It was then producing 42,800 h.p. and exporting 10,000 h.p. The Canadian Niagara Power Company enjoyed the right to develop 100,000 h.p. on the Canadian side. Its plants were capable of

producing 46,000 h.p., all of which was being exported.' Grauer, 'The Export of Electricity from Canada,' p. 250, n. 3.

41 Nelles, *The Politics of Development*, p. 35.
42 Ibid., p. 237.
43 Ibid., p. 242.
44 Ibid., p. 244.
45 Ibid., p. 245.
46 Ibid., pp. 245-6.
47 Ibid., p. 263.
48 Ibid., pp. 287-8. Grauer indicates that 'on April 12, 1917, all the assets in Canada of the Ontario Power Company were purchased by the Hydro-Electric Power Commission of Ontario [Ontario Power Commission].' Grauer, 'The Export of Electricity from Canada,' p. 250.
49 Nelles, *The Politics of Development*, p. 301.
50 Neil Bradford and Glen Williams, 1989, 'What Went Wrong? Explaining Canadian Industrialization,' ed. Clement, *The New Canadian Political Economy*, pp. 54-76.
51 Christopher Armstrong and H.V. Nelles, *Monopoly's Moment: The Organization and Regulation of Canadian Utilities, 1830-1930* (Toronto: University of Toronto Press, 1986), pp. 90-1.
52 Clement found that at the time 'industrialization was undertaken by smaller Canadian entrepreneurs and already developed firms from the United States.' Clement, *The Canadian Corporate Elite*, p. 66.
53 Tom Naylor, *The History of Canadian Business, 1897-1914*, Vol. 2, *Industrial Development* (Toronto: James Lorimer, 1975), pp. 152-4.
54 Naylor, *The History of Canadian Business*, Vol. 2, p. 150. Whether or not generating equipment was produced at this Westinghouse location requires further research.
55 Ibid., p. 61.
56 Clement describes the history of Canadian General Electric as 'having three features common among manufacturing industries at the turn of the [twentieth] century; consolidation by a financier (Nicholls), technological dependence on a US company, and eventual takeover by the US parent to integrate the branch into its worldwide operations.' Wallace Clement, *Continental Corporate Power: Economic Linkages between Canada and the United States* (Toronto: McClelland and Stewart), p. 53.
57 Dales observes that the excellence of the final installation of the Niagara Falls Power Company was due to 'the best of both European and American engineering talent. Swiss firms were chosen to design the turbines.' John H. Dales, *Hydroelectricity and Industrial Development: Québec 1898-1940* (Cambridge: Harvard University Press, 1957), p. 19. See also Belfield, 'The Niagara Frontier,' p. 87 and Davenport, *The Niagara Falls Electrical Handbook*, pp. 163-4.
58 Belfield, 'The Niagara Frontier,' pp. 87-8.
59 Ibid., p. 118.
60 For a general description of the Ontario Power Company's generating station, see Davenport, *The Niagara Falls Electrical Handbook*, pp. 185-7.
61 Belfield, 'The Niagara Frontier,' pp. 115-7; Davenport, *The Niagara Falls Electrical Handbook*, p. 175.
62 Belfield, 'The Niagara Frontier,' p. 117.
63 Nelles, *The Politics of Development*, pp. 293-4.
64 Between 1900 and 1986, Ontario bought all its large hydro turbines, a total of 60, from Canadian General Electric and 48 of its 55 generators from Canadian Westinghouse, the exception being 7 from Western Electric. Canada, Statistics Canada, *Electric Power Statistics: Inventory of Prime Movers and Electrical Generating Equipment as of December 31, 1986*, vol. 3, cat. 57-206.
65 Belfield, 'The Niagara Frontier,' pp. 143-4. For information about the Shawinigan plant equipment, see Dales, *Hydroelectricity and Industrial Development*, pp. 230-1 n19.
66 Official Report of the Ontario Power Commission, dated 28 March 1906, p. 27 (Toronto: The Monetary Times Printing Company; source: Ontario Hydro Archives, Beck Papers, file: ORR-104.11-11).

67 These observations were made about the Baldwin Locomotive Works and generally applied to operations employing no more than 100 horsepower. Official Report of the Ontario Power Commission, dated 28 March 1906, p. 29 (Toronto: The Monetary Times Printing Company; source: Ontario Hydro Archives, Beck Papers, file: ORR-104.11-11).

68 Chief Engineer [unsigned], Hydro-Electric Power Commission, 1911 Power Survey, for the Hon. Adam Beck, 29 August 1911 (Toronto: Ontario Hydro Archives, Archival Folder 3, Sir Adam Beck, General [1911]).

69 Ibid,. pp. 2-5.

70 Ibid., pp. 5-7.

71 Beck papers, 'Corporate Relations Municipal Customers General,' List of Industrial customers who applied for public power in southern Ontario municipalities in 1919, ten pages (Toronto: Ontario Hydro Archives, file OR-510, year 1930).

72 Ibid.

73 *Niagara Falls, Canada: A History of the City and the World Famous Beauty Spot* (no place, no printer, A Centennial Volume, 1967), p. 81.

74 Ibid., p. 81.

75 Ibid., p. 83.

76 Ibid., p. 84.

77 T.C. Keefer's address (in 1899) to the Royal Society of Canada entitled, 'Canadian Water Power and Its Electrical Product in Relation to the Underdeveloped Resources of the Dominion.' Cited by Nelles, *The Politics of Development*, p. 216.

78 Ibid.

79 Nelles, *The Politics of Development*, p. 108.

80 Mines, *Annual Report*, 1894, pp. 8-9. Cited by Nelles, *The Politics of Development*, p. 146.

81 Mark Fram, *Ontario Hydro Ontario Heritage: A Study of Strategies for the Conservation of the Heritage of Ontario Hydro* (Toronto: Ministry of Culture and Recreation, 1980), p. 31.

82 Ibid., p. 32.

83 Merril Denison, *The People's Power: A History of Ontario Hydro* (Toronto: McClelland and Stewart, 1960), p. 131.

84 A report entitled 'Errors and Misrepresentations made by The Hydro-Electric Inquiry Commission (Known as the Gregory Commission) Respecting the Publicly Owned and Operated Hydro-Electric Power Undertaking of Municipalities in the Province of Ontario,' Toronto, Ontario, 1925, p. 31 (Toronto: Ontario Hydro Archives, archival folder #292).

85 Nelles describes the members and influence of the Timber Ring thus: 'Between 1911 and 1920 no one cut anything in the northwestern part of the province without first doing business with Col. J.A. Little or some member of what came to be known as "the old Tory Timber Ring." The colonel's associates were Gen. Don Hogarth, provincial organizer of the Conservative party, banker, mining promoter, and timber speculator; W.H. Russel, a young Detroit lawyer turned pulpwood exporter; and J.J. Carrick, a former mayor of Port Arthur, Conservative MP, MPP, real estate promoter, and mining speculator.' Nelles, *The Politics of Development*, pp. 376-7.

86 'Errors and Misrepresentations made by The Hydro-Electric Inquiry Commission,' Toronto, Ontario, 1925, p. 31 (Toronto: Ontario Hydro Archives, archival folder #292).

87 Notes of the meeting between members of the Canadian Manufacturers Association and the Hydro-Electric Power Commission, 4 May 1920, pp. 4-6 (Toronto: Ontario Hydro Archives, Archival Folder #7, Adam Beck General [1920-2]).

88 Ibid., p. 5.

89 Ibid., pp. 7-8.

90 Ibid., p. 21.

91 'Memorandum Respecting the Contracts of the Goodyear Tire and Rubber Company of Canada, Limited. With The Hydro Electric Commission and The Toronto Power Company. For the Supply of Electric Power to its Factory at New Toronto. Ontario.' [original capitalization and punctuation], p. 4; covering letter of memorandum from C.H. Carlisle, vice-president and general manager, to Lt. Col. Hon. D. Carmichael, Parliament Bldg., Toronto, 9 February 1922; attachments are letters from Carlisle to Beck (8 May 1922) and

to Fleming (General Manager of Toronto Power Company, 5 July 1922); Toronto Power Company Contract with Goodyear Tire and Rubber Goods Company of Canada, 28 July 1920 (Toronto: Archives of Ontario, Collection Id. RG35 1-4, file 1922 Go-Gy-11).

92 'Memorandum Goodyear Tire,' p. 4.

93 Ibid., p. 2.

94 Letter on Goodyear letterhead, 23 October 1922, to Lt. Col. Hon. D. Carmichael, Commissioner of the Hydro-Electric Power Commission of Ontario and C.H. Carlisle, Vice-President and General Manager, 3 pages (Toronto: Archives of Ontario, collection Id. RG35 1-4, file 1922 Go-Gy-11), p. 1.

95 'Memorandum Goodyear Tire,' p. 2.

96 Letter, Carmichael/Carlisle, 23 October 1922, p. 2.

97 'Memorandum Goodyear Tire,' p. 4.

98 Grauer, 'The Export of Electricity from Canada.'

99 Ibid., p. 251. An alternative source for the history of federal regulation by the National Energy Board can be found in 'Regulations of Electricity Exports: Report of an Inquiry By a Panel of the Energy Board Following Hearing in November and December 1986' (Ottawa: Minister of Supply and Services Canada, 1987), pp. 5-6.

100 Commission of Conservation, *Water Powers of Canada* (Ottawa: Mortimer Company, 1911); cited by Grauer, 'The Export of Electricity from Canada,' p. 251, n. 4.

101 Arthur V. White, Memorandum Respecting Exportation of Electricity, Commission of Conservation, 5 May 1914, p. 13; cited by Grauer, 'The Export of Electricity from Canada,' pp. 252, n. 7, 255.

102 Grauer, 'The Export of Electricity from Canada,' p. 255.

103 Ibid., p. 251.

104 Ibid.

105 Ibid., p. 256.

106 Ibid.

107 Hydro-Electric Power Commission of Ontario, *Murray Report on Electric Utilities: Refutation of Unjust Statements*, 'The Allied War Needs as a Governing Factor' (Toronto: Hydro-Electric Power Commission of Ontario, 1922), p. 34 (Toronto: Ontario Hydro Archives, archival folder 7, Adam Beck General [1920-2]).

108 Beck correspondence with T.J. Hannigan, Secretary, Ontario Municipal Electrical Association, Guelph, and S.R.A. Clement, Secretary, Association of Municipal Electrical Utilities, Toronto, dated, 23 June 1925, pp. 3-4 (Toronto: Ontario Hydro Archives, Archival Folder #8, Sir Adam Beck, General Correspondence [1923-5]).

109 Grauer, 'The Export of Electricity from Canada,' p. 260.

110 Rt. Hon. W.L. Mackenzie King, House of Commons, Official Report of Debates, 1929, p. 415; cited by Grauer, 'The Export of Electricity from Canada,' pp. 261-2.

111 Grauer, 'The Export of Electricity from Canada,' p. 260.

112 Christopher Armstrong, *The Politics of Federalism: Ontario's Relations with the Federal Government, 1867-1942* (Toronto: University of Toronto Press, 1981), p. 166.

113 Armstrong, 'Water Power and the Constitution,' and 'The Battle of the St Lawrence,' both in *Politics of Federalism* (University of Toronto Press, 1981).

Chapter 4: Power from the North and Neighbour

1 David Cass-Beggs, 'Economic Feasibility of Trans-Canada Electrical Interconnection,' Paper presented at the Western Zone Meeting of the Canadian Electrical Association, Edmonton, Alberta, 21-23 March 1960, p. 9. Also see two reports prepared 'for the confidential use of the Federal-Provincial Working Committee on Long-Distance Transmission' by T. Ingledow and Associates, Consulting Engineers, 'National Power Network Stage II: Assessment' (Vancouver: T. Ingledow), Interim Report (March 1966), Volume 1 (February 1967).

2 Cass-Beggs, 'Economic Feasibility of Trans-Canada Electrical Interconnection,' p. 11; T. Ingledow, 'National Power Network,' vol. 1, p. 8.

3 Ingledow, 'National Power Network,' vol. 1, p. 25.

4 Robert Bourassa, *Deux Fois La Baie James* (Ottawa: Les Éditions La Presse, 1981), p. 8.

5 Hydro-Québec, *Strategic Plan 1998-2002* (Montréal: Hydro-Québec, October 1997), p. 2. André Caillé, President, Hydro-Québec, 'Industry Review: Hydro-Québec,' *Connections: 1999 Electricity Industry Review* (Montreal: Canadian Electricity Association, January 1999), p. 18.
6 'Innu disrupt power talks in Labrador,' *Vancouver Sun*, 10 March 1998, p. A5. See also the related issue of transmission-line access for Labrador-generated electricity discussed by Konrad Yakabuski, 'U.S. energy regulators push Québec on Churchill Falls,' *Globe and Mail*, 6 June 1997.
7 Roland Parenteau, 'Hydro-Québec: Les Relations Entre l'Etat et Son Entreprise' (Ottawa: Conseil économique du Canada, Sept. 1986), Document no. 312, pp. 7-9; John Dales, *Hydroelectricity and Industrial Development: Québec 1898-1940* (Cambridge, MA: Harvard University Press, 1957), p. 122.
8 Roland Parenteau, 'Hydro-Québec,' p. 7.
9 André Bolduc and Daniel Larouche, *Québec: Un Siècle d'Électricité* (no place: Libre Expression, 1979), pp. 268-9.
10 René Durocher, 'The Quiet Revolution,' in *The Canadian Encyclopedia* (Edmonton: Hurtig Publishers, 1988), 2nd ed., vol. 3, p. 1813.
11 Bolduc et al., *Québec: Un Siècle d'Électricité*, pp. 260, 268.
12 For example, on 12 February 1962 Lévesque pointed out the shortcomings of the private power corporations to 500 representatives from the electrical industry during la semaine nationale de l'électricité. Ibid., pp. 264-5.
13 Ibid., p. 268.
14 Of course, private hydro companies such as the Shawinigan Water and Power Company tried to resist nationalization. They raised questions about who actually would be the masters. In Shawinigan's publication, 'Les Faits,' dated June 1962, they queried: 'Dans leur désir d'être "maîtres chez eux," les francophones ne risquent-ils pas d'engendrer une vaste bureaucratie qui ferait de l'Etat le véritable maître?' Ibid., pp. 268-9.
15 Ibid., p. 265.
16 On 15 December 1961, Michel Bélanger, director of planning in the Ministry of Natural Resources, stated to the Conseil d'orientation économique, which advised the Lesage government on economic matters, 'mais on éliminera ainsi un handicap au développement industriel des régions démunies de sources d'énergie d'importance.' Bélanger, thereby indicated that the power supply was needed in those regions not supplied by profit-oriented hydro corporations. Ibid., pp. 264-5.
17 Ibid., pp. 280, 283. For a more comprehensive account of the nationalization, see Carol Jobin, *Les enjeux économiques de la nationalisation de l'électricité* (Montréal: Editions coopératives Albert St-Martin, 1978).
18 Ibid., p. 282.
19 Ibid., p. 287.
20 The Manic-Outardes complex furnished a total of 5,517 MW of which the capacity of the twenty-five generators of the four powerhouses at Manicouagan totalled 3,675 MW by 1976; the three powerhouses on the Outardes River had a capacity of 1,842 MW by 1978. Ibid., pp. 306-11, 313.
21 Ibid., p. 306.
22 The case of Churchill Falls is more fully described in the next chapter, 'The Churchill Power Trap.'
23 Bourassa states that as a result of the Churchill Falls contract 'the Manicouagan Outardes power development was stopped; and the Outardes II and Manic III projects were postponed by five years.' Robert Bourassa, *Power from the North* (Scarborough: Prentice-Hall, 1985), p. 87.
24 Hélène Connord-Lajambe's MA thesis draws attention to the similarities between la Société d'energie and the British Newfoundland Company's (Brinco's) La Société Churchill Falls. This similarity did not extend to getting firm electricity contracts before building projects during the 1970s. Hélène Connord-Lajambe, 1977, 'Coûts et Externalités de l'Aménagement Hydroélectrique de la Baie James' (Montréal: MA dissertation, Department of Economics, McGill University, 1977).
25 Robert Bourassa, *James Bay* (Montreal: Harvest House, 1973), p. 10.

26 The JBDC (SDBJ) had only five ·executive administrators (Pierre Nadeau, Fred Ernst, Roland Giroux, Raymond Primeau, and Lucien Cliche), who started their first meeting in the Hydro-Québec building at 75 West Dorchester on 13 September 1971 in the presence of Premier Robert Bourassa. The money for JBDC was raised by issuing shares which were bought by the Québec minister of finance. In other words, public money was used to subsidize the participation of foreign industry. *Procès-Verbal d'un Conseil d'Administration de la Société de Développement de la Baie James*, 13 September 1971, and 28 October 1971 (Montréal: Centre d'Archives Hydro-Québec).

27 Bourassa, *James Bay*, p. 71. Pierre Fournier, *Les sociétés d'Etat et les objectifs economiques du Québec: une evaluation préliminaire*, p. 92.

28 Robert Bourassa, *Power from the North*, p. 20.

29 Ibid, p. 21.

30 'The first subsidiary set up was the James Bay Energy Corporation. The partner is the Québec Hydro-Electric Commission (Québec Hydro), which holds a minimum of 51 percent of the Energy Corporation's shares,' Ibid., p. 21. See also, Robert Bourassa, *James Bay*, p. 71.

31 Derek Hill and John Kalbfleisch, 'Bourassa, Rockefeller discuss James Bay.' *The Gazette* (Montreal), 10 March 1972.

32 Robert Bourassa, *Power from the North*, p. 91.

33 Bourassa, *Deux Fois La Baie James,* pp. 84-5.

34 Ibid.

35 Ibid., p. 85.

36 Industrial linkages that can be related to infrastructure and Canada's economic growth are conceptualized by Mel Watkins, 'A Political Economy of Growth,' pp. 18-9, and by Neil Bradford and Glen Williams, 'What Went Wrong? Explaining Canadian Industrialization,' p. 60; both in *The New Canadian Political Economy*, edited by Wallace Clement (Kingston: McGill-Queen's University Press, 1989).

37 The idea of 'nationalistic' procurement of hydro equipment stems from Philippe Faucher and Kevin Fitzgibbons, 'The Political Economy of Electrical Power Generation: Procurement Policy and Technological Development in Québec, Ontario and British Columbia' (Montréal: Université de Montréal and Centre for Research on the Development of Industry and Technology, 1990), p. 9. The paper was first presented at the conference entitled 'The Political Economy of Electrical Power in Canada,' held at the University of Western Ontario, 24 March 1990. The transfer of Amercian technique to Canadian enterprise is discussed by Mel Watkins in 'The Political Economy of Growth,' p. 29.

38 Philippe Faucher and Kevin Fitzgibbons, 'The Political Economy of Electrical Power Generation,' p. 10.

39 Jorge Niosi and Philippe Faucher, 'Public Enterprises Procurements and Industrial Development,' monograph (Montréal: Centre de recherche en développement industriel et technologique, 1986), p. 15.

40 Philippe Faucher, 'Procurement by State-Owned Enterprises: Potential and Limits for Industrial Development,' monograph (Montréal: Centre de recherche en développement industriel et technologique, 1988), p. 19.

41 Niosi and Faucher, 'Public Enterprises Procurements,' p. 15.

42 Ibid., 10.

43 Interview with Jorge Niosi, 21 March 1989, at the *Centre de recherche en développement industriel et technologique (CREDIT)* in Montréal. Export restrictions as part of import substitution industrialization by branch plants are examined by Glen Williams, *Not for Export: Toward a Political Economy of Canada's Arrested Industrialization* (Toronto: McClelland and Stewart, 1986).

44 Interview with Jorge Niosi, 21 March 1989.

45 Gouvernement du Québec, Développement Économique, *L'Électricité, Facteur de Developpement Industriel au Québec* (Québec: Éditeur officiel du Québec, 1980), p. 25.

46 Québec, *l'Électricité, Facteur de Developpement Industriel au Québec*, p. 137.

47 Québec, Government of Québec. Request by Hydro-Québec to the Lieutenant-Governor-in-Council to grant approval, in the 'Resolution of June 3, 1968,' for its intention to sign

the Churchill Falls power contract (Ottawa: National Archives of Canada, Churchill Falls (Labrador) Corporation, MG II 73, vol. 58, file 11).

48 The 1970 October Crisis, with bombings and political kidnappings by extremist groups in the separatist movement, and the 1980 referendum to separate Québec from Canada contributed to the flight of capital from Québec. But political events are in themselves not sufficient to explain Québec's low industrial growth: as will be shown, the structure of industry in other provinces changed little in the wake of large-scale hydro projects during the last thirty years.

49 Canada, Natural Resources Canada, 'Canada's Largest Hydro Stations, 1994,' *Electric Power in Canada, 1994* (Montreal: Canadian Electricity Assocation, no date), Table 7.4, p. 74.

50 Total consumption in 1981 of 'Industries manufact.' = 22,130 GWh; the sum of Chimie, Fonte d'affinage, Sidérurgie, and Pâtes et papiers = 18,538, these four sectors having consumed 84 percent of power supplied to all manufacturing industries in 1981. Total consumption in 1984 of 'Industries manufact.' = 22,862 GWh; sum of Chimie, Fonte d'affinage, Sidérurgie, and Pâtes et papiers = 19,351; four sectors constitute 85 percent of all manufacturing industries in 1984. These industrial electricity consumption data originate from two applications to the National Energy Board carrying a different date but the same title, Hydro-Québec, 'Demande à l'Office National de l'Énergie Pour l'Obtention de Licences Visant l'Exportation d'Électricité,' vol. 1.1, décembre 1982, exhibit B-2, sec. 6 (2) o, p. 3 de 4; juillet 1985, exhibit B-3, sec. 6 (2) o, p 3 de 3 (Ottawa: Library of the National Energy Board).

51 Bourassa, *James Bay*, pp. 68, 71. For more detailed information consult the *Procès-Verbal de la Première Assemblée du Conseil d'Administration de la Société de Développement de la Baie James*, 13 September 1971, and the *Procès-Verbal du Conseil d'Administration de la Société de Développement de la Baie James*, 28 October 1971 (Montréal: Centre d'Archives Hydro-Québec).

52 Pierre Fournier, 1979, *Les Sociétés d'Etat*, p. 102.

53 Ibid.

54 André Raynauld, *La Propriété des entreprises au Québec* (Montréal: Les Presses de l'Université de Montréal, 1974), p. 60; cited by Pierre Fournier in *The Québec Establishment: The Ruling Class and the State* (Montreal: Black Rose Books, 1976), 2nd ed., pp. 27, 37.

55 *Procès-Verbal d'un Conseil d'Administration de la Société de Développement de la Baie James*, 16 August 1973 (Montréal: Centre d'Archives Hydro-Québec), p. 13.

56 Ibid., 13 December 1973, p. 14.

57 Ibid., 25 April 1974, pp. 2-3.

58 Ibid., 29 November 1974, p. 16; and 13 December 1974, p. 3.

59 Canadian Reynolds Metal Company, Contract #4407-75, date: 1 October 1975. Documents: Conseil Exécutif, Autres Documents, Hydro Québec Articles 159 et 160 for the period 1969-76, Instrument de Recherche 300034 (Sainte-Foy: Archives nationales du Québec).

60 Québec, Commission de la construction du Québec, *Les Affaires*, Saturday, 25 March 1989, p. 35. Claude Turcotte, 'Contrats à partage des risques et de bénéfices: Hydro mise sur le long terme,' *Le Devoir*, 13 March 1994, p. B1.

61 Ibid.

62 Ibid.

63 Québec, Order in Council, *An Act to Establish the Québec Hydro-Electric Commission*, 1944, Chapter 22, Division V, paragraph 29.

64 Bolduc et al., *Québec: Un Siècle d'Électricité*, p. 232.

65 Canada, NEB, 'Hydro-Québec,' Hearings Transcript, 27 November 1974 (Ottawa: Library of the National Energy Board), p. 125.

66 This answer was not in response to Zanis, but to questions by Robert Cousino, who raised these questions on behalf of the National Energy Board about planning for export. Ibid., p. 105.

67 Bourassa, *James Bay*, p. 16.

68 Ibid., pp. 16-17.

69 Bourassa, *Deux Fois La Baie James*, p. 90.

70 The figure of 176 MW is derived by calculating the difference between the total industrial

manufacturing capacity needed between 1981 (3,218 MW) and 1984 (3,394 MW) (Appendix 6).

71 The National Energy Board issued 'Reasons for Decisions' on Hydro-Québec licence applications during the following months: December 1974, September 1976, November 1977, July 1978, August 1980, September 1983, January 1984, August 1984, January 1985, June 1985, July 1985, November 1985, May 1987, and February 1988.

72 'Lesage: 'Aucune entente avec Brinco si l'énergie ne nous appartient pas',' *Le Devoir,* 5 May 1965, p. 1. Philip Smith, *Brinco: The Story of Churchill Falls* (Toronto: McClelland and Stewart, 1975), p. 204. Letter from Jean Lesage to John Diefenbaker, prime minister of Canada, 8 February 1962, in which he declined Québec's participation in the national power network discussions. Source: Sessional Paper No. 199. Canada. Department of the Secretary of State, Friday, 16 February 1962. Canada. Unpublished Sessional Paper, 24th Parliament, Session 5, vol. 912-919 #81-238, 1962; Papers 199 and 1,990.

73 Canada, National Energy Board, 'Report to the Governor in Council in the Matter of the Application Under the National Energy Board Act of Québec Hydro-Electric Commission,' September 1976 (Ottawa: National Energy Board, 1976), p. 5.

74 Ibid., pp. 16, 83.

75 Ibid., p. 3.

76 Ibid., p. 81.

77 Ibid., p. 24.

78 Ibid., p. 22.

79 Ibid., pp. 41, 45.

80 Ibid., p. 41.

81 Canada, National Energy Board, 'Inter-Utility Trade Review' (Calgary: National Energy Board, 1992), pp. 2-6 to 2-11.

82 Canada, National Energy Board, 'Report to the Governor in Council in the Matter of the Application Under the National Energy Board Act of Québec Hydro-Electric Commission,' September 1976 (Ottawa: National Energy Board, 1976), p. 42.

83 Ibid., p. 43.

84 Ibid., p. 55.

85 Ibid., p. 45.

86 Ibid.

87 Ibid., p. 46.

88 In the 'first 5 years, 2.77734 mill per kilowatthour [sic]' was the base paid for power from the 'Power Contract between Québec Hydro-Electric Commission and Churchill Falls (Labrador) Corporation Limited, May 12, 1969,' p. 16. This copy of the power contract was submitted as part of the Hydro-Québec Licence application entitled 'l'exportation d'électricité à Citizens Utilities Company, Contrats et Conventions, Volume 1 [originally dated 'juillet 1985']' during NEB Hearings in Montréal, 24-25 September 1985, Ordonnance EH-1-85, exhibit #B-4 (Ottawa: National Energy Board Library). Canada, NEB, *Report: Hydro Québec, September 1976,* pp. 58, 20.

89 Ibid., p. 58.

90 Canada, NEB, *Report: Hydro Québec, September 1976,* p. 83.

91 Ibid., pp. 70-7, 20-2, 46.

92 Ibid., pp. 68-71.

93 Ibid., p. 80.

94 House of Commons Debates, First Session: 23rd Parliament, vol. IV, 1963, 8 October 1963, 3,301.

95 Canada, NEB, *Report: Hydro Québec, September 1976,* p. 58.

96 Canada, NEB, *Reasons for Decision: Hydro-Québec, January 1984,* Appendices: VI, VII.

97 Canada, NEB, *Reasons for Decision: Hydro-Québec, for Export to the New-England Utilities, May 1987,* p. 3.

98 Ibid.

99 Ibid.

100 Canada, NEB, *Reasons for Decision Hydro-Québec, for Exports to the New England Utilities, February 1988,* p. 23.

101 Ibid., p. 1.
102 Ibid.
103 Canada, NEB, *Reasons for Decision: Hydro-Québec, for Export to the New-England Utilities, May 1987*, pp. 8, 12.
104 Ibid., p. 11.
105 Ibid., pp. 11-12.
106 Ibid., p. 20.
107 Canada, NEB, *Reasons for Decision Hydro-Québec, for Exports to the New England Utilities, February 1988*, pp. 3, 27-29.
108 Ibid., p. 37.
109 Ibid., pp. 37, 40.
110 Bourassa, *Power from the North*, pp. 92, 153.
111 Ibid., pp. 77-8, Table 5-1, 81. The evaluation of the size of Québec's energy resources is an excerpt from the investment banking firm of Kidder, Peabody and Company, Inc., in a report cited by Bourassa.
112 Ibid., pp. 133-55.
113 Ibid.
114 Canada, Natural Resources Canada, *Electric Power in Canada, 1994* (Montreal: Canadian Electricity Association, no date), Table 8.13, p. 90, Table 8.6, p. 86, Table 7.2, *Electric Power in Canada, 1986*, p. 30.
115 Jack Aubry, 'Quebec cancels $13-billion project,' *Vancouver Sun*, 19 November 1994. Réal Séguin, et al., 'Quebec shelves Great-Whale project,' *Globe and Mail*, 19 November 1994. United States, US Department of Energy, Energy Information Administration, 'Hydroelectric Trade and Canada,' excerpts from *International Energy Outlook 1995*, by Linda Doman.
116 Ibid., US Department of Energy.
117 Hydro-Québec, *Strategic Plan 1998-2002*, pp. 18, 55. Export revenue amounted to $488 million in 1995 and $537 million in 1996, and has tended to fluctuate over the years; see Canada, Natural Resources Canada, *Electric Power in Canada, 1996*, p. 83.
118 Hydro-Québec, *Strategic Plan, 1998-2002*, p. 32. The covering letter, dated 21 October 1997, is addressed to Guy Chevrette, Minister of State for Natural Resources and signed by L. Jacques Menard, Chairman of the Board and André Caillé, President and Chief Executive Officer. This plan was submitted to the Québec government for review by 1 November 1997. André Caillé, 'Industry Review: Hydro-Québec,' p. 18.
119 Ibid.
120 Ibid., p. 1.
121 Ibid., pp. 9, 10.
122 Ibid., p. 35.
123 Ibid., p. 17.
124 Ibid.
125 Ibid., p. 10.
126 Ibid., p. 11.
127 Ibid.
128 Ibid.
129 Michael Gourdeau, director of the Board of Hydro-Québec Energy Services (US), Hydro-Québec Press Release with Canada News Wire, November 1997.
130 Canada, National Energy Board, *Regulatory Agenda* (Calgary: National Energy Board), Issue No. 51, 1 January 1995, p. 4.
131 Matthew Coon Come, Grand Chief of the Cree, Grand Council of the Cree, 'Decision on Hydro Québec Puts Québec Energy Supply at Risk,' Press Release by Canada New Wire, Montreal, 13 November 1997.
132 Ibid.
133 Ibid.
134 Hydro-Québec, *Strategic Plan 1998-2002*, p. 7.
135 Ibid., p. 57.
136 Michell MacAffee (Canadian Press), St. John's, 'Aboriginals plan protest during Lower Churchill event,' *Evening Telegram*, 6 March 1998.

137 Ibid.; Hydro-Québec, *Strategic Plan 1998-2002,* pp. 2, 44.
138 Daniel Salée and William Coleman, 'The Challenges of the Québec Question: Paradigm, Counter-paradigm and ...?' *Understanding Canada,* edited by Wallace Clement (Montreal: McGill-Queen's University Press, 1997), pp. 262-85.
139 Ibid., p. 267.
140 André Bolduc et al., *Québec: Un Siècle d'Électricité,* p. 265.
141 Robert Bourassa, *Power from the North,* p. ix.

Chapter 5: The Churchill Power Trap

1 'The case for Brinco,' *St. John's Daily News,* 3 July 1968.
2 Edward Greenspon and Barrie McKenna, 'Tobin warns he'll pull Churchill plug,' *Globe and Mail,* 21 September 1996, p. A1.
3 'Innu disrupt power talks in Labrador,' *Vancouver Sun,* 10 March 1998, p. A5.
4 Churchill Falls (Labrador) Corporation, *An Energy Giant: The Churchill Falls Power Development* (Churchill Falls, Labrador, no date, received from CFLCo in March 1998), p. 2.
5 Newfoundland, Supreme Court of Newfoundland Trial Division, Between Her Majesty's Attorney General of Newfoundland and Churchill Falls (Labrador) Corporation Limited, Québec Hydro-Electric Commission, the Royal Trust Company, and General Trust Company, Judgment of Goodridge, J., 6 July 1982, 'Introduction,' pp. 4-5.
6 Smith, *Brinco,* p. 9.
7 'Brinco in Newfoundland: A Summary,' Confidential memorandum by Brinco, approx. date 1972 (Ottawa: National Archives, CFLCo, Document location: MG 28 III 73, vol. 1, file 7), p. 1.
8 Smith, *Brinco,* pp. 17-23.
9 Ibid., p. 9.
10 Newfoundland, Supreme Court, ... Attorney General ... and CFLCo ... Judgment of Goodridge, J., 6 July 1982, p. 5.
11 Ibid., pp. 5-6.
12 Ibid., p. 8.
13 Ibid., pp. 9, 34-5, 38.
14 W. Gillies Ross, 'Power on Hamilton River, Labrador,' *Canadian Geographical Journal* 66 (1963): 78.
15 'Brinco in Newfoundland: A Summary,' pp. 5, 14.
16 Newfoundland, Supreme Court, ... Attorney General ... and CFLCo ... Judgment of Goodridge, J., 6 July 1982, p. 8.
17 The Québec engineer and president of CFLCo, Don McParland, was subordinate in command to the directives from Brinco and the Bechtel Corporation in San Francisco during the construction of Churchill Falls; Smith, *Brinco,* pp. 164-5.
18 By 1968, $25 million and 'in 1968 and 1969 a further $25 million and $75 million of General Mortgage Bonds were purchased by Hydro-Québec.' 'Brinco in Newfoundland: A Summary,' p. 4.
19 See 'Brinco in Newfoundland: A Summary,' pp. 4, 7. The National Energy Board indicates, without providing references or specifying its use for the construction of Churchill Falls, that 'CFLCo's financing came from Brinco (fifty-eight percent), Hydro-Québec (thirty-four percent), and Newfoundland (eight percent);' see Canada, National Energy Board, 'Inter-Utility Trade Review: Inter-Utility Cooperation' (Calgary: National Energy Board, 1992), pp. 3-32.
20 Duplessis, cited by Philip Smith, *Brinco: The Story of Churchill Falls* (Toronto: McClelland and Stewart, 1975), p. 62.
21 Alec McEwen, 'The Labrador Boundary,' *Canadian Surveyor* 36, 2 (June 1982); cited in *The Canadian Encyclopedia* (Edmonton: Hurtig Publishers, 1988), 2nd. ed., vol. 2, p. 1160.
22 During the 1974 Energy Conference, attended by provincial premiers, Newfoundland further proposed to develop the remaining power sites in Labrador to supply the national power grid; a further 6,000 MW (2,000 MW at Lower Churchill site and 4,000 MW at other sites) was to be developed in Labrador, and the surplus made available to consumers in Canada over a national power grid. See Newfoundland, 'Position on A National

Power Grid' (Ottawa, Federal-Provincial First Ministers' Conference on Energy, 22-23 January 1974), Document No. FP-4119, pp. 3, 5.

23 'Power Contract between Québec Hydro-Electric Commission [Hydro-Québec] and Churchill Falls (Labrador) Corporation Limited [CFLCo], May 12, 1969' (source in Ottawa: National Energy Board Library). This copy of the contract was submitted by Hydro-Québec for the licence application, 24-25 September 1985, ordonnance EH-1-85, as exhibit B-4, pp. 16, 17.

24 The Churchill Falls (Labrador) Corporation (CFLCo) has become part of a group of smaller utilities run by Newfoundland and Labrador Hydro (NLH), a provincial Crown corporation established by an act of the provincial legislature in 1954 and incorporated in 1975 with a mandate to generate and transmit electricity in the province. This Newfoundland utility owns CFLCo, which operates the Churchill Falls plant, one of the largest power facilities in the world. Canada, Natural Resources Canada, *Electric Power in Canada, 1994* (Montreal: Canadian Electricity Association, 1995), p. 2.

25 Newfoundland, Supreme Court, ... Attorney General ... and CFLCo ... Judgment of Goodridge, J., 6 July 1982, p. 376.

26 See also, Newfoundland, Government of Newfoundland, Gordon F. Pushie, Commission Chairman, *Report on the Royal Commission on the Economic State and Prospects of Newfoundland and Labrador* (St. John's: Queen's Printer, Dec. 1987), pp. 279-97.

27 Canada, NEB, 'Inter-Utility Trade Review: Transmission Access and Wheeling' (Calgary: National Energy Board, 1992), p. 1-6.

28 Ibid.

29 Ibid., p. 1-7. Canada, NEB, 'The Regulation of Electricity Exports: Report of an Inquiry by a Panel of the National Energy Board Following a Hearing in November and December 1986' (Ottawa: Minister of Supply and Services Canada, 1987), p. 8.

30 Ibid.

31 Ibid.

32 About export to foreign markets, see Canada, NEB, Hearing, Order No. 1-86, 'In the Matter of the National Energy Board Act and the Regulations made thereunder; and in the Matter of an inquiry to review and report on the federal regulation of electricity exports,' held in Ottawa, 27 November 1986, Transcript, vol. 4-4, pp. 876, 886.

33 Ibid., p. 886.

34 Ibid., pp. 884-6

35 Ibid.

36 Ibid., p. 876.

37 Newfoundland, 'Position on A National Power Grid,' p. 2.

38 For electricity as 'manufactured product' see hearing day, 25 November 1986, Transcript, vol. 2-1, p. 161. Québec wanted the NEB to support a 'total absence of regulation in the case of the transportation of electricity' interprovincially (hearing day 25 November 1986, Transcript, vol. 2-1, p. 164). M. Richards, representing Hydro-Québec, added: 'que l'électricité est un produit manufacturable et qu'en conséquence toutes les companies et toutes les provinces au Canada peuvent produire leur électricité,' hearing day, 24 November 1986, Transcript, vol. 1-1, p. 23. Canada, NEB, Hearing, '... Regulations ... Exports,' held in Ottawa, Order No. 1-86, 25 November 1986, Transcript, vol. 2-1, pp., 161-5; see also Hearing, 24 Nov. 1986, Transcript, vol. 1-2, p. 24.

39 'Lesage: Aucune entente avec Brinco si l'énergie ne nous appartient pas,' *Le Devoir*, 5 May 1965, p. 1. Smith, *Brinco*, p. 204.

40 *Le Devoir*, 5 May 1965; Smith, *Brinco*, p. 217.

41 Canada, NEB, 'Inter-Utility Trade Review: Transmission Access and Wheeling' (Calgary: National Energy Board, 1992), pp. 1-6.

42 Canada, Cabinet Conclusions, Meeting Day 6 May 1965, 'Hamilton Falls Power Project: Discussion with Premier Lesage and Premier Smallwood' (source in Ottawa: National Archives, RG2, vol. # 6271).

43 Smith, *Brinco*, p. 262.

44 Canada, NEB, Hearing, Order No. 1-86, '... Regulations ... Exports,' held in Ottawa, 24 November 1986, Transcript, vol. 1-2, p. 81.

45 For instance, when Noel Clarke (representing Newfoundland) probed further during the 1986 NEB hearings, he asked, 'Can you think of any other kinds of obstacles to having a wheeling arrangement [renting a transmission-line capacity from Hydro-Québec] with the Lower Churchill?' Guevremont (representing Hydro-Québec) replied, 'Cette question contient je ne sais pas combien de pièges, et je ne sourais répondre à une question aussi piégée.' Ibid., p. 145.

46 Jacques Guevremont, Executive Vice President of External Markets, Hydro-Québec, Canada, NEB, Hearing, '... Regulations ... Exports,' held in Ottawa, Order No. 1-86, 25 November 1986, Transcript, vol. 2-1, pp. 191-2.

47 See Québec, Québec Hydro-Electric Commission, 3 October 1966, regarding a resolution to be approved by the Lieutenant-Governor-in-Council to make legal the '... the draft Power Contract to be signed between CFLCo and Hydro-Québec ... to purchase electric energy from CFLCo ...' (source in Ottawa: National Archives, MG III 73, vol. 58, file 11), pp. 1, 2. This was a draft version of the order-in-council to which was attached a letter discussing changes to this order-in-council from the law firm Ogilvy, Cope, Porteous, Hansard, Marler, Montgomery, and Renault, 15 May 1968, addressed to the Law Department, Québec Hydro-Electric Commission, Montreal.

48 Canada, NEB, *Report to the Governor in Council: In the Matter of an Application under the National Energy Board Act of Québec Hydro-Electric Commission, September 1976* (Ottawa: National Energy Board Library), p. 5. For transmission-line capability, see Energy, Mines and Resources Canada, *Electric Power in Canada, 1988* (Ottawa: Minister of Supply and Services Canada, 1989), Table 7.3, p. 52. For turbine and generator installation period, see 'CFLCo, Churchill Falls,' Statistics Canada, Electric Power Statistics: Inventory of Prime Mover and Electric Generating Equipment (Ottawa: Minister of Supply and Services, 1987), cat. no. 57-206, p. 22.

49 Hydro-Québec projected that its power surpluses during the most critical October periods would increase from 3,342 MW in 1977, to 3,694 MW in 1987, to 5,224 MW in 1992 (entire capacity of Churchill Falls), and to 8,239 MW in 1998; see Canada, NEB, *Report ... of an Application under the NEB Act of Québec Hydro-Electric Commission [Hydro-Québec], Sept. 1976*, Appendix 5.

50 Canada, NEB, *Reasons for Decision: In the Matter of an Application under the National Energy Board Act of Hydro-Québec, January 1984* (Ottawa: Minister of Supply and Services, 1984), pp. 1, 7.

51 Ibid., p. 91.

52 Ibid.

53 Ibid., p. 40.

54 Noel Clarke, representing Churchill Falls (Labrador) Corporation, questions Guevremont, representing Hydro-Québec, 'What you are saying is: "Let's leave it all to free negotiations between utilities." But, it appears that it would have to be entirely up to you [Hydro-Québec] in terms of whether access [of Labrador electricity to provincial or US networks] would be provided. You might not be prepared to provide access on any terms.' Canada, NEB, Hearing, '... Regulations ... Exports,' held in Ottawa, Order No. 1-86, 24 November 1986, Transcript, vol. 1-2, pp. 136-7.

55 Edward Greenspon and Barrie McKenna, 'Tobin warns he'll pull Churchill plug,' *Globe and Mail*, 21 September 1996, front page.

56 The Foreword by James Schlesinger states: 'No higher pragmatism can be imagined than the joint use – for the benefit of both the American and Canadian economies – of this unutilized and renewable resource [hydro power from the north]'; in Robert Bourassa, *Power from the North* (Scarborough: Prentice-Hall, 1985), pp. ix, x.

57 Konrad Yakabuski, 'U.S. energy regulators push Québec on Churchill Falls,' *Globe and Mail*, 6 June 1997, p. B6.

58 Ibid.

59 Ibid.

60 'Power Contract Hydro-Québec and CFLCo, 12 May 1969,' p. 15.

61 Newfoundland, Supreme Court, ... Attorney General ... and CFLCo ... Judgment of Goodridge, J., 6 July 1982, p. 376. The contract extends over forty years and is automatically

renewed for another twenty-five years. For renewal terms, see 'Power Contract Hydro-Québec and CFLCo, 12 May 1989,' pp. 1, 7; the year 2041 is from Greenspon, et al., 'Tobin warns he'll pull Churchill plug,' *Globe and Mail*, 21 September 1996, p. A1.

62 Smith, *Brinco*, pp. 132-3.

63 'Power Contract Hydro-Québec and CFLCo, 12 May 1989,' Delivery Point, p. 16.

64 'Brinco in Newfoundland: A Summary,' p. 15.

65 Smith, *Brinco*, p. 376.

66 Ibid., p. 375.

67 Ibid.

68 Ibid.

69 Brinco arranged $840 million financing in New York and Hydro-Québec bought $125 million in mortgage bonds to finance the $946 million Churchill Falls project. See 'Brinco in Newfoundland: A Summary,' pp. 4, 7.

70 'Power Contract Hydro-Québec and CFLCo, 12 May 1969,' p. 27.

71 'Brinco in Newfoundland: A Summary,' p. 10.

72 CFLCo, see 'Remarks by Roy C. Legge, General Manager, Head Office Brinco Limited, at the 59th Annual Convention of the General Accountants Association of Canada, Friday, July 7, 1972' (source in Ottawa: National Archives, MG 28, III 73, vol. 1, file 7).

73 CFLCo, 'The Churchill Falls Power Development: Its Transport and Transportation Information System.' This report contains portions of papers by H.L. Snyder, R.D. Boivin, P.A.T. Keeping, J.S. Howard, J.A. Harrower, and T.W. Kierans (source in Ottawa: National Archives, MG 28, III, 73, vol.1, file 9).

74 CFLCo, 'Turbine and Generator Contract, Press Release Churchill Falls,' 30 June 1967 (source in Ottawa: National Archives, CFLCo, MG 28, III 73, vol. 59, file 7). See transcribed press release in Appendix 9, this book.

75 Smith, *Brinco*, pp. 341-2.

76 CFLCo, Minutes of a Meeting, 'Confidential Turbine and Generators,' held at CFLCo office, Friday, 9 June 1967 at 2:30 p.m. Present were representatives from CFLCo (McParland, Wermelinger), from Acres Canada Bechtel (Ressegieu, Lazier, Thomson), and the Churchill Falls (Machinery) Consortium (Lewis, Lunan, Pontbriand), see subsection 'Québec Manufacture' (source in Ottawa: National Archives, CFLCo, MG 28, III 73, vol. 59, file 7), p. 7.

77 See CFLCo, Donald McParland, president, letter of intent regarding the 'Churchill Falls (Labrador) Power Project + Supply of Turbines and Generators,' dated 28 June 1967 (Ottawa: National Archives, CFLCo Manuscripts, MG 28, III 73, vol. 59, file 7).

78 'Giant powerhouse start soon,' *Montreal Star*, 25 April 1968.

79 CFLCo, 'Turbine and Generator Contract, Press Release Churchill Falls,' 30 June 1967, p. 2. See transcribed press release in Appendix 9, this book.

80 CFLCo, see letter from Acres Canadian Bechtel, signed for F.E. Ressegieu, to Mr. McParland of Hamilton Falls Power Corporation [renamed CFLCo], 6 November 1969 (source in Ottawa: National Archives, MG 28 III 73, vol. 54, file 4). See also Churchill Falls (Labrador) Corporation, *An Energy Giant: The Churchill Falls Power Development* (Labrador: Churchill Falls, no date), a seventeen-page brochure available from the CFLCo in March 1998.

81 This is an indirect linkage to a development sector such as the expansions in the hydro-electric infrastructure – the linkage is merely fiscal rather than physical. The Churchill Falls project in Labrador was not connected by transmission lines to the Island of Newfoundland.

82 'Brinco in Newfoundland: A Summary,' p. 4.

83 Ibid.

84 Ibid.

85 Canada, Science Council of Canada, 43rd Meeting, Science Council of Canada: Visits to Churchill Falls Hydro Electric Development and Iron Ore Company's Developments at Wabush City and Sept-Iles. 2-5 October 1973, p. G-3. See also Smith, *Brinco*, p. 160.

86 Smith, *Brinco*, p. 160.

87 Ibid., p. 104.

88 Newfoundland, *Report of the Royal Commission on Electrical Energy, February 1966* (Ottawa: National Energy Board Library), pp. 60-2.

89 Huet Massue, 'Highlights of Electric Power in Canada' (Montreal: Shawinigan Water and Power Company, 1954); electricity cost as a percentage of production in finished goods and semi-processed goods manufacturing was cited in Newfoundland, *Report of the Royal Commission on Electric Energy, February 1966*, pp. 61-2.

90 Ibid.

91 'A New Industrial Horizon presented by Newfoundland,' *Daily Commercial News and Building Record* [Toronto], 27 March 1968. A thirty-page-page tabloid section of the paper promoted Newfoundland as an industrial province.

92 Newfoundland, Government of Newfoundland, Gordon F. Pushie, Commission Chairman, *Report on the Royal Commission on the Economic State and Prospects of Newfoundland and Labrador* (St. John's: Queen's Printer, Dec. 1987), pp. 275-8.

93 'Churchill Control by Britain, U.S.' *Electrical Business*, 26 August 1968. Note also that in 1967, the company [CFLCo] was owned 72% by BRINCO, with 17% of the shares held by Hydro-Québec. The balance was held by the Newfoundland Department of Resources and by Rio Algom Mines, each with approximately 5% of the shares.

94 'Agreements are signed,' *Evening Telegram*, St. John's, Newfoundland, 11 June 1968. For a critique of the loan guarantees Newfoundland provided for this project, see Patrick Fellows, 'All Newfie really needs is an economic miracle,' *Toronto Star*, 31 May 1968.

95 Ibid.

96 Don McParland, President of CFLCo, 'Brinco and the Churchill Falls Development,' Speech to the Canadian Club of Toronto, 23 January 1967 (source in Ottawa: National Archives of Canada, MG 28, III 73).

97 Churchill Falls (Labrador) Corporation, *An Energy Giant: The Churchill Falls Power Development* (Churchill Falls, Labrador, no date), a glossy seventeen-page brochure, received in March 1998 from CFLCo in St. John's, Newfoundland, p. 7.

98 Lower Churchill Development Corporation, *1996 Annual Report*, St. John's, Newfoundland, 1996, p. 3.

99 Bob Benson, 'The deal that never was,' *Evening Telegram*, St. John's, Newfoundland, 22 February 1998. In 1998, former Newfoundland Premier Frank Moores recalled that 'he had worked out a deal' between his province and Québec with Québec's Premier René Lévesque in the 1970s 'to jointly harness the hydroelectric potential of the Lower Churchill River,' but because of intense opposition in Moores's Cabinet 'that deal was never inked.' This deal with Lévesque needs further study. Bob Benson, 'Lower Churchill revisited,' *Evening Telegram*, St. John's, Newfoundland, 25 January 1998.

100 Canadian Press, 'Brian Tobin will meet his Quebec counterpart ...,' *Evening Telegram*, St. John's, 4 March 1998, p. 1. Vancouver *Sun*, 'Innu disrupt power talks in Labrador,' Vancouver, 10 March 1998, p. A5.

101 Bob Benson, 'The deal that never was,' *Evening Telegram*, St. John's, 22 February 1998. Tobin on 27 June 1997 reasoned in a similar way when he said 'God bless FERC.' Konrad Yakabuski, 'U.S. energy regulators push Québec on Churchill Falls: Developments in Washington give impetus to break the deadlock with Newfoundland over power projects in Labrador,' *Globe and Mail*, 6 June 1997, p. B6.

102 Canada, National Energy Board, 'Review of Inter-Utility Trade in Electricity,' Ottawa: Minister of Public Works and Government Services Canada, 1994, p. 18.

103 Hydro-Québec, *Strategic Plan 1998-2002*, p. 1.

104 Chris Flanagan, 'Project stands on own: Furey,' *Evening Telegram*, St. John's, Newfoundland, 23 February 1998.

105 'No progress yet in Lower Churchill talks, *Globe and Mail*, 20 January 1998.

106 Peter Fenwick, 'Innu owed a debt in new power deal,' *Sunday Telegram*, 25 January 1998.

107 Glen Whiffen, 'Innu Nation wants a say in talks on Lower Churchill development,' *Evening Telegram*, St. John's, 17 January 1998.

108 NTV TV *Evening News*, 6:00 p.m., Newfoundland, 3 March 1998.

109 'Innu disrupt power talks in Labrador,' *Vancouver Sun*, 10 March 1998, p. A5.

110 Ibid.

111 Canada, Natural Resources Canada, *Electric Power in Canada, 1996* (Montreal: Canadian Electricity Association, 1997), Table 5.4, p. 51.
112 Chris Flanagan, 'No easy task for negotiators to anticipate future needs,' *Evening Telegram,* St. John's, Newfoundland, 4 March 1998.
113 Ibid.
114 The price of 30.6 mills per kWh was published by Natural Resources Canada, *Electric Power in Canada, 1994,* Table 8.7, p. 87. The one-quarter of a cent is from the 'Power Contract Hydro-Québec and CFLCo, 12 May 1969,' p. 16.
115 Jean-Thomas Bernard is cited by Edward Greenspon and Barrie McKenna, in 'Tobin warns he'll pull Churchill plug,' *Globe and Mail,* 21 September 1996, p. A3.
116 Canada, NEB, 'The Regulation of Electricity Exports: Report ... 1986;' see section 'The Statement of National Power Policy,' pp. 8-9.
117 For export to foreign markets see Canada, NEB, Hearing, Order No. 1-86, '... Regulations ... Exports,' held in Ottawa, 27 November 1986, Transcript, vol. 4-4, pp. 876, 886.
118 Hydro-Québec, *Strategic Plan 1998 to 2002,* p. 1.

Chapter 6: Nelson River Power

1 Canada, Energy Mines and Resources, Federal-Provincial Nelson River Review Committee, G.M. MacNabb, chairman of Review Committee and Assistant Deputy Minister (Energy Development) of Energy, Mines and Resources, *First Annual Report of Federal-Provincial Nelson River Review Committee, Canada, Manitoba Agreement, Development of Hydroelectric Potential of the Nelson River, March 31, 1970* (Ottawa: Energy Mines and Resources, 1970; source in Winnipeg: Provincial Archives of Manitoba, reference, 550.1), p. 6.
2 Manitoba, Commission of Inquiry into Manitoba Hydro, letter from B.E. Maxfield, Chief Engineer of Inco Metals Company, Thompson, to G.E. Tritschler, Manitoba Hydro Inquiry, Winnipeg, 30 November 1978 (source in Winnipeg: Legislative Library, Information Services). See also Manitoba Hydro, Public Affairs, *A History of Hydro-Electric Power in Manitoba,* May 1986, p. 14.
3 This term is modified from the term 'unplanned oversupply' used by Gatt-Fly, *Power to Choose* (Toronto: Between the Lines, 1981) pp. 66-7.
4 H.V. Nelles, 'Public Ownership of Electrical Utilities in Manitoba and Ontario, 1906-30,' *Canadian Historical Review* 57, 4 (1976): 473.
5 Manitoba, Public Utilities Commissioner, 'Report on the Projected Hydro-Electric System for the Province of Manitoba by the Public Utilities Commissioner, 1913,' by H.A. Robson, Commissioner, Winnipeg, 3 February 1914, to Hon. J.H. Howden, Attorney-General of Manitoba, Winnipeg (source in Winnipeg: Provincial Archives of Manitoba, RG 22, box 6).
6 Robson, 'Report on the Projected Hydro-Electric System, 1913,' p. 1. Nelles claims Robson's report was significant in firmly squelching the idea' of the 'provincially owned hydroelectric system' when Premier R.P. Roblin referred this matter to him for an opinion in 1914 (probably in 1913, since Robson on 3 February 1914 mentions that a year has passed since the resolution was approved by the Legislature), see Nelles, 'Public Ownership,' p. 471. Furthermore, Nelles regards Robson as a commissioner who was appointed by Premier R.P. Roblin because he shared Roblin's view of being 'privately hostile and publicly lukewarm toward public power,' in that Robson shared his concern to protect the private property of the [Winnipeg Electric Railway Company]; see Nelles, 'Public Ownership,' p. 471. A closer reading of Robson's 1914 memo reveals that he may have respected private property, but called for repossession of dormant water rights and a halt to granting any further rights to the private sector so that they could be held for public uses (p. 3). In contrast to Nelles, I value different aspects of Robson's memorandum, such as its early historic vision of public hydro-electric development on northern and southern rivers, which became a reality in subsequent decades.
7 Robson, 'Report on the Projected Hydro-Electric System, 1913,' pp. 3-4.
8 Ibid., pp. 3-4.
9 Ibid., p. 4.
10 Ibid.
11 Ibid., pp. 7-9.

12 Ibid., p. 2
13 Ibid., p. 5.
14 Ibid.
15 Ibid., p. 11.
16 Ibid.
17 Ibid., p. 12.
18 Ibid.
19 The L.E. Myers Company, Construction and Management of Public Utilities, prepared this survey 'for distributing power over a portion of Manitoba,' on behalf of Mr. Samuel Insull, President, Middle West Utilities Company, Chicago, Illinois, 12 July 1916 (source in Winnipeg: Provincial Archives of Manitoba, RG 22, box 6), pp. 22, 50.
20 Ibid., p. 50.
21 The Myers survey reports that the 'The Brandon Electric Light Co. is the only power producer. There is some uncertainty as to the ownership of the property. It was originally owned by Judge Walker of Winnipeg and Messrs. Geo. Patterson, who is the Manager, and E.L. Christie of Brandon, Manitoba.' [possible co-ownership with Abbott-Eaton Company], p. 9.
22 One of many case studies in the Myers survey describes how small enterprises had modern electric motors, but often did not use them because of inflated electricity prices: 'The Brandon Creamery & Supply Co. has 25 H.p. connected and uses steam power in its plant as much as possible owing to the high power rate charged. It also has a 50 H.p. motor for its refrigerating plant but does not use the motor for the same reason. An 85 H.p. producer gas unit has been used for the compressor but at present natural ice is used. As much as $500.00 per month has been paid for power when running the motor driven compressor. At present the daily use is only about 15 K.W. hours and for this the electric company charges 10 cents net per K.W. Hr.' Ibid., p. 17.
23 Not only entrepreneurs, but also the municipality of Brandon could not afford the unreasonable cost of privately produced electricity: 'For instance the City is now about to buy a 500 H.p. pump to meet its increased demand and will probably install steam power, instead of electricity, which it would use if given a reasonable rate. The City's coal bill now runs from $17,000 to $20,000 per year for pumping only.' Ibid., p. 14.
24 Manitoba Hydro, Public Affairs, 'Formation of Manitoba Hydro,' a one-page chronology of the formation of Manitoba Hydro, revised August 1989.
25 The City of Winnipeg manager, J.G. Glassco, noted that the private sector had acquired the more lucrative plants. The Myers survey paraphrases Glassco as saying: 'the privately owned plant in Winnipeg had very much the best of the situation there; indeed, [Glassco] felt it would continue to have it until its franchise expired.' The L.E. Myers Company Survey, p. 2.
26 Nelles, 'Public Ownership,' pp. 464-5.
27 Ibid., p. 465.
28 Manitoba Hydro, 'Formation of Manitoba Hydro.'
29 Manitoba Hydro, Public Affairs, *A History of Hydro-Electric Power in Manitoba*, May 1986, p. 6.
30 Ibid.
31 Nelles, 'Public Ownership,' p. 466.
32 Robson, 'Report of the Projected Hydro-Electric System,' p. 13.
33 Nelles, 'Public Ownership,' p. 471.
34 Manitoba Hydro, Public Affairs, *A History of Hydro-Electric Power in Manitoba*, May 1986, p. 8.
35 Nelles, 'Public Ownership,' p. 474.
36 Robson, 'Report of the Projected Hydro-Electric System,' p. 3.
37 Nelles, 'Public Ownership,' pp. 475-6.
38 Ibid., pp. 481, 482.
39 Manitoba Hydro, 'The Formation of Manitoba Hydro.'
40 Manitoba Hydro, Public Affairs, 'What is Manitoba Hydro?' a nine-page brochure, dated 87 11 [November 1987], p. 1.

41 Ibid., pp. 2-3.
42 Ibid., p. 6.
43 Duff Roblin, at the Provincial Premiers' Conference, Victoria, British Columbia, 6-7 August 1962. Proceedings, 6 August 1962 (source in Victoria: Legislative Library, CAN ZC 1962), p. 164.
44 Ibid.
45 Canada, Energy Mines and Resources, Federal-Provincial Nelson River Review Committee, *First Annual Report*, 1970, p. 6.
46 Ibid.
47 Ibid. pp. 7-8.
48 Manitoba Hydro, Marc Eliesen, 'Chairperson's Message,' *Annual Report, 1987*, p. 6.
49 Clark-Jones found that the Paley Report singled out what she called the 'strategic' nature of nickel 'being what the American military termed "true war metal."' It became part of the stockpile program between 1956 and 1961. The demand for Inco's nickel was artificial: 'Production diversification had preceded to its logical (profitable) extreme in Canada with exploitation of both Sudbury and Manitoba ores on the basis of the artificially induced levels of high demand.' Melissa Clark-Jones, *A Staple State: Canadian Industrial Resources in Cold War* (Toronto: University of Toronto Press, 1987), pp. 122, 142, 151-2.
50 Because this was a control structure regulating the Nelson River as outflow from Lake Winnipeg, special turbines and generators were needed. See L.A. Bateman, chair of Manitoba Hydro, 'The Nelson River: An Energy Source,' paper presented at the 89th Annual Congress of the Engineering Institute of Canada, Winnipeg, Manitoba, 30 September to 3 October, 1975, pp. 9, 11. The prices are discussed, and warnings about the Leningrad manufacturer's history of inability to meet delivery dates and manufacturing standards are given. See Manitoba, Commission of Inquiry into Manitoba Hydro, Final Report, by G.E. Tritschler, Commissioner, December 1979 (Winnipeg), pp. 156, 455-6.
51 Manitoba, The Bertram Study Group, chair P.C. Hensman, Consultant, Manitoba Hydro, 'The Pas Machinery Manufacturing Project: Report of The Bertram Study Group on the Feasibility of Manufacturing Generating and Associated Hydraulic Equipment for Manitoba Hydro at the Bertram Plant, The Pas, Manitoba' (Winnipeg: Bertram Study Group, December 1973), pp. 2-1 to 2-3.
52 Ibid., p. 2-1.
53 Ibid., pp. 2-1 to 2-3.
54 Letter from L.A. Bateman, chair and chief executive officer, Manitoba Hydro, addressed to W.R.C. Blundel, executive vice-president, Apparatus and Heavy Machinery, Business Division, Canadian General Electric Company Limited, Montreal, 17 December 1973; included in The Bertram Study Group, 'The Pas Machinery Manufacturing Project,' as Appendix IV.
55 Manitoba Hydro, *Annual Report 1987*, p. 12; *A History*, p. 25 and 'Supplement.'
56 Manitoba Hydro, *Annual Report 1986*, p. 12.
57 Manitoba Hydro, *A History*, p. 25 and 'Supplement;' *Annual Report 1987*, pp. 11-12.
58 Interview with E.R. (Ruth) Kristjanson, senior economic consultant, Economic Analysis Department, Manitoba Hydro, Winnipeg, 24 August 1989.
59 Manitoba Hydro, *Hydro Gram*, A weekly newsletter about Manitoba Hydro's activities, 24 August 1989, vol. 29, no. 21, photo caption on front page.
60 For a content listing of the Canadian General Electric benefits package, see Manitoba, Manitoba Energy Authority, *Annual Report Year Ended March 31, 1985*, p. 15.
61 Ibid.
62 In industrial offset agreements, the purchaser invites the tenderer to propose activities producing economic benefits to Manitoba that are not directly related to the work on hydro projects; for example, activities such as employment, research and development, transfer of technology, business investment, and starting affirmative action ventures for target groups (i.e., northern Native populations). See Manitoba Hydro, Purchasing Department, 'Instructions to Tenderers,' Specification No. 1950, paragraph 10.2 *Industrial Offsets* (Winnipeg: Manitoba Hydro, received 23 August 1989), pp. 6-7.
63 Manitoba, Manitoba Energy Authority, *Annual Report, 1985*, p. 15.

64 Such manufacturing would take place in a local business. 'A "Manitoba business" is a business which is registered to do business in the Province of Manitoba, and the firm, or its principals, maintains in Manitoba on a continuous basis, the facilities, equipment and staff necessary to perform the work required, or to manufacture, or assemble, or supply, the work.' Al Kazima, 'Manitoba Business Involvement' (Winnipeg: Manitoba Hydro, August 1989).
65 Manitoba Hydro, *Annual Report*, 1987, p. 6.
66 Canada, National Energy Board, 'Application to the National Energy Board for a Certificate of Public Convenience for an International Power Line and a Licence to Export Power,' vol. 1, by the Manitoba Hydro Electric Board, Winnipeg Manitoba, 1975 (source in Ottawa: National Energy Board Library, 1975); see Exhibit 3 entered during the public NEB hearings in Winnipeg, 13 January 1976, p. 11.
67 Manitoba Hydro, *What is Manitoba Hydro?*, p. 6.
68 Clark-Jones, *A Staple State*, pp. 122, 142, 152.
69 Manitoba Hydro, Manitoba Hydro Systems Load Forecast, 1983/84 to 2003/04, p. 41.
70 Ibid., p. 42.
71 Interview with John F. Funnel, Q.C., general counsel and secretary, in his office at Manitoba Hydro in Winnipeg, 23 August 1989.
72 John E.G. Giesbrecht, the librarian of the Business Resource Centre, Manitoba's Department of Industry, Trade and Tourism, informed me in August 1989 that only the most current business resource material is kept in the library, and it is unlikely that 1970s industrial promotion documentation was kept.
73 Manitoba, Department of Industry and Commerce, 'An Assessment of the Impact of Nelson River Electric Power on Industrial Development in Manitoba,' by Arthur D. Little, Inc., of Cambridge MA, June 1966 (source in Winnipeg: Manitoba Hydro Corporate Library).
74 Ibid., p. 28.
75 Ibid., p. 41.
76 Ibid.
77 Walter C. Newman, *What Happened When Dr. Kasser Came to Northern Manitoba* (Winnipeg: Newmac Publishing Company, 1976), p. 1. See also Philip Mathias, '"Good Deal for Manitoba": the Churchill Forest Industries Project,' in his book *Forced Growth: Five Studies of Government Involvement in the Development of Canada* (Toronto: James Lewis and Samuel, 1971).
78 Newman, *What Happened When*, p. 1.
79 Edward Schreyer, Premier of Manitoba, 'Address to Canadian Electrical Week Luncheon,' 14 February 1974, p. 5. See also Schreyer's 'Opening Remarks for the First Ministers Conference on Energy,' Ottawa, 22 January 1974 (source in Winnipeg: Provincial Archives of Manitoba).
80 Manitoba, Manitoba Energy Authority, 'Chairperson and Executive Director's Message,' by Marc Eliesen, chair and executive director, Manitoba Energy Authority, *Annual Report 1985/86*, p. 4.
81 Interview with Garry R.S. Hastings and Henryk Mardarski, Winnipeg, 21 August 1989.
82 Manitoba, Manitoba Energy Authority, *Annual Report, 1987*, p. 17.
83 Manitoba, Manitoba Hydro, *Annual Report 1986*, pp. 14-15.
84 Ibid., p. 15.
85 Manitoba, Manitoba Energy Authority, *Annual Report, 1988*, p. 5.
86 Interview with Garry R.S. Hastings, executive officer, and Henryk Mardarski, information access officer, in the Lombard Ave. offices of the Manitoba Energy Authority, Winnipeg, 21 August 1989.
87 Ibid.
88 Manitoba, Manitoba Energy Authority, *Annual Report, 1985*, p. 16.
89 Alcoa and an official from the Manitoba Energy Authority agreed to the 'gag order,' which I understand means that negotiating partners cannot release details about their negotiations. Ron W. Pritchard, Director of the Energy Policy Branch of Manitoba Energy Mines and Resources, stated that another aluminum company, Alcan, asked for part-ownership

of the Limestone plant in return for locating in Manitoba; to the Pawley government, this was an unacceptable demand. Interview with Ron. W. Pritchard, 24 August 1989, the Offices of Manitoba Energy Mines and Resources, in Winnipeg.
90 Manitoba Hydro, *Annual Report*, 1986, p. 12.
91 The task force consisted of L.A. Bateman, Hydro chair; J.S. Anderson, deputy minister of finance and hydro; H.L. Briggs, of Ottawa, a consulting engineer; Eric W. Kierans, then a professor with the Department of Economics at McGill University; and Marc Eliesen, then secretary of the planning secretariat of the Manitoba Cabinet. In 1973, this task force submitted to the Schreyer government the 'Report of the Examination of a Proposal for the Long Term Export of 1000 MW to Northern States Power Company.' This controversial report was reviewed in 1976 by the Commission of Inquiry into Manitoba Hydro, and the critical comments about Manitoba's over-investment in hydroelectric facilities by Kierans (contained in the 1973 task force report) and by Commissioner Tritschler were published in the commission's final report to D.W. Craik, minister charged with the administration of the Manitoba Hydro Act, in December 1979. Manitoba, Commission of Inquiry into Manitoba Hydro, Final Report, by G.E. Tritschler, Commissioner, December 1979, Winnipeg, pp. 223-7.
92 Manitoba, Manitoba Hydro Task Force, 'Report on the Examination of a Proposal for the Long Term Export of 1000 MW to Northern States Power Company (1976 Sessional Paper # 94, 1973), p. 3.
93 Manitoba Hydro Task Force, 'Report on the Examination,' Subsection 'Manitoba Hydro: Commentary Relating to December Task Force Report on Examination of a Proposal for the Long Term Export of 1000 MW to Northern States Power Co.,' p. 1.
94 Ibid.
95 Ibid., subsection 'V. Conclusions and Recommendations,' no page numbers.
96 Ibid., subsection VII Addendum, 'Comments by Mr. E. Kierans on Hydro Task Force,' p. 10. See also Manitoba, Commission of Inquiry into Manitoba Hydro, Final Report, by G.E. Tritschler, pp. 224, 227.
97 Ibid., subsection VII Addendum, 'Comments by Mr. E. Kierans on Hydro Task Force,' p. 10.
98 Ibid.
99 Ibid.
100 Ibid.
101 Letter from Edward Schreyer, premier of Manitoba, to Donald S. Macdonald, minister of energy, mines and resources, 28 November 1973 (Winnipeg: Provincial Archives of Manitoba, Schreyer Papers). Messrs. Bateman, Fraser, and Aaranson of Manitoba Hydro met in Winnipeg with Messrs. MacNabb and Humphrys of the Ministry of Energy, Mines and Resources to discuss the acceleration of 'at least two sites on the lower Nelson.' Manitoba, Premier's Office, letter from Donald S. Macdonald, minister of energy, mines and resources, to Edward Schreyer, premier of Manitoba, 7 January 1974 (Winnipeg: Provincial Archives of Manitoba, Schreyer Papers).
102 Manitoba, Premier's Office, letter from Macdonald to Schreyer, 7 January 1974.
103 Ibid.
104 Manitoba Hydro Task Force, 'Report on the Examination,' Subsection 'Manitoba Hydro: Commentary Relating to December 1973 Task Force Report on Examination of a Proposal for the Long Term Export of 1000 MW to Northern States Power Co.,' pp. 1-2.
105 Ibid. p. 2
106 Canada, Office of the Prime Minister, Press Release, 22 January 1974, 'Opening Statement by the Prime Minister of Canada at First Ministers Conference on Energy, Ottawa, January 22, 1974' (source in Winnipeg: Provincial Archives of Manitoba, Schreyer Papers), p. 5.
107 In the same press release of 22 January 1974 Trudeau explained that 'this unusual arrangement was worked out some 12 years ago, in Canadian fashion, as a compromise,' whereby 'Canada is roughly self-sufficient in oil. About half our oil serves Canada west of the Ottawa Valley, and about half is exported to the United States. We replace that half by imports from various parts of the world including Venezuela and the Middle East.' Ibid., pp. 4-5.

108 Schreyer, 'Address to Canadian Electrical Week,' 14 February 1974.
109 In the Mid-continental Area Power Pool, electricity was 72 percent coal-based, with the remainder based on nuclear power, gas, oil, and hydro. Mid-continental Area Power Pool, 'Paths of Power: Discussion Guide' (Minneapolis: Mid-continental Area Power Pool), a pamphlet.
110 Manitoba Hydro, 'Manitoba Hydro Systems Load Forecast, 1983/84 to 2003/04,' p. 48.
111 Canada, National Energy Board, 'Application to the National Energy Board for a Certificate of Public Convenience for an International Power Line and a Licence to Export Power by Manitoba Hydro, Winnipeg, 1969,' entered as Exhibit C during the NEB hearings in Winnipeg, 4 November 1969, p. 9.
112 Manitoba, Commission of Inquiry, Final Report, Tritschler, December 1979, pp. 337-8.
113 Canada, Energy Mines and Resources, Federal-Provincial Nelson River Review Committee, *First Annual Report*, 31 March 1970, p. 6.
114 The way to begin a national power grid was 'by closer interconnection between four Western provinces.' Schreyer, 'Address to Canadian Electrical Week,' 14 February 1974, pp. 4, 7-8.
115 Edward Schreyer, 'Opening Remarks for the First Ministers Conference on Energy,' Ottawa, 22 January 1974, pp. 22-23.
116 Canada, NEB, 'National Energy Board Report to the Governor in Council in the Matter of the Application under the National Energy Board Act of the Manitoba Hydro-Electric Board [Manitoba Hydro], March 1976,' p. 57.
117 Ibid., p. 55.
118 A second hearing in Winnipeg on 13 September 1976 was called to resolve the routing of the 230 kV international power line. Canada, NEB, 'National Energy Board Reasons for Decision in the Matter of the Application under Part IV of the National Energy Board Act of The Manitoba Hydro-Electric Board [Manitoba Hydro], September 1976,' p. 4.
119 Canada, NEB, Report, Manitoba Hydro, March 1976, p. 47.
120 Ibid., p. 48.
121 Ibid., p. 52.
122 Ibid., p. 55.
123 Ibid., p. 59.
124 Ibid. pp. 6-10.
125 Ibid., p. 54.
126 Ibid.
127 Ibid., pp. 64, 3-5. Brooks did not allow firm power (100 MW) and 2,000 GWh of energy, but agreed with three other licences for the period of 1976-86: interruptible power, short-term firm power, and import/export transfer arrangements between Manitoba Hydro and MP&L.
128 Ibid. p. 60.
129 Canada, NEB, Reasons for Decision, Manitoba Hydro, September 1976, pp. 9-11.
130 Canada, *Electric Power in Canada, 1996,* Table A5, p. 149. See also Appendix 16 in this book.
131 Canada, Natural Resources Canada, *Electric Power in Canada, 1994*, Tables 9.2 and 9.3, pp. 96-7.
132 Canada, NEB, 'National Energy Board Reason for Decision in the Matter of the Application under the National Energy Board Act of the Manitoba Hydro-Electric Board [Manitoba Hydro], August 1987,' pp. 17-18.
133 Manitoba Hydro, *A History*, p. 25.
134 Canada, Natural Resources Canada, *Electric Power in Canada, 1993*, p. 5.
135 Canada, *NEB Reasons for Decision, Manitoba Hydro, August 1987*, p. 18.
136 Ibid., Table 2, p. 13.
137 I thank one of the reviewers of this book for pointing out this ex post facto approach to National Energy Board regulation.
138 Manitoba Hydro, 'The Hydro Province,' five pages of description on the website of Manitoba Hydro, 5 April 1998.
139 Hellmut Schroeder-Lanz, 'The Mega-Hydropower Project of "Hydro Manitoba,"' Unpublished essay, Department of Geography, University of Trier, Germany, 1996.

140 Electricity Forum Home Page, 'Manitoba Hydro: Bad News in Ontario May Be Good News for Manitobans,' 1 August 1997. See also, Debate on Bill 55, The Manitoba Hydro Amendment Act, in the Legislature of Manitoba, 11 June 1997, statements by the Leader of the Opposition Mr. Doer.
141 Manitoba Hydro, Manitoba Hydro, 'The Hydro Province,' five pages of description on the website of Manitoba Hydro, 5 April 1998, pp. 2-3.
142 Interview with the Manager of Power Planning and Export Marketing, Manitoba Hydro, 24 March 1998.
143 Manitoba Hydro, 'News Release' (Winnipeg: Manitoba Hydro 10 April 1996), No. 02-96.
144 Bob Brennan, excerpts from his speech on 'utility convergence,' reprinted in *Insights* vol. 7, issue 1, May 1997.
145 *Insights*, 'Changes to the Manitoba Hydro Act Passed,' August 1997, vol. 7, issue 2. See also Manitoba Hydro, *Hydro Gram*, 18 September 1997, vol. 27, no. 25.
146 Manitoba Hydro, 'Changes to the Manitoba Hydro Act' (Winnipeg: Manitoba Hydro, 1997) a three-fold pamphlet summarizing key sections in the Manitoba Hydro Amendment Act (Bill 55).
147 Ibid.
148 Manitoba, Second Reading Debate in the Legislature of Manitoba of 'Bill 55: The Manitoba Hydro Amendment Act,' Wednesday, 11 June 1997. Legislative Assembly of Manitoba, The Standing Committee on Law Amendments, Thursday 12 June 1997.
149 Ibid., Legislature, Mr. Doer, 11 June 1997.
150 Ibid.
151 Legislative Assembly of Manitoba, The Standing Committee on Law Amendments, Thursday 12 June 1997.
152 Manitoba, Steering Committee, Western Electric Power Grid Study Agreement, *Western Electric Power Grid*, vol. I, executive summary, p. 5. Source: Energy Policy Branch of Energy, Mines and Resources, Manitoba. I was made aware of the 'job issue' related to the western grid by Ron W. Pritchard, director of the Energy Policy Branch of Energy, Mines and Resources, Manitoba, in an interview in his Winnipeg office, 24 August 1989.

Chapter 7: Peace, Pulp, and Power Hunger

1 T.L. Sturgess and R.W. Bonner, Department of Industrial Development, Trade, and Commerce, Bureau of Economics and Statistics, Government of British Columbia, *British Columbia Facts and Statistics*, vol. X, 1956, pp. 25-6. R.W. Bonner later became chairman of BC Hydro. In the same publication, the 'Manufacturing' section's opening paragraph began, for seven consecutive years (1956-62), with the same claim: 'Secondary industries are rapidly accounting for a large portion of British Columbia's total production values.'
2 British Columbia, 'Memorandum of Intention made this 16th day of November, 1956, between Birger Strind, director, of Stockholm, Sweden ... acting on behalf of Axel Wenner-Gren of Stockholm, and The Government of the Province of British Columbia, represented by the Minister of Finance,' tabled in the BC legislature, 12 February 1957. Cited in 'Text of Wenner-Gren Memorandum of Intent,' *Western Business and Industry*, vol. 31, March 1957, p. 70.
3 Cecil Maiden, *Lighted Journey: The Story of the B.C. Electric* (Vancouver: BC Electric, Public Information Dept., printed by Keystone Press, 1948), pp. 101-2.
4 Maiden, *Lighted Journey*, pp. 101-13. The turbine installation date is provided in Canada, Statistics Canada, *Electric Power Statistics: Inventory of Prime Mover and Electrical Generating Equipment as of December 31, 1986*, vol. 3, cat. no. 57-206, p. 63.
5 Maiden, *Lighted Journey*, p. 109. Statistics Canada, *Electric Power Statistics*, 1986, cat. no. 57-206, p. 63.
6 Maiden, *Lighted Journey*, p. 109.
7 Ibid., pp. 109-10.
8 BC Government, 'Electric Power: An Act to Provide for Improving the Availability and Supply of Electrical Power,' assented to 28 March 1945, chapter 27, p. 85.
9 Swainson (1979) indicates that, as many small utilities were taken over, the Power Commission's capacity increased from 14,700 to 324,735 KW [14.7 to 325 MW] between

March 1948 and March 1958. Neil Swainson, *Conflict over the Columbia* (Montreal: McGill-Queen's University Press, 1979), p. 33.

10 British Columbia, BC Power Commission, its Journal entitled *Progress* (Aug. 1957), pp. 7, 8.

11 Gordon Shrum, Peter Stursberg, and Clive Cocking, *Gordon Shrum: An Autobiography* (Vancouver: UBC Press, 1986), edited by Clive Cocking, pp. 74-5.

12 Shrum, et al., *Gordon Shrum*, p. 77.

13 Ibid.

14 Ibid., p. 78.

15 Ibid.

16 *Western Business*, 'Text of Wenner-Gren Memorandum,' p. 71.

17 Patricia Marchak, *Green Gold: The Forest Industry in British Columbia* (Vancouver: UBC Press, 1983), p. 308. The group's proposal also included building railroads to tap these resources in northern British Columbia, 'Memorandum of Intention ... between ... Axel Wenner Gren ... B.C. Minister of Finance,' reprint in *Western Business*, p. 70.

18 Shrum, et al., *Gordon Shrum*, p. 79.

19 British Columbia, BC Energy Board, *Report on the Columbia and Peace Power Projects*, Victoria, 31 July 1961, written by chairman Gordon Shrum. W.A.C. Bennett argued that 'opening the North' was necessary because the 'the rich resources of the northern and central regions' were needed for the further development of British Columbia (Budget Speech 1954), cited by Marchak in *Green Gold*, p. 39.

20 BC Energy Board, Shrum, *Report on the Columbia and Peace*, p. 28.

21 Shrum, et al., *Gordon Shrum*, p. 79.

22 Ibid., p. 81.

23 BC Laws, Statutes, etc. An act to provide for the reorganization of the British Columbia Electric Company Limited and the development of power resources. *Statutes of B.C.* 1961, 2nd Session, chapter 4. An Act to Amend the Power Development Act, 1961 (British Columbia: A. Sutton Queen's Printer, 1962), chapter 50.

24 Ronald Worley, *The Wonderful World of W.A.C. Bennett* (Toronto: McClelland and Stewart, 1971), p. 233.

25 Worley, *The Wonderful World*, pp. 234-5.

26 Shrum, et al., *Gordon Shrum*, p. 85.

27 Ibid., p. 86.

28 Canada, Cabinet Conclusion, 'CR [Columbia River] terms of reference,' date of Cabinet meeting 9 March 1962 (Ottawa: National Archives, RG2 A5a, Cabinet Conclusions, vol. 6192). See also Shrum, et al., *Gordon Shrum*, p. 93.

29 Cited by R. Worley in *The Wonderful World*, p. 235.

30 Paddy Sherman, *Bennett* (Toronto: McClelland and Stewart, 1966), p. 255.

31 Dal Grauer was a political economist who taught social science at the University of Toronto; he became the general secretary of 'B.C. Electric ... which gave him charge of the Public Utilities Commission Data, and of the rate and wage hearings.' Maiden, *Lighted Journey*, p. 156.

32 Ibid., p. 160.

33 Ibid.

34 Glen Williams, 'Canada in the International Political Economy,' in *The New Canadian Political Economy*, edited by Wallace Clement and Glen Williams (Kingston: McGill-Queen's University Press, 1989), p. 116.

35 BC Hydro, Industrial Development Department, 'The Pulp and Paper Industry,' by John Raybould; a two-page reprint from this report entitled 'Paper for the World,' in BC Hydro's magazine *Progress*, fall issue, 1966.

36 British Columbia, Ministry of Industry and Small Business Development, Program Directory 1981, vol. No. 12, April 1981, Program Title: 'Incoming Buyers,' p. 10. The 'Incoming Buyers' program description outlined the terms of assistance and purpose: '1. Designed to assist firms in attracting potential buyers from outside the province. Often bringing potential purchasers to view a plant is a valuable sales tool for the firm concerned. 2. Financial support offered consists of return economy excursion air fare for the potential buyer, to a maximum of $2,000. 3. The ministry retains the right to encourage

and assist the prospective purchaser in visiting as many local companies as is deemed appropriate.'

37 BC Hydro, Statistical Analysis Department, T.D. Buchanan, *Pulp and Paper Prospects in B.C.* (Vancouver: BC Hydro, 1964), p. 2.

38 Philippe Faucher and Kevin Fitzgibbins, 'Political Economy of Electrical Power Generation' (Montréal: Université de Montréal and Centre for Research on the Development of Industry and Technology, 1990), pp. 24-5.

39 Faucher, et al., 'Political ... Power Generation,' p. 11.

40 Stephen Tang, the senior electrical equipment inspection engineer of BC Hydro's Quality Control and Inspection Department, interview, 17 July 1986, at his office on Burrard and Nelson streets in Vancouver.

41 Ibid.

42 BC Hydro, Statistical Analysis Department, T.D. Buchanan, *Pulp and Paper*, p. 7. Buchanan interpreted these changes with reference to the Sloan Report, 'The Forest Resources of British Columbia, 1956' (Victoria: Queen's Printer, 1957), p. 306.

43 Ibid., p. 7.

44 Ibid.

45 Ibid.

46 Marchak, *Green Gold*, p. 40.

47 Ibid.

48 BC Power Commission, *Annual Report*, 1956, p. 63.

49 BC Power Commission, *Progress*, 'New Industry Already Promised Arrow Lakes,' July 1951, p. 10. Other references: 'Power Means Progress' (August 1952); Elk Falls Co. Ltd newsprint mill, Alberni Plywood (August 1952).

50 British Columbia, BC Power Commission, C.W. Nash and H.L. Briggs, *Pulp and Paper Opportunities in Central British Columbia,* June 1956, p. ii.

51 BC Power Commission, Nash, *Pulp and Paper*, pp. 15, 17.

52 BC Energy Board, Shrum, *Report on the Columbia and Peace*, p. 28.

53 Marchak, *Green Gold*, p. 309.

54 When I interviewed Mr. B. Crooks (manager of Finlay Forest Products in Mackenzie) and Mr. Karl Baker (logging manager of BC Forest Products in Mackenzie) personally, they stressed the importance of the Williston Lake reservoir on 22 July 1986.

55 B. Crooks; the importance of the reservoir was stressed in an interview on 22 July 1986.

56 Marchak, *Green Gold*, p. 309.

57 British Columbia, BC Utilities Commission, BC Hydro, Table 1, 'Bulk Power,' exhibit 97, *Site C Hearings*, 10 February 1982, sheet 2 of 4.

58 Information from Mr. W. Hurd of the Corporate Communications Department of BC Forest Product's Vancouver office, and the company's logging manager in Mackenzie, Mr. Karl Baker, and a man (who identified himself only as Murray) in the Accounting Department, July and 14 August 1986. Marchak's analysis of the town of Mackenzie is in 'The Instant Town,' *Green Gold*, pp. 303-22.

59 'Peace River still big puzzle: Wenner-Gren project experts' battleground,' *Vancouver Sun*, 8 June 1959, p. 1; an article in a series published by the *Vancouver Sun* on the theme of 'Development of Power' (6-11 June 1959).

60 BC Hydro, Industrial Development Department, *Power Intensive Industries for Peace River At-Site Power* (Vancouver: BC Hydro, August 1963), p. i.

61 Ibid., p. 1.

62 Ibid.

63 Mary Doreen Taylor, 'Development of the Electricity Industry in British Columbia,' MA thesis, Department of Geography, UBC, 1965, p. 177. Taylor took some of this information from a paper by J.V. Rogers, 'Power, the Pathway to Progress.'

64 British Columbia led the way in the country in 1891 by prohibiting raw log exports – an early imposition of a manufacturing condition; in Hugh Aitken, 'Defensive Expansionism: The State and Economic Growth in Canada,' p. 211.

65 The initial function of the British Columbia Energy Board was to enforce cooperation

between the BC Power Commission (a rural public utility) and BC Electric (an urban util-ity) in British Columbia. Shrum et al., *Gordon Shrum*, p. 77.

66 BC Energy Board, Shrum, *Report on the Columbia and Peace*, pp. 28-30.
67 Cited by Shrum, Ibid., p. 29.
68 H.F. Angus, 'Statement by the Chairman of the Public Utilities Commission,' an adden-dum to BC Energy Board, Shrum, *Report on the Columbia and Peace*, p. 31.
69 British Columbia, BC Energy Board (an advisory committee on the power market), *Elec-tric Power Requirements: in British Columbia Projections to 1990* (Victoria: BC Energy Board, 1 May 1971).
70 Ibid., p. 11.
71 Russel Kelly, 'B.C. Has Too Much Dam Power,' 14 February 1984, photocopied article.
72 Albert Sigurdson, 'B.C. Hydro foresees further cuts in capital projects as growth slows,' *Globe and Mail*, 26 September 1983, B1.
73 British Columbia, BC Utilities Commission, *Site C Hearings*, 10 February 1982, Exhibit 97, Table 1; 'B.C. Hydro Bulk Power (1800 Series) Accounts Detail,' Probable Electrical Pro-jected Sales; Exhibit 98, Table 2, 'B.C. Hydro 1981 Bulk Load Projections – "Other Assumed Staging"'; and exhibit 96, Statement of Evidence of Richard Overstall (SPEC)'; these exhibits were presented to and filed by the BC Utilities Commission. See Site C Hearing Transcript, p. 4,331.
74 BC Utilities Commission (BCUC) *Site C Hearings*, 10 February 1982, Fort St. John, Mr. Richard Overstall, vol. 26, p. 4,339.
75 British Columbia, BC Utilities Commission, *Site C Report: Report and Recommendations to the Lieutenant Governor-in-Council* (Vancouver: BC Utilities Commission, May 1983), p. 300.
76 BC Utilities Commission, *Site C Report*. See also, Mark Jaccard, John Nyboer, and Timo Makinen, 'Managing Instead of Building: B.C. Hydro's Role in the 1990s,' *BC Studies* (autumn and winter 1991-92): 104-5.
77 BC Hydro, *1997 Annual Report*. This report indicates BC Hydro's thermal capacity as 1,084 MW.
78 John Sheehan, President, BC Hydro, in *Electricity Today*, Canadian Electricity Forum, no date, but after 1995.
79 Tim Newton, principal engineer at Powerex, the export subsidiary of BC Hydro, indicated that an agreement had been reached in the fall of 1997 that the Columbia power could be returned over existing international power lines or sold in the United States. Personal interview with Tim Newton on 24 March 1998.
80 Canada, National Energy Board, *Reasons for Decision in the Matter of an Application under the NEB Act of B.C. Hydro and Power Authority Part A: Licences*, March 1980, pp. 1-2.
81 Ibid., p. 23
82 Ibid., p. 25.
83 Ibid., p. 32.
84 Canada, NEB, *Reasons ... B.C. Hydro*, March 1980, Appendix 7, p. 1 of 1.
85 Ibid.
86 Ibid., Appendix 6, 1 of 4.
87 Canada, NEB, *Reasons ... B.C. Hydro*, March 1980, p. 10.
88 Gordon Bell, *Victoria Times*, '"Power Hunger" Claims for U.S. Misleading?' 12 August 1961.
89 Ibid.
90 US Department of Energy, Bonneville Power Administration, *Issue Alert*, 'Selling South: BPA Seeks Ways of Marketing Surplus Power to the Pacific Southwest.' Source: BC Utilities Commission, file 14, no date (possibly early 1980s), p. 5.
91 Brian Lewis, business reporter for the *Vancouver Sun*, on CBC AM Radio, 12:50 h, 3 Sep-tember 1985.
92 US Department of Energy, Bonneville Power Administration, *Near Term Intertie Access Pol-icy*, 1 June 1985, p. 10. A copy of this policy was obtained from the BC Utilities Commis-sion, exhibit 21, hearing 4, entered by BC Hydro on 9 January 1986.

93 National Energy Board, *Review of Inter-Utility Trade in Electricity* (Ottawa: Minister of Public Works and Government Services Canada, 1994), p. 19.

94 Canada, National Energy Board, *The Regulatory Agenda* (Calgary: National Energy Board), issue no. 44 (1 March 1993), p. 3; issue no. 62 (1 October 1997), p. 7.

95 British Columbia, Powerex, *Electricity Trade* (Vancouver: Powerex, Director of Communications, no date), an eleven-page pamphlet explaining electricity trade and its benefits for British Columbia.

96 British Columbia, British Columbia Utilities Commission, *Annual Report 1995*, p. 11; see also BCUC's 'The British Columbia Electricity Market Review: Report and Recommendations to the Lieutenant Governor in Council,' September 1995.

97 Marjorie Cohen, 'Public Power and the Political Economy of Electricity Competition,' *Power Grab: The Future of BC Hydro*, a conference co-sponsored by the Canadian Centre for Policy Alternatives, BC Office and the Institute for Governance Studies (Vancouver: Simon Fraser University, Harbour Centre Campus, 10 January 1998), p. 11.

98 Elliot Roseman and Anil Malbotra, 'The Dynamics of Independent Power,' *Public Policy for the Private Sector*, note no. 3 (Washington, DC: World Bank, June 1996). Cited by Marjorie Cohen in 'Public Power,' p. 11.

99 *Globe and Mail*, Oct. 10, 1997. Cited by Marjorie Cohen, 'Public Power,' p. 26.

100 British Columbia Utilities Commission, 'In The Matter of British Columbia Hydro and Power Authority: Wholesale Transmission Service Application, Decision, June 25, 1996' (Vancouver: BCUC), p. 2.

101 Ibid., p. 1.

102 Ibid.

103 Ibid., 'Executive Summary,' p. 1 of 3.

104 BC Hydro, *1997 Annual Report*, p. 1.

105 Rod Nutt, 'Powerex given green light for exports to U.S.,' *Vancouver Sun*, 25 September 1997, section D.

106 Edward Alden, 'Hydro clears export hurdle,' *Vancouver Sun*, 4 July 1997, section D.

107 Ibid.

108 Ibid.

109 Ibid.

110 Ibid.

111 Ibid.

112 Ibid.

113 Rod Nutt, 'Powerex given green light for exports to U.S.,' *Vancouver Sun*, 25 September 1997, section D.

114 Ibid.

115 BC Hydro, Powerex, information sheet provided by Sheena Macdonald, Powerex Information Officer in Vancouver, 30 March 1998.

116 Rod Nutt, 'Powerex given green light for exports to U.S.,' *Vancouver Sun*, 25 September 1997, section D.

117 Edward Alden, 'Hydro clears export hurdle,' *Vancouver Sun*, 4 July 1997, section D.

118 Ibid.

119 Rod Nutt, 'Powerex given green light for exports to U.S.,' *Vancouver Sun*, 25 September 1997, section D.

120 Ibid.

121 Brian Smith with the host of the *Early Edition*, CBC Radio One, Vancouver, 25 March 1998, 7:45 a.m.

122 BC Hydro, Power Smart Marketing, *The Power Is Yours: BC Hydro the Power Company of Choice* (Burnaby: A Publication for the Customers and Employees of BC Hydro, spring 1998), a seventeen-page brochure.

123 For an account that favours a market approach, see Dr. Mark Jaccard, Advisor to the Minister of Employment and Investment, British Columbia Task Force on Electricity Market Reform, who wrote 'Reforming British Columbia's Electricity Market: A Way Forward,' Second Interim Report, December 1997 and Final Report, January 1998 (Vancouver: British Columbia Task Force on Electricity Market Reform).

124 Brian Smith with the host of the *Early Edition*, CBC Radio One, Vancouver, 25 March 1998, 7:45 a.m.
125 Ibid. See also BC Hydro, *Current Directions: The Power Is Yours*, a publication for BC Hydro customers, a two-page insert mailed with hydro bills (Vancouver: BC Hydro, April/May 1998).
126 Ibid.
127 Ibid.

Chapter 8: Conclusion
 1 Cass-Beggs, 'Economic Feasibility of Trans-Canada Electrical Interconnection.' This paper was presented to the Canadian Electrical Association, Western Zone Meeting, Edmonton, Alberta, 21-23 March 1960 (New Westminster: BC Hydro Information Centre, Retrieval System No. AK.138), p. 16.
 2 Janine Brodie, 'The New Political Economy of Regions,' *Understanding Canada: Building on the New Canadian Political Economy* (Montreal: McGill-Queen's University Press, 1997), p. 254.
 3 The Cabinet discussed the premier of BC's intent to finance the Peace River scheme by selling the extra power generated at US plants because of storing water in Columbia River reservoirs in BC. Canada, Cabinet Conclusions, 9 March 1962 (Ottawa: National Archives, RG2 A5a, vol. # 6192)
 4 Canada, Cabinet Conclusions, Meeting Date 22 August 1964, 'National Energy Policy' (source in Ottawa: National Archives, RG2 A5a, Cabinet Conclusions, vol. # 6264).
 5 Canada, National Energy Board, *Inter-Utility Trade Review: Inter-Utility Cooperation* (Calgary: National Energy Board, 1992), p. 2-12.
 6 Ibid., pp. 3-1 to 3-24.
 7 Claus Offe, 'The Theory of the Capitalist State and the Problem of Policy Formation,' *Stress and Contradiction in Modern Capitalism: Public Policy and the Theory of the State*, edited by Leon Lindberg, et al. (Lexington, MA: Lexington Books, 1975), p. 128.
 8 Claus Offe, 'Tauschverhältnis und Politische Steuerung: Zur Aktualität des Legitimationsproblems,' in *Strukturprobleme des kapitalistischen Staates: Aufsätze zur Politischen Soziologie* (Frankfurt: Suhrkamp, 1972), p. 54.
 9 Bolduc et al., *Québec: Un Siècle d'Électricité* (no place: Libre Expression, 1979), p. 265.
10 For more details about Brinco, see Chapter 5.
11 Philip Smith, *Brinco: The Story of Churchill Falls* (Toronto: McClelland and Stewart, 1975), p. 376.
12 Wallace Clement and Glen Williams, 'Resources and Manufacturing in Canada's Political Economy,' *Understanding Canada: Building on the New Canadian Political Economy* (Montreal: McGill-Queen's University Press, 1997), p. 54.
13 Ibid., p. 55.
14 Gordon Laxer, 'Foreign Ownership and Myths about Canadian Development,' *Canadian Review of Sociology and Anthropology*, 22, 3 (August 1985): 333. Tom Naylor, *The History of Canadian Business,1867-1914*, Vol. 2, *Industrial Development* (Toronto: James Lorimer, 1975), p. 279.
15 National Resources Canada, *Electric Power in Canada, 1996*, Table 1.4, p. 10. See Chapter 7 for more detail on BC Hydro.
16 Natural Resources Canada, *Electric Power in Canada [Reports of], 1992, 1993, 1994*, Table 1.4, on p. 11 in all three reports. CBC Radio AM, 23 October 1996, repeated a news report of Radio Canada that 6,400 Hydro-Québec employees would be laid off in the next three years. Under the guise of needed efficiency, proposals for privatizing Hydro-Québec's transmission and distribution system were announced together with the layoffs in the same news report. Hydro-Québec *Strategic Plan 1998-2002*, p. 44.
17 Dales, on the other hand, argued that building power projects in advance and then creating a local market has the greatest effect on the growth of diversified manufacturing. John Dales, *Hydroelectricity and Industrial Development, Québec 1898-1940* (Cambridge, MA: Harvard University Press, 1957), pp. 156-7.
18 See Chapter 7 for more details.
19 Natural Resources Canada, *Electric Power in Canada, 1994*, p. 48.

20 As defined in the introduction, load is the amount of electric power or energy consumed by a particular customer or group of customers. A load factor is the ratio of average load during a designated period to peak, or maximum, load in that same period (usually expressed in percent). The energy consumption [demand] in Québec in 1994 was 172,172,000 MWh (*Electric Power in Canada, 1994*, Table 5.1, p. 50), and the peak demand in Québec in the same year was 33,755 MW (ibid., Table 5.6, p. 52). One megawatt is 1,000 kilowatts (ibid, p. 173 , one year has 8,760 hours; the load factor is defined (ibid., p. 48) as the 'energy demand in kilowatt hours divided by the product of the number of hours in the period, multiplied by peak demand in kilowatts;' thus 172,172,000/(33,755 x 8,760) = 58.2 percent load factor (ibid., Table 5.7, p. 53) in the province of Québec for 1994. Natural Resources Canada, *Electric Power in Canada, 1994*, Table 5.7, p. 53.

21 Ibid., p. 53.

22 Canada, House of Commons Debates, First Session: Twenty Sixth Parliament, Vol. IV, 1963, 8 October 1963, p. 3,301.

23 From the reprint of Bill Best's statement in *Intercom*, a BC Hydro employee news bulletin, 16 May 1980 issue. This claim is repeated by C.W.J. Boatman, vice-president Corporate and Environmental Affairs, BC Hydro, in 'B.C. dams are not built with power export in mind,' *Vancouver Sun*, 19 April 1991, p. A-10.

24 British Columbia, Ministry of Energy, Mines and Petroleum Resources, 'Reasons for Decision: in the Matter of an Application for an Energy Certificate under the Utilities Commission Act made jointly by BC Hydro and Powerex [BC Hydro's export subsidiary],' ERC-EOW (9207), 16 September 1992, p. 2.

25 The province of Québec during the mid-1980s changed this mandate to allow building power plants for exports. A similar change in mandate may also have been necessary in Manitoba to build the Limestone project solely for export. See also Section 16 of Manitoba's 1997 Hydro Amendment Act wherein building hydroelectric projects for export to the US is clarified.

26 By April 1992, two major contracts that Hydro-Québec counted on for its further James Bay expansion, the $15 billion Maine contract and the $17 billion New York contract, fell apart. 'Hydro-Québec may cut back,' *Ottawa Citizen*, 28 March 1989, p. C3. Peter Maser and Ian Austen, 'Lost contract a blow for Bourassa,' *Vancouver Sun*, 2 April 1992.

27 Canada, Energy Mines and Resources, 'The Canada-U.S. Free Trade Agreement and Energy: An Assessment,' c. 1988, p. 44.

28 Ibid., p. 25.

29 Ibid., p. 44. See also Chapter 9 of the Free Trade Agreement which refers to the Bonneville Power Transmission issue.

30 Laura Macdonald, 'Going Global: The Politics of Canada's Foreign Economic Relations,' p. 189.

31 George Chuchman, 'Electric Utility Deregulation and Competition Implications for Manitoba,' a paper prepared for CUPE (Canadian Union of Public Employees), Manitoba Regional Office, June 1997, p. 18.

32 Ibid.

33 United States, Federal Energy Regulatory Commission, 'Promoting Wholesale Competition Though Open Access: Non-discriminatory Transmission Service by Public Utilities,' Docket No. RM95-8-001 (Washington: Federal Energy Regulatory Commission, Order No. 888-A, issued 3 March 1997), p. 1.

34 Ronald Osborne, President and Chief Executive Officer, Ontario Hydro, 'Industry Review: Ontario Hydro,' *Connections: 1999 Electricity Industry Review* (Montréal: Canadian Electricity Association, January 1999), p. 26.

35 Ibid.

36 André Callé, President Hydro-Québec, 'Industry Review: Hydro Québec,' *Connections: 1999 Electricity Industry Review* (Montréal: Canadian Electricity Association, January 1999), p. 18.

37 Hydro-Québec, *Strategic Plan 1998-2002*, pp. 9, 10.

38 Chris Flanagan, 'Project stands on own: Furey' *Evening Telegram*, St. John's, Newfoundland, 23 February 1998.

39 Konrad Yakabuski, 'U.S. energy regulators push Québec on Churchill Falls,' *Globe and Mail,* 6 June 1997, B6.
40 Brian Smith with the host of the *Early Edition,* CBC Radio One, Vancouver, 25 March 1998, 7:45 a.m.
41 Canada, Department of Indian Affairs and Northern Development Research Division, 'Negotiating a Way of Life: Initial Cree Experience with Administrative Structure Arising from the James Bay Agreement,' Ignatius La Rusic, research director (Ottawa: Research Division in the Department of Indian Affairs, 1979).
42 Ibid.
43 Boyce Richardson, *James Bay: The Plot to Drown the North* (Toronto: Clarke, Irwin and Company and Sierra Club, 1972).
44 Ibid., p. 17; and see Robert Bourassa, *James Bay* (Montreal: Harvest House, 1973), pp. 83-5.
45 Glen Williams, 'Greening the New Canadian Political Economy,' *Studies in Political Economy* 37, 5, (Spring 1992): 26. Dieter Soyez, 'Industrial Resource Use and Transnational Conflict Patterns: Geographical Implications of the James Bay Hydro Power Schemes,' a lecture given at Simon Fraser University, Department of Geography, 19 September 1996. James Waldran, *As Long as the Rivers Run: Hydro-Electric Development in Native Communities in Western Canada.*
46 Canada, Natural Resources Canada, *Electric Power in Canada, 1994* (Montreal: Canadian Electricity Association, no date), p. 33.
47 Matthew Coon Come, Grand Chief of the Cree, 'Decision on Hydro-Québec Puts Québec Energy Supply at Risk,' press release by Canada News Wire, Montreal, 13 November 1997.
48 Edward Alden, 'Hydro clears export hurdle,' *Vancouver Sun,* 4 July 1997, section D.

Glossary of Technical Terms

The following technical terms are selectively transcribed, with minor modifications, from *Electric Power in Canada, 1995*, pp. 171-3, written and published by Natural Resources Canada and the Canadian Electricity Association.

capacity In the electric power industry, capacity has two meanings:
(1) System Capacity: The maximum power capability of a system. For example, a utility system might have a rated capacity of 5,000 MW (capacity of Churchill Falls is 5,429 MW for example), or might sell 50 MW (the capacity needed by a large pulp mill) of capacity (i.e., power).
(2) Equipment Capacity: The maximum power capability of a piece of equipment. For example, a generating unit might have a rated capacity of 50 MW.

electrical energy The quantity of electricity delivered over a period of time. The commonly used unit of electrical energy is the kilowatt-hour (kWh)

energy source The primary source that provides the power that is converted to electricity. Energy sources include water, coal, petroleum products, gas, uranium, wind, geothermal [energy], and so on.

export conversion To convert from GWh of export to MW of plant capacity needed, the following calculation is necessary: an export amount of 10,000 GWh divided by 8,760 (hours per year) x 1.25 (for load factor of approximately 80 percent) would result in the size of plant needed in GW. Thereafter GW can be converted into MW.

firm energy of power Electrical energy or power intended to be available at all times during the period of the agreement for its sale.

gigawatt (GW) One million kilowatts.

gigawatt hours (GWh) A unit of bulk energy. A million kilowatt hours. A billion watt hours.

horse power One horsepower is equivalent to 746 watts.

hydraulic sources The falling water that can rotate turbines connected to generators.

hydro infrastructure The power system of one utility or several interconnected utilities into a provincial or national power system. State intervention can lead to an expansion of the hydro infrastructure.

hydroelectric power station A power station which generates electricity from hydraulic sources.

interruptible energy or power Energy or power made available under an agreement that permits curtailment or interruption of delivery at the option of the supplier.

kilowatt (kW) The commercial unit of electric power; 1,000 watts. A kilowatt can be visualized as the total amount of power needed to light ten 100-watt light bulbs.

kilowatt hour (kWh): The commercial unit of electric energy; 1,000 watt hours. A kilowatt hour can best be visualized as the amount of electricity consumed by ten 100-watt light bulbs burning for an hour.

load The amount of electric power or energy consumed by a particular customer or group of customers.

load factor The ratio of the average load during a designated period to the peak or maximum load in that same period (usually expressed in percent).

megawatt (MW) A unit of bulk power; 1,000 kilowatts.

megawatt hours (MWh) A unit of bulk power; 1,000 kilowatt hours.

mill 1/1000 of a dollar or 1/10 of a cent.

power system All interconnected facilities of an electrical utility. A power system includes all the generation, transmission, distribution, transformation, and protective components necessary to provide service to customers.

standard industrial classification The 1970 Revision of the Standard Industrial Classification includes: (1) food-and-beverage industries, (2) tobacco products industries, (3) rubber and plastic products industries, (4) leather industries, (5) textile industries, (6) knitting mills, (7) clothing industries, (8) wood industries, (9) furniture and fixture industries, (10) paper and allied industries, (11) printing and publishing industries, (12) primary metal industries, (13) metal fabricating industries, (14) machinery industries, (15) transportation equipment industries, (16) electrical products industries, (17) non-metallic mineral products industries, (18) petroleum and coal products industries, (19) chemical and chemical products industries, and (20) miscellaneous manufacturing industries. Canada, Statistics Canada, *Consumption of Purchased Fuel and Electricity*, cat. 57-208, 1982, 62-8.

terawatt hours (TWh): One billion kilowatt hours.

voltage The electrical force or potential that causes a current to flow in a circuit (just as pressure causes water to flow in a pipe). Voltage is measured in volts (V) or kilovolts (kV). 1 kV = 1000 V.

watt The scientific unit of electric power. A typical light bulb is rated 25, 40, 60, or 100 watts, meaning that it consumes that amount of power when illuminated.

Bibliography

Archival Sources

British Columbia. BC Hydro Library. Cass-Beggs Papers. Industry studies and internal reports especially prepared by and for BC Hydro. Vancouver.

–. BC Utilities Commission. Submissions to hearings for the proposed Site C dam on the Peace River. Vancouver.

–. Legislative Library. Third Provincial Premiers Conference, Proceedings, 6 August 1962. Victoria.

Canada. Library of the National Energy Board. Transcripts of Public Export Licence Hearings, 1960s-80s. Transcripts of Hearings into the Regulation of Electricity held November-December 1986. Original export licence documents submitted by British Columbia Hydro, Manitoba Hydro, and Hydro-Québec, Ottawa (NEB library at present in Calgary).

–. National Archives of Canada. Churchill Falls (Labrador) Corporation Limited [CFLCo] (MG 28 III 73), corporate files from CFLCo's Montreal office. Ottawa. Cabinet Conclusions about the National Power Grid, miscellaneous years 1961 to 1965 (RG 2, B2, RG2 A5a).

Manitoba. Provincial Archives of Manitoba. Schreyer Papers (PAM Range 42, 550 files, box 147). Winnipeg.

–. Provincial Archives of Manitoba. Access to Senator Duff Roblin's papers in the Provincial Archives of Manitoba was refused by Roblin's Winnipeg office in August 1989.

–. Legislative Library. Briefs to the Tritschler *Commission of Inquiry into Manitoba Hydro.* Winnipeg.

Ontario. Ontario Hydro Archives. Sir Adam Beck, General and correspondence files from years: 1911, 1920-22, 1923-5. Toronto.

–. Archives of Ontario. The Ontario Hydro-Electric Commission. Hon. D. Carmichael (Commissioner) correspondence, 1920-2. Toronto.

Québec. Archives nationales du Québec. Industrial Power Contracts, Documents: Conseil Exécutif, Autres Documents, Hydro-Québec Articles (159 et 160 for the period 1969-76, Instrument de Recherche 300034). Sainte-Foy.

–. Centre d'Archives Hydro-Québec. Procès-Verbal d'une Conseil d'Administration de la Société de Développement de la Baie James, 1971-76. Société d'Energie de la Baie James. Le Consortium La Grande. Montréal.

Secondary Sources

Acres, H.G., and Company. 'National Power Network Stage I Assessment,' 24 January 1964. Cited by Canada, National Energy Board, 'Inter-Utility Trade Review: Inter-Utility Cooperation' (Calgary: National Energy Board, 1992), pp. 2-4.

Aitken, Hugh G.J. 'Defensive Expansionism: The State and Economic Growth in Canada.' *Approaches to Canadian Economic History.* Edited by W.T. Easterbrook and M.H. Watkins. Toronto: McClelland and Stewart, 1969.

Angus, H.F. 'Statement by the Chairman of the Public Utilities Commission.' *Report on the Columbia and Peace Power Projects*. By Gordon Shrum. Victoria: BC Energy Board, 1961.

Armstrong, Christopher. 'Water-Power and the Constitution.' *Politics of Federalism: Ontario's Relations with the Federal Government, 1987-1942*. Toronto: University of Toronto Press, 1981, pp. 160-77.

Armstrong, Christopher, and H.V. Nelles. *Monopoly's Moment: The Organization and Regulation of Canadian Utilities, 1830-1930*. Toronto: University of Toronto Press, 1986.

Arthur D. Little, Inc. 'An Assessment of the Impact of Nelson River Electric Power on Industrial Development in Manitoba.' Report to the Department of Industry and Commerce, Government of Manitoba. Cambridge, MA: Arthur D. Little, Inc., June 1966, report # 67191. Source in Winnipeg: Manitoba Hydro Library.

Bateman, L.A., Chairman of Manitoba Hydro. 'The Nelson River: An Energy Source.' Paper presented at the 89th Annual Congress of the Engineering Institute of Canada, Winnipeg, Manitoba, 30 September to 3 October, 1975.

Belfield, Robert B. 'The Niagara Frontier: The Evolution of Electric Power Systems in New York and Ontario, 1880-1955.' PhD dissertation, University of Pennsylvania, 1981. Source located in Ottawa: National Research Council, CISTI Library.

Bolduc, André, and Daniel Larouche. *Québec: Un Siècle d'Électricité*. No place: Libre Expression, 1979.

Bourassa, Robert. *Le Défi Technologique*. Montréal: Québec/Amérique, 1985.

–. *Deux Fois La Baie James*. Ottawa: Les Editions de la Presse, 1981.

–. *James Bay*. Montreal: Harvest House, 1973.

–. *Power from the North*. Scarborough: Prentice-Hall, 1985.

Bradford, Neil, and Glen Williams. 'What Went Wrong? Explaining Canadian Industrialization.' *The New Canadian Political Economy*. Edited by Wallace Clement and Glen Williams. Kingston: McGill-Queen's University Press, 1989.

Brennan, Bob. 'Utility Convergence.' *Insights*. Winnipeg: Manitoba Hydro, May 1997, vol. 7, issue 1.

British Columbia. BC Energy Board. *Report on the Columbia and Peace Power Projects*. By Gordon Shrum, as chairman of the BC Energy Board. Victoria. 31 July 1961.

–. BC Energy Board. 'Electric Power Requirements in British Columbia Projection to 1990.' Victoria: BC Energy Board, 1 May 1971.

–. BC Power Commission. 'New Industry Already Promised Arrow Lakes.' *Progress*. Victoria: BC Power Commission, July 1951.

–. BC Power Commission. *Pulp and Paper Opportunities in Central British Columbia*. By C.W. Nash and H.L. Briggs. No place [likely Victoria]: June 1956.

–. BC Utilities Commission [BCUC]. *Annual Report 1995*.

–. BCUC. 'The British Columbia Electricity Market Review: Report and Recommendations to the Lieutenant Governor in Council.' Vancouver: BCUC, September 1995.

–. BCUC. BC Hydro. Table 1, 'Bulk Power.' Exhibit 97. *Site C Hearings*, Fort St. John, 10 February 1982.

–. BCUC. BC Hydro. Table 2, 'B.C. Hydro Bulk Power (1800 Series) Accounts Detail: Probable Electrical Projected Sales.' Exhibit 98. *Site C Hearings*, Fort St. John, 10 February 1982.

–. BCUC. 'B.C. Hydro 1981 Bulk Load Projections: Other Assumed Staging.' Exhibit 96. Statement of Evidence of Richard Overstall (SPEC). Site C. *Hearings*, Fort St. John, 10 February 1982.

–. BCUC. BC Hydro. *Site C Hearings*. Testimony by Mr. Overstall and Mr. Brassington. Fort St. John, 10 February 1982. Transcript, vol. 26.

–. BCUC. Decision in the Matter of 'British Columbia Hydro and Power Authority: Wholesale Transmission Services Application.' Vancouver: BCUC, 25 June 1996.

–. BCUC. 'In the Matter of the Application of British Columbia Hydro and Power Authority for an Energy Project Certificate for the Peace River Site C Project.' *Site C Report: Report and Recommendations to the Lieutenant Governor in Council*. Vancouver: BCUC, May 1983.

–. BCUC. 'In The Matter of British Columbia Hydro and Power Authority: Wholesale Transmission Service Application, Decision, June 25, 1996.' Vancouver: BCUC, 1996.

–. Government of British Columbia. 'Electric Power: An Act to Provide for Improving the Availability and Supply of Electrical Power.' *B.C. Statutes*. Chapter 27. Assented to 28 March 1945.

–. Government of British Columbia. An Act to Povide for the Reorganization of the British Columbia Electric Company Limited and the Development of Power Resources. An Act to Amend the Power Development Act, 1961. *B.C. Statutes*. Chapter 50. Assented to 29 March 1962.

–. Ministry of Industry and Small Business Development [MISBD]. 'Incoming Buyers.' *Program Directory*. Victoria: MISBD, April 1981.

–. MISBD. *British Columbia Facts and Statistics*. Annual Review 1983 and 1984. Victoria: MISBD.

–. MISBD. 'Introduction.' *B. C. Manufacturers' Directory 1985: A Directory of Manufacturing Activity in British Columbia*. Victoria: MISBD, 1985.

–. Rural Electrification Committee. *Progress Report*. Victoria: King's Printer, 1944. Cited by Mary D. Taylor. 'Development of the Electricity Industry in British Columbia.' MA thesis, Department of Geography, UBC, April 1965.

BC Hydro. *Annual Reports 1995-97*.

–. 'Biography of David Cass-Beggs.' Approved by Cass-Beggs on 23 December 1974. New Westminster: BC Hydro Information Centre, Retrieval System No. AK. 138.

–. Industrial Development Department. 'Power Intensive Industries for Peace River at-Site Power.' Vancouver: BC Hydro, August 1963.

–. Industrial Development Department. *The Mining Industry of British Columbia and the Yukon*. By J.C. Dawson. Vancouver: BC Hydro, January 1968.

–. *Intercom*. BC Hydro's employee newspaper. Reprint from the 16 May 1980 issue.

–. Load Forecast Department. Actual Electrical Sales Statistics. 6 May 1968.

–. 'Paper for the World.' By John Raybould. A two-page reprint from the BC Hydro journal *Progress*. Vancouver: B.C. Hydro, 1966.

–. Power Smart Marketing. *The Power is Yours: BC Hydro: the Power Company of Choice*. A seventeen-page brochure. New Westminster: A Publication for the Customers and Employees of BC Hydro, spring 1998.

–. Statistical Analysis Department, Commercial Services Division. *British Columbia as an Investment Prospect*. By T.D. Buchanan. Vancouver, 26 June 1967.

–. Statistical Analysis Department, Commercial Services Division. 'Electric Consumption in the B.C. Mining Industry.' Vancouver: BC Hydro, June 1964.

–. Statistical Analysis Department, Commercial Services Division. *Industrial Classification of KWh Sold to Primary Potentials in the B.C. Hydro Service Area For Fiscal 1964/65*. By T.D. Buchanan. Vancouver: BC Hydro, 12 November 1965.

–. Statistical Analysis Department, Commercial Services Division. 'Pulp and Paper Prospects in B.C.' By T.D. Buchanan. Vancouver: BC Hydro, 21 December 1964.

Brodie, Janine. 'The New Political Economy of Regions.' *Understanding Canada: Building on the New Canadian Political Economy*. Edited by Wallace Clement. Montreal: McGill-Queen's University Press, 1997.

Caillé, André. 'Industry Review: Hydro-Québec,' *Connections: 1999 Electricity Industry Review*. Montreal: Canadian Electricity Association, January 1999, p. 18.

Canada. The Constitution Act, 1982, amended by the Constitution Amendment Proclamation, 1983. Ottawa: Minister of Supply and Services, 1986.

–. Energy, Mines and Resources. *The Canada-U.S. Free Trade Agreement and Energy: An Assessment*. No date, c. 1988.

–. Energy, Mines and Resources. *Electric Power in Canada, 1986 to 1991*. Ottawa: Minister of Supply and Services, 1987 to 1992.

–. Energy, Mines and Resources. Federal-Provincial Nelson River Review Committee. 'First Annual Report of Federal-Provincial Nelson River Review Committee, Canada, Manitoba Agreement, Development of Hydroelectric Potential of the Nelson River, March 31, 1970.' By G.M. MacNapp, Chair of Review Committee and Assistant Deputy Minister (Energy Development) of Energy, Mines and Resources. Ottawa: Energy Mines and Resources, 1970. Source in Winnipeg: Provincial Archives of Manitoba, reference, 550.1.

–. National Energy Board of Canada [NEB].'Application to the National Energy Board for a Certificate of Public Convenience for an International Power Line and a Licence to Export Power by Manitoba Hydro, Winnipeg, 1969.' Exhibit C entered during the NEB hearings in Winnipeg, 4 November 1969.

–. NEB. 'Application to the National Energy Board for a Certificate of Public Convenience for an International Power Line and a Licence to Export Power, Vol. 1, by The Manitoba Hydro Electric Board, Winnipeg Manitoba, 1975.' Source in Ottawa: National Energy Board Library, 1975.

–. NEB. 'Hydro-Québec.' Hearings, Transcript, 27 November 1974. Ottawa: Library of the National Energy Board.

–. NEB. 'Inter-Utility Trade Review: Inter-Utility Cooperation.' Calgary: National Energy Board, 1992.

–. NEB. 'Inter-Utility Trade Review: Transmission Access and Wheeling.' Calgary: National Energy Board, 1992.

–. NEB. 'Inter-Utility Trade in Electricity.' Ottawa: Minster of Public Works and Government Services. Calgary: National Energy Board, 1994.

–. NEB. 'Inter-Utility Trade in Electricity: Analyses of Submissions, April 1994.' Ottawa: Minister of Public Works and Government Services Canada 1994.

–. NEB. 'National Energy Board Report to the Governor in Council in the Matter of the Application under the National Energy Board Act of the Manitoba Hydro-Electric Board [Manitoba Hydro], March 1976.' Ottawa: National Energy Board, 1976.

–. NEB. 'National Energy Board Reasons for Decision in the Matter of the Application under Part IV of the National Energy Board Act of The Manitoba Hydro-Electric Board [Manitoba Hydro], September 1976.'

–. NEB. 'National Energy Board Reason for Decision in the Matter of the Application under the National Energy Board Act of the Manitoba Hydro-Electric Board [Manitoba Hydro], August 1987.'

–. *Reasons for Decisions in the Matter of an Application under the National Energy Board Act of British Columbia and Power Authority Part A: Licences, March 1980.* Ottawa: National Energy Board, 1980.

–. NEB. *Reasons for Decision: In the Matter of an Application under the National Energy Board Act of Hydro-Québec, January 1984.* Ottawa: Minister of Supply and Services, 1984.

–. NEB. 'Regulation of Electricity Exports: Report of an Inquiry by a Panel of the National Energy Board Following a Hearing in November and December 1986, National Energy Board, June 1987.' Ottawa: Minister of Supply and Services, 1987.

–. NEB. *Regulatory Agenda.* A quarterly publication containing brief summaries of hearings, applications, decisions, and appeals before the National Energy Board. Calgary: National Energy Board, years 1992-8.

–. NEB. *Report to the Governor in Council in the Matter of the Application Under the National Energy Board Act of Québec Hydro-Electric Commission, September 1976.* Ottawa: National Energy Board, 1976.

–. Natural Resources Canada. *Electric Power in Canada, 1992, 1993.* Ottawa: Minister of Supply and Services, 1993, 1994.

–. Natural Resources Canada. *Electric Power in Canada, 1994, 1996.* Montreal: Canadian Electricity Association, 1995, 1998.

–. *North American Free Trade Agreement between the Government of Canada, the Government of the United Mexican States, and the Government of the United States of America.* Ottawa: Minister of Supply and Services Canada, 1992.

–. Office of the Prime Minister. 'Opening Statement by the Prime Minister of Canada at First Ministers' Conference on Energy, Ottawa, January 22, 1974.' Press Release, 22 January 1974. Source in Winnipeg: Provincial Archives of Manitoba, Schreyer Papers.

–. Revised Statutes of Canada, 1985. The Constitution Act 1867. Ottawa: Queen's Printer for Canada.

–. Science Council of Canada. '43rd Meeting, Science Council of Canada: Visits to Churchill Falls Hydro Electric Development and Iron Ore Company's Developments at Wabush City and Sept-Iles. 2-5 Oct. 1973.' Ottawa: National Archives, CFLCo MG 28 III 73, vol. 31, file 4.

–. Secretary of State, *Sessional Paper 199;* 'A copy of all correspondence, telegrams, and other documents exchanged between the government and each province since January 1, 1961 [to 16 February 1962], regarding the establishment of a national power grid system,' thirty-five pages. Vancouver: UBC Library, Unpublished Sessional Papers, 24th Parliament, Session 5, vols. 912-19 #81-238, 1962, reel 22/22.

–. Statistics Canada. *Electric Power Statistics III: Inventory of Prime Mover and Electric Generating Equipment as of December 31, 1986.* Cat. 57-206. Ottawa: Minister of Supply and Services, 1987.

–. Statistics Canada. *Consumption of Purchased Fuel and Electricity: By the Manufacturing, Mining and Electrical Power Industries.* Years 1962 to 1974. Cat. 57-206. For the Minister of Industry, Trade and Commerce. Ottawa: Supply and Services, 1977.

–. Statistics Canada. *Consumption of Purchased Fuel and Electricity: By the Manufacturing, Mining and Electrical Power Industries.* Years 1975 to 1984. Cat. 57-208. Ottawa: *Supply and Services.*

Canadian Encyclopedia. Edmonton: Hurtig, 1988.

Cass-Beggs, David. 'Economic Feasibility of Trans-Canada Electrical Interconnection.' Paper presented at the Canadian Electrical Association, Western Zone Meeting, Edmonton, Alberta, 21-23 March 1960. New Westminster: BC Hydro Information Centre, Retrieval System No. AK. 138.

Chuchman, George. 'Electric Utility Deregulation and Competition Implications for Manitoba.' A paper prepared for CUPE (Canadian Union of Public Employees), Manitoba Regional Office, June 1997.

Churchill Falls (Labrador) Corporation [CFLCo]. *An Energy Giant: The Churchill Falls Power Development.* A glossy seventeen-page brochure, received in March 1998 from CFLCo in St. John's, Newfoundland. Churchill Falls, Labrador: CFLCo, no date.

Clark-Jones, Melissa. *A Staple State: Canadian Industrial Resources in Cold War.* Toronto: University of Toronto Press, 1987.

Clement, Wallace. *The Canadian Corporate Elite: An Analysis of Economic Power.* Toronto: McClelland and Stewart, 1975.

–. *Continental Corporate Power: Economic Linkages between Canada and the United States.* Toronto: McClelland and Stewart, 1977.

–. 'Debates and Directions: A Political Economy of Resources.' *The New Canadian Political Economy.* Edited by Wallace Clement and Glen Williams. Kingston: McGill-Queen's University Press, 1989.

–. 'Introduction: Whither the New Canadian Political Economy?' *Understanding Canada: Building on the New Canadian Political Economy.* Edited by Wallace Clement. Montreal: McGill-Queen's University Press, 1997.

– (ed.). *Understanding Canada: Building on the New Canadian Political Economy.* Montreal: McGill-Queen's University Press, 1997.

Clement, Wallace, and Glen Williams. 'Resources and Manufacturing in Canada's Political Economy.' *Understanding Canada: Building on the New Canadian Political Economy.* Montreal: McGill-Queen's University Press, 1997.

Cohen, Marjorie Griffin. 'Public Power and the Political Economy of Electricity Competition.' Presented at the conference on Deregulation and Competition in the Electrical Industry, co-sponsored by the Canadian Centre for Policy Alternatives, BC office and the Institute for Governance Studies. Vancouver, Simon Fraser University, Harbour Centre Campus, 10 January 1998.

Connord-Lajambe, Hélène. 'Couts et externalités de l'aménagement hydroélectrique de la baie James.' MA thesis, Département d'Economique, Université McGill, Montréal, 1977.

Daily Commercial News and Building Record. 'A New Industrial Horizon presented by Newfoundland.' [Toronto] 27 March 1968. A thirty-two-page tabloid section of the paper which promoted Newfoundland as an industrial province.

Dales, John. *Hydro-Electricity and Industrial Development: Québec 1898-1940.* Cambridge: Harvard University Press, 1957.

Davenport, George W., and the American Institute of Electrical Engineers. *The Niagara Falls Electrical Handbook.* Syracuse: Mason Press, 1904.

Denison, Merril. *Niagara's Pioneers*. Niagara Falls Power Company. No date, place, or publisher.

–. *The People's Power: A History of Ontario Hydro*. Toronto: McClelland and Stewart, 1960.

Dewar, Kenneth. C. 'State Ownership in Canada: Origins of Ontario Hydro.' PhD dissertation, University of Toronto, 1975.

Dominion Engineering Works Limited. 'Proposal for a Joint Agreement for the operation of the Bertram Plant at The Pas, Manitoba, for the Manufacture and Supply of Hydraulic Turbine Components: Manitoba Hydro Specification number 896 Long Spruce Generating Station, 17 April 1973.' Source in Winnipeg: Manitoba Hydro Corporate Library.

Duquette, Michel. 'Conflicting Trends in Canadian Federalism: The Case of Energy Policy.' *New Trends in Canadian Federalism*. Edited by François Rocher and Miriam Smith. Peterborough: Broadview Press, 1995, pp. 391-413.

Durocher, René. 'Quiet Revolution.' *The Canadian Encyclopedia*, vol. 3. Edmonton: Hurtig Publishers, 1988.

'Errors and Misrepresentations made by The Hydro-Electric Inquiry Commission (Known as the Gregory Commission) Respecting the Publicly Owned and Operated Hydro-Electric Power Undertaking of Municipalities in the Province of Ontario. 1925.' Toronto, 1925. Source in Toronto: Ontario Hydro Archives, Archival Folder #292.

Faucher, Philippe, and Kevin Fitzgibbons. 'The Political Economy of Electrical Power Generation: Procurement Policy and Technological Development in Québec, Ontario and British Columbia.' Montréal: Université de Montréal and Centre for Research on Development of Industry and Technology, 1990.

–. 'Procurement by State-Owned Enterprises: Potential and Limits for Industrial Development.' Montréal: Université du Québec à Montréal, Centre de recherche en développement industriel et technologiques (CREDIT), 1988.

Fournier, Pierre. *The Québec Establishment: The Ruling Class and the State*. Montreal: Black Rose, 1976.

–. Les societé s d'Etat et les objectifs économiques du Québec: une evaluation préliminaire. Coproducteur, Office de planification et de developpement du Québec. Québec: Editeur officiel du Québec, 1979.

Fram, Mark. *Ontario Hydro Ontario Heritage: A Study of Strategies for the Conservation of the Heritage of Ontario Hydro*. Toronto: Ministry of Culture and Recreation, 1980.

Froschauer, Karl. 'B.C. Hydro Is a Major Institutional Force in Extending and Intensifying Staples Dependence.' MA thesis, Department of Anthropology and Sociology, UBC, Vancouver, 1986.

–. 'Provincial Hydro Expansions: Required to Serve Industrial Development in Canada and Continental Integration.' PhD dissertation, Department of Anthropology and Sociology, Carleton University, Ottawa, 1993.

Gatt-Fly. *Power to Choose*. Toronto: Between the Lines, 1981.

Gerschenkron, Alexander. *Economic Backwardness in Historical Perspective*. Cambridge, MA: Belknap Press, 1966. Chapters 1 and 2.

Glassco, J.G. 'Water-Power Development: Public or Private Development.' *The Commercial Manchester* [England], 13 June 1929.

Grauer, Dal. 'The Export of Electricity from Canada.' *Canadian Issues: Essays in Honour of Henry F. Angus*. For UBC in Vancouver by the University of Toronto Press, 1961.

Hirschman, Albert. 'A Generalized Linkage Approach to Development, with Special Reference to Staples.' *Essays in Trespassing: Economics to Politics and Beyond*. Cambridge: Cambridge University Press, 1981.

Hughes, Thomas. *Networks of Power: Electrification in Western Society, 1880-1930*. Baltimore: Johns Hopkins University Press, 1983.

–. 'Technology as a Force for Change in History: The Effort to Form a Unified Electric Power System in Weimar Germany.' *Industrielles System und Politische Entwicklung in der Weimar Republik*. Düsseldorf: Droste Verlag, 1974.

Hydro-Québec. *Hydro-Québec Development Plan 1988-1990, Horizon 1997*. Montréal: Hydro-Québec, 1988.

–. Ministère de l'Industrie et du Commerce. 'L'Energie au Québec et son influence sur la localisation des entreprises manufacturières.' By L. Marechal. 1975. Source in Montréal: Library of Hydro-Québec.

–. 'Power Contract between Québec Hydro-Electric Commission and Churchill Falls (Labrador) Corporation Limited, May 12, 1969.' Source: Exhibit 8-4, Hydro-Québec. Des licences d'exportation d'électricité et d'énergie. Montréal, P.Q., les 24-5 septembre. Ordonnace EH-1-85. Juillet 1985. Source in Ottawa: Library of the National Energy Board.

–. *Strategic Plan 1998-2002*. Montréal: Hydro-Québec, October 1997.

Ingledow, T., and Associates. 'National Power Network Stage II Assessment, Interim Report.' Prepared for the Confidential Use of the Federal-Provincial Working Committee on Long Distance Transmission. Vancouver: T. Ingledow and Associates. Interim Report, March 1966. Source in Vancouver: BC Hydro Library, Vancouver.

–. 'National Power Network Stage II Assessment.' Prepared for the Confidential Use of the Federal-Provincial Working Committee on Long Distance Transmission. Vancouver: T. Ingledow and Associates, Volume I, February 1967. Source in Ottawa: National Energy Board Library.

Innis, Harold. *The Problems of Staple Production in Canada*. Toronto: Ryerson, 1933.

Interprovincial Advisory Council on Energy (IPACE) Networks Study Group. 'An Evaluation of Strengthened Interprovincial Interconnections of Electric Power Systems,' October 1978, vol. I and I. Cited by Canada, National Energy Board, 'Inter-Utility Trade Review,' pp. 2-7 to 2-11.

Jaccard, Mark. *Reforming British Columbia's Electricity Market: A Way Forward*. Vancouver: British Columbia Task Force on Electricity Market Reform, Interim Report dated December 1997 and Final Report dated January 1998.

Jaccard, Mark, John Nyboer, and Timo Makinen. 'Managing Instead of Building: B.C. Hydro's Role in the 1990s.' *BC Studies* (autumn and winter 1991-2): 99-126.

Keefer, T.C. 1899. 'Canadian Water Power and Its Electrical Product in Relation to the Undeveloped Resources of the Dominion.' *Proceedings and Transactions*, Royal Society of Canada, 23 May 1899. Source in Ottawa: National Archives of Canada, 2nd Series, vol. V., 3-40. Also cited by H.V. Nelles, *The Politics of Development: Forests Mines and Hydro-Electric Power in Ontario, 1849-1941*. Toronto: Macmillan, 1974; and Hamden, CT: Archon, 1974.

Kierans, Eric. *Report on Natural Resources Policy in Manitoba*. Winnipeg: Queen's Printer, 1973.

La Rusic, Ignatius. *Negotiating a Way of Life*. Montreal: Department of Indian Affairs, October 1979.

Laxer, Gordon. 'Foreign Ownership and Myths about Canadian Development.' *Canadian Review of Sociology and Anthropology* 22 (1985): 3.

–. *Open for Business: The Roots of Foreign Ownership in Canada*. Toronto: Oxford University Press, 1989.

Levitt, K. *Silent Surrender: The Multinational Corporation in Canada*. Toronto: Macmillan, 1970.

Lower Churchill Development Corporation. *1996 Annual Report*. St. John's, Newfoundland: Lower Churchill Development Corporation, 1996

Macdonald, Donald S. (Minister of Energy, Mines and Resources). 'Notes for Statement of Other Elements of Energy Policy.' First Ministers' Conference on Energy, 22-23 January 1974, Ottawa. Document No. FP-3133.

Macdonald, Laura. 'Going Global: The Politics of Canada's Foreign Economic Relations.' *Understanding Canada: Building on the New Political Economy*. Edited by Wallace Clement. Montreal: McGill-Queen's University Press, 1997.

Mahon, Rianne. 'Canadian Public Policy: Unequal Structure of Representation.' *The Canadian State*. Edited by Leo Panitch. Toronto: University of Toronto Press, 1977.

Maiden, Cecil. *Lighted Journey: The Study of B.C. Electric*. Vancouver: Keystone, 1948.

Manitoba. *Commission of Inquiry into Manitoba Hydro: Final Report, December, 1979*. G.E. Tritschler, Commissioner. Winnipeg: Government of Manitoba.

–. Committee on Manitoba's Economic Future. 'Community Development as a Factor in Industrial Growth in Manitoba.' By the Midwest Research Institute, Economics Division, Study conducted by Bruce W. Macy and Donald L. Vogelsang, Final Report, dated 1 June 1962. Source in Winnipeg: Provincial Archives of Manitoba, RG6 D1.

–. Department of Industry and Commerce. The Bertram Study Group. 'The Pas Machinery Manufacturing Project: Report of the Bertram Study Group on the Feasibility of Manufacturing Generating and Associated Hydraulic Equipment for Manitoba Hydro at Bertram Plant: The Pas Manitoba, December 1973.' Source in Winnipeg: Manitoba Hydro Corporate Library.

–. Energy Policy Branch of Energy, Mines and Resources. Study Directorate. *Western Electric Power Grid. Vol. I, Executive Summary.* Steering Committee, Western Electric Power Grid Study Agreement, December 1980. Winnipeg: Energy Policy Branch of Energy Mines and Resources.

–. Legislative Assembly of Manitoba. Second Reading Debate in the Legislature of Manitoba of 'Bill 55: The Manitoba Hydro Amendment Act,' Transcripts of Wednesday, 11 June 1997.

–. Manitoba Energy Authority. *Annual Reports.* Years 1984 to 1989.

–. Public Utilities Commissioner. 'Report on the Projected Hydro-Electric System for the Province of Manitoba by the Public Utilities Commissioner, 1913.' By H.A. Robson, Commissioner, Winnipeg, 3 February 1914 to Hon. J.H. Howden, Attorney-General of Manitoba, Winnipeg. Source in Winnipeg: Provincial Archives of Manitoba, RG 22, box 6.

Manitoba Hydro. *Annual Reports.* Years 1984 to 1989.

–. 'Changes to the Manitoba Hydro Act.' A three-fold pamphlet summarizing key sections in the Manitoba Hydro Amendment Act (Bill 55). Winnipeg: Manitoba Hydro, 1997.

–. 'Formation of Manitoba Hydro.' A Manitoba Hydro information pamphlet. Winnipeg: Manitoba Hydro, Public Affairs. Revised: PA D14 F9 89 08 [August 1989].

–. *A History of Hydro-Electric Power in Manitoba.* Brochure. Winnipeg: Manitoba Hydro, Public Affairs, 86 05 [May 1986].

–. *Hydro Gram.* A small internal newspaper. Vol. 19, no. 21, 1989. Vol. 27, no. 25, 1997. Winnipeg: Manitoba Hydro, Public Affairs.

–. 'Instructions to Tenderers.' An internal document used by Al Kazima, Purchasing Service Supervisor, in the procurement process. Specification No. 1950, paragraph 10.2, *Industrial Offsets.* Winnipeg: Manitoba Hydro, 1989.

–. *Manitoba Hydro System Load Forecast 1983/84 to 2003/04.* Winnipeg: Manitoba Hydro, 1984. [Submitted as exhibit B-4, Manitoba Hydro-Electric Board ... for an export licence ..., Winnipeg, Man., Nov. 5-16, 1984. Order no. EH- 6-84.] Source in Ottawa: National Energy Board Library.

–. 'What Is Manitoba Hydro?' Pamphlet. Winnipeg: Manitoba Hydro, Public Affairs. D10-F10, 87 11 [November 1987].

Marchak, Patricia. *Green Gold: The Forest Industry in British Columbia.* Vancouver: University of British Columbia Press, 1983.

Massue, Huet. 'Highlights of Electric Power in Canada.' Montreal: Shawinigan Water and Power Company, 1954. Cited in the *Report of the Royal Commission on Electric Energy, February 1966.* Government of Newfoundland and Labrador. Source in Ottawa: National Energy Board Library.

Mathias, Philip. 'A Good Deal for Manitoba:' The Churchill Forest Industries Project. *Forced Growth: Five Studies of Government Involvement in the Development of Canada.* Toronto: James Lewis, 1971.

Mavor, James. *Niagara in Politics: A Critical Account of the Ontario Hydro-Electric Commission.* New York: Dutton, 1925.

Maxfield, B.E., Chief Engineer, Inco Metal Company. A brief dated 30 November 1978 addressed to the Honourable G.E. Tritschler, Commissioner of the Commission of Inquiry into Manitoba Hydro. Source in Winnipeg: Legislative Library.

Mid-continental Area Power Pool. 'Paths of Power: Discussion Guide.' Minneapolis: Mid-continental Area Power Pool. A pamphlet.

Myers, L.E. Company, Construction and Management of Public Utilities. 'For Distributing Power over a Portion of Manitoba,' survey on behalf of Mr. Samuel Insull, President, Middle West Utilities Company, Chicago, IL, 12 July 1916. Source in Winnipeg: Provincial Archives of Manitoba, RG 22, box 6.

Nash, C.W., and H.L. Briggs. 'Pulp and Paper Opportunities in Central British Columbia.' Victoria: British Columbia Power Commission, June 1956.

Naylor, R.T. 'Canada in the European Age.' *Canadian Journal of Political and Social Theory*. Fall/automne (1983): 97-127.

–. *The History of Canadian Business, 1867-1914*, Vol. 1, *The Banks and Finance Capital*. Toronto: James Lorimer, 1975.

–. *The History of Canadian Business, 1867-1914*, Vol. 2, *Industrial Development*. Toronto: James Lorimer, 1975.

Nelles, H.V. *The Politics of Development: Forests Mines and Hydro-Electric Power in Ontario, 1849-1941*. Toronto: Macmillan, 1974.

–. 'Public Ownership of Electrical Utilities in Manitoba and Ontario, 1906-30.' *Canadian Historical Review* 57 (1976): 461-85.

Netherton, Alex. 'International Electricity Trade and Paradigm Shift.' A paper presented at *Power Grab: The Future of BC Hydro*, a conference on Deregulation and Competition in the Electrical Industry, co-sponsored by the Canadian Centre for Policy Alternatives, BC office and the Institute for Governance Studies, Simon Fraser University, Harbour Centre Campus, 10 January 1998.

Newfoundland. Government of Newfoundland. 'Position on a National Power Grid.' Federal-Provincial First Ministers' Conference on Energy, 22-23 January, 1974, in Ottawa. Document FP- 4119. Schreyer Papers, PAM, Range 42, 550 files, box 159. Source in Winnipeg: Provincial Archives of Manitoba.

–. *Report on the Royal Commission on the Economic State and Prospects of Newfoundland and Labrador*. Chairman, Gordon F. Pushie. St. John's: Queen's Printer, 1967.

–. 'Report of the Royal Commission on Electrical Energy.' February 1966. Ottawa: National Energy Board Library.

–. Supreme Court of Newfoundland Trial Division. Between Her Majesty's Attorney General of Newfoundland and Churchill Falls (Labrador) Corporation Limited, Québec Hydro-Electric Commission, the Royal Trust Co., and General Trust Co. Judgment of Goodridge, J., 6 July 1982.

Newman, Walter C. *What Happened When Dr. Kasser Came to Northern Manitoba*. Winnipeg: Newmac, 1976.

Niagara Falls, Canada: A History of the City and the World Famous Beauty Spot. An Anthology. A Centennial Volume 1967. [Niagara Falls]: no author or publishers listed.

Niosi, Jorge, and Philippe Faucher. 'Public Enterprises[,] Procurements[,] and Industrial Development: The Case of Hydro-Québec.' Montréal: Université du Québec à Montréal, Centre de recherche en développement industriel et technologiques (CREDIT), 1986.

Offe, Claus. 'Demokratische Legitimation der Planung.' *Strukturprobleme des Kapitalistischen Staates*. Frankfurt: Suhrkamp Verlag, 1972.

–.'Tauschverhältnis und Politische Steuerung: Zur Aktualität des Legitimationsproblems.' *Strukturprobleme des Kapitalistischen Staates*. Frankfurt: Suhrkamp Verlag, 1972.

–. 'The Theory of the Capitalist State and the Problem of Policy Formation.' *Stress and Contradiction in Modern Capitalism: Public Policy and The Theory of the State*. Edited by Leon Lindberg, et al. Lexington: Heath, 1975.

Ontario. Hydro-Electric Power Commission of Ontario [Ontario Hydro]. 'The Allied War Needs as a Governing Factor.' *Murray Report on Electric Utilities: Refutation of Unjust Statements*. Toronto: Hydro-Electric Power Commission of Ontario, 1922. Source in Toronto: Ontario Hydro Archives, archival folder 7, Adam Beck General [1920-2].

–. *Official Report of the Ontario Power Commission*. 28 March 1906. Sir Adam Beck papers. Toronto: Ontario Hydro Archives.

Osborne, Ronald. 'Industry Review: Ontario Hydro,' *Connections: 1999 Electricity Industry Review*. Montréal: Canadian Electricity Association, January 1999.

Parenteau, Roland. 'Hydro-Québec: Les relations entre l'état et son entreprise.' Document 312. Edited by the directeur de l'information Conseil économique du Canada. Ottawa: Economic Council of Canada, 1986.

Plourde, André. 'Energy and the NAFTA.' Commentary, No. 46. Ottawa: C.D. Howe Institute, 1993 [May 1993].

Power Grab: The Future of B.C. Hydro. A conference on Deregulation and Competition in the Electrical Industry, co-sponsored by the Canadian Centre for Policy Alternatives, BC office, and the Institute for Governance Studies, Simon Fraser University, Harbour Centre Campus, 10 January 1998.

Priddle, Roland. 'Regulation of Canadian Energy Exports in the Free Trade Era.' Notes from a speech presented to the Twenty-First Annual Conference of the Institute of Public Utilities, entitled 'Emerging Markets and Regulatory Reform: An Agenda for the 1990s,' Michigan State University, Colonial Williamsburg, Virginia, 11 December 1989. Calgary: National Energy Board Library.

Raybould, John. 'Paper for the World.' A two-page reprint from the BC Hydro magazine *Progress*. Vancouver: BC Hydro, 1966.

Québec. Direction générale de l'Energie, et al. *l'Électricite, Facteur de Developpement industriel au Québec*. Éditeur officiel du Québec, 1980.

–. Order in Council, An Act to Establish the Québec Hydro-Electric Commission, 1944, Chapter 22, Division V, paragraph 29.

Richardson, Boyce. *James Bay: The Plot to Drown the North Woods*. Toronto: Clark Irwin, 1972.

Ross, W. Gillies. 'Power on Hamilton River, Labrador.' *Canadian Geographical Journal* 66 (1963):78.

Salée, Daniel, and William Coleman. 'The Challenges of the Québec Question: Paradigm, Counter-paradigm and ...?' *Understanding Canada: Building on the New Canadian Political Economy*. Edited by Wallace Clement. Montreal: McGill-Queen's University Press, 1997.

Schlesinger, James. 'Foreword.' *Power from the North*. By Robert Bourassa. Scarborough: Prentice-Hall, 1985.

Schreyer, Edward. 'Address to Canadian Electrical Week Luncheon.' Winnipeg, 14 February 1974. Source in Winnipeg: Provincial Archives of Manitoba, Schreyer Papers.

–. 'Opening Remarks for the First Ministers Conference on Energy.' Ottawa, 22 January 1974. Source in Winnipeg: Provincial Archives of Manitoba, Schreyer Papers.

Schroeder-Lanz, Hellmut. 'The Mega-Hydropower Project of "Hydro Manitoba."' Unpublished essay. Trier, Germany: Department of Geography, University of Trier, 1996.

Sharp, Mitchell. 'Power, Announcement of National Policy October 8, 1963.' *House of Commons Debates: Official Report*. Vol. IV, 1963, First Session, 26th Parliament, pp. 3,299-301. Ottawa: Queen's Printer, 1963.

Sherman, Paddy. *Bennett*. Toronto: McClelland and Stewart,1966.

Shrum, Gordon. *Gordon Shrum: An Autobiography*. Edited by Peter Stursberg and Clive Cocking. Vancouver: University of British Columbia Press, 1986.

Sigurdson, Albert. 'B.C. Hydro foresees further cuts in capital projects as growth slows.' *Globe and Mail*, 26 September 1983.

Smith, Philip. *Brinco: The Story of Churchill Falls*. Toronto: McClelland and Stewart, 1975.

Stevenson, Garth. 'Federalism and the Political Economy of the Canadian State.' *The Canadian State*. Edited by Leo Panitch. Toronto: University of Toronto Press, 1983.

Sturgess T.L., and R.W. Bonner. Department of Industrial Development, Trade, and Commerce. Bureau of Economics and Statistics, Government of British Columbia. *British Columbia Facts and Statistics*, years 1956 to 1962.

Swainson, Neil. *Conflict over the Columbia*. Montreal: McGill-Queen's University Press, 1979.

Tieleman, William. 'Political Economy of Nationalization: Social Credit and the Takeover of the British Columbia Electric Company.' MA thesis, Department of Political Science, University of British Columbia, 1984.

United States. Federal Energy Regulatory Commission, 'Promoting Wholesale Competition Though Open Access: Non-discriminatory Transmission Service by Public Utilities.'

Docket No. RM95-8-001. Washington, DC: Federal Energy Regulatory Commission, Order No. 888-A, issued 3 March 1997.

–. US Department of Energy. Bonneville Power Administration. 'Near Term Intertie Access Policy,' 1 June 1985. A copy of this policy was obtained from the BC Utilities Commission, exhibit 21, hearing 4, entered by BC Hydro on 9 January 1986.

–. US Department of Energy. Bonneville Power Administration. 'Selling South: BPA Seeks Ways of Marketing Surplus Power to the Pacific Southwest.' *Issue Alert,* no date. Source: BC Utilities Commission, file 14.

Watkins, Mel. 'Canadian Capitalism in Transition.' *Understanding Canada: Building on the New Canadian Political Economy.* Edited by Wallace Clement. Montreal: McGill-Queen's University Press, 1997.

–. 'The Political Economy of Growth.' *New Canadian Political Economy.* Kingston: McGill-Queen's University Press, 1989.

–. 'A Staples Theory of Economic Growth.' *Approaches to Canadian Economic History: A Selection of Essays.* Toronto: McClelland and Stewart, 1967.

–. 'The Staple Theory Revisited.' *Journal of Canadian Studies* 12 (1977): 25.

Western Business and Industry. 'Text of Wenner-Gren Memorandum of Intent [16 November 1956].' 31 (1957): 71-72.

Williams, Glen. 'Canada in the International Political Economy.' *New Canadian Political Economy.* Edited by Wallace Clement and Glen Williams. Kingston: McGill-Queens University Press, 1989.

–. 'Greening the New Canadian Political Economy,' *Studies in Political Economy* 37 (1992): 5.

–. *Not for Export: Toward a Political Economy of Canada's Arrested Industrialization.* Toronto: McClelland and Stewart, 1986.

Worley, Ronald. *The Wonderful World of W.A.C. Bennett.* Toronto: McClelland and Stewart, 1971.

Index